干旱瘠薄土地作物
需水量与灌溉制度研究

（以新疆阿勒泰地区为例）

（汉英对照版本）

A Case Study on Crop Water Requirement and Irrigation Scheduling in Arid and Barren Land

(Altay Prefecture, Xinjiang)

(Chinese-English Bilingual Edition)

中 文 作 者：程发林　陈亚宁

Chief　Editor：Cheng Falin　Chen Yaning

译　　　者：王立成　金　鹏　张　猛　杨科特　徐　丽
　　　　　　刘春锋　赵　健　李浩瑾　李　娅　翟　政
　　　　　　张　扬　郑慧洋　邱海岭

**Translators：Wang Licheng　Jin Peng　Zhang Meng
　　　　　　　Yang Kete　Xu Li　Liu Chunfeng　Zhao Jian
　　　　　　　Li Haojin　Li Ya　Zhai Zheng　Zhang Yang
　　　　　　　Zheng Huiyang　Qiu Hailing**

黄河水利出版社

·郑州·

图书在版编目(CIP)数据

干旱瘠薄土地作物需水量与灌溉制度研究:以新疆阿勒泰地区为例=A Case Study on Crop Water Requirement and Irrigation Scheduling in Arid and Barren Land (Altay Prefecture, Xinjiang):英文/程发林,陈亚宁主编;王立成等译.—郑州:黄河水利出版社,2019.6
ISBN 978-7-5509-2418-5

Ⅰ.①干… Ⅱ.①程…②陈…③王 Ⅲ.①作物需水量-研究-阿勒泰地区-英文 ②灌溉制度-研究-阿勒泰地区-英文 Ⅳ.①S274

中国版本图书馆 CIP 数据核字(2019)第 126053 号

出 版 社:黄河水利出版社
地址:河南省郑州市顺河路黄委会综合楼 14 层　邮政编码:450003
发行单位:黄河水利出版社
发行部电话:0371-66026940、66020550、66028024、66022620(传真)
E-mail:hhslcbs@126.com
承印单位:河南新华印刷集团有限公司
开本:787 mm×1 092 mm　1/16
印张:22.25
字数:668 千字　　　　　　　　　印数:1—1 000
版次:2019 年 6 月第 1 版　　　　印次:2019 年 6 月第 1 次印刷
定价:88.00 元

主　　　　编：程发林　陈亚宁
Chief　　　Editor：Cheng Falin　Chen Yaning
副　主　　编：巴　新　欧阳军　哈飞章
Associate　Editor：Ba Xin　Ouyang Jun　Ha Feizhang
技　术　顾　问：陈宝奎　杨文亮
Technical　Advisor：Chen Baokui　Yang Wenliang

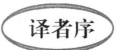

Preface by Translators

作物在生长发育过程中,是否供水越多越好?有人认为"水大,浇透才能高产""水越多越好,作物才能快长",这些认知缺乏科学依据,不仅会造成水资源的浪费,还会破坏土壤结构及通气性,反而造成作物产量降低。

Themore water, the better for crops. It is right? Some people believe a thorough watering helps increase the yield of crops. Actually, this kind of belief is scientifically groundless. To practice it would not only waste water, but also damage soil texture and aeration, reducing the yield of crops.

在长期生产实践中,广大干旱灌区干部、职工总结出根据季节时令、降水多少、气温高低、风力的大小等气候情况确定灌水时间的经验;根据田间水分状况、土壤质地、肥瘦、地温及地下水位高低等情况确定灌水量的经验,但经验的总结不能作为指导干旱地区灌溉工作的科学依据。

During long time practice, people from the arid areas have summed up some experience on irrigation. They determine the timing of irrigation based on such factors as season, rainfall, temperature, and wind; and they determine the amount of irrigation water based on such soil conditions as humidity, texture, fertility, temperature, and groundwater table. However, the experience summarized from past practice cannot science to replace guide irrigation in arid areas.

科学灌溉用水一直是农田水利管理运用和科研工作的一个重要内容。阿勒泰地区的农业生产主要在荒漠戈壁地上开发利用的,荒漠戈壁地的灌溉面积占阿勒泰全地区灌溉面积的80%以上。这种戈壁地与一般土地在土质结构、性状等方面有着很大差别,土壤有机质含量、田间持水率等也不同于其他土质。

Irrigation and wateringin a scientific manner has always been a significant part in farmland water conservancy. Altay Prefecture reclaims farmlands from Gobi Desert for crop production; and these farmlands account for over 80% of the total irrigation area in Altay Prefecture. They differ greatly from other lands in soil texture, soil property, organic content, field capacity, and others.

《干旱瘠薄土地作物需水量与灌溉制度研究(以新疆阿勒泰地区为例)》这本书中对阿勒泰地区春小麦、玉米、苜蓿及甜菜等作物的需水量、需水规律、生理需水量、潜在腾发量、作物需水量数学模型、作物需水量与产量的关系以及作物灌溉制度和灌水方法进行了详细地分析和总结,对于指导阿勒泰地区农业生产具有很大的实际意义,其研究方法和结论对全国干旱瘠薄地区也具有很大的指导意义。

A Case Study on Crop Water Requirement and Irrigation Scheduling in Arid and Barren Land (Altay Prefecture, Xinjiang) analyzes water demand, physiological water demand, potential evapotranspiration of crops planted in Altay Prefecturesuch as spring wheat, corn, alfalfa, and beet. It establishes mathematical models of water demand for these plants, defines the law governing water demand, and exposes the correlation between water demand and crop yield. Besides, this book also gives a detailed account of the irrigation regime and method in Altay Prefecture. In a word, this book promotes crop production in Altay Prefecture; and the research method and conclusion in this book also serve as an important guideline for irrigation in other arid and barren areas in China.

近年来,随着国外灌溉工程项目的增加,和国外同行交流日益频繁,我们感觉国内外灌溉方法殊途同归,无论采用何种灌溉方式,最终都归结到基本灌溉制度上。因此,我们选择翻译此书,希望借着这本极具代表性的著作,能够给相关行业的从业人员提供一些借鉴和参考。限于个人水平和文笔习惯不同,可能存在一些疏漏,本书如有不当之处,尚请同行专家和读者批评指正。

In recent years, we participated in more and more overseas irrigation projects. Through communication with our foreigner counterparts, we found the different irrigation methods we may adopt boil down to the fundamental irrigation regime. Therefore, we chose this book to translate, hoping that this professional work could provide some insight to our counterparts in other countries. However, limited by our ability in this profession and writing, this translated version must have some flaws and omissions. In view of this, the good suggestions and corrections from our readers are warmly welcomed.

全书编写分工如下：全书章节安排及统稿由王立成、杨科特负责，第一部分翻译由徐丽、邱海岭、李娅执笔；第二部分翻译由张扬、张猛、刘春锋、赵健、李浩瑾执笔；第三部分翻译由金鹏、翟政、郑慧洋执笔。杨科特负责全书的英文稿审定。另外，在此书出版过程中，承蒙中水北方勘测设计研究有限责任公司编辑部王晓红、于荣海两位编辑给予的大力支持，谨致以衷心的感谢：

This translated edition is completed thanks to the work of a diligent team. Mr. Wang Lichen and Yang Kete are responsible for the general chapter arrangement and edition of thewhole book. The book consists of three parts, of which Part I is translated by Xu Li, Qiu Hailing, and Li Ya, Part II is translated by Zhang Yang, Zhang Meng, Liu Chunfeng, Zhao Jian, and Li Haojin, and Part III is translated by Jin Peng, Zhai Zheng, and Zheng Huiyang. The draft English version is examined and reviewed by Yang Kete. In addition, heartfelt thanks shall also be extended to editor Wang Xiaohong and Yu Ronghai from the editorial department of BIDR, without whosestrong support, this book would not come out.

译者

2018 年 12 月

Translators

December 2018

readers are warmly welcomed.

本书翻译的完成还要感谢一支辛勤的翻译团队。王辰和杨可参与了本书的整体策划和各章节安排；第Ⅰ部分由徐海玲、李娅翻译；第Ⅱ部分由张阳、张萌、吕春晨、肖健和马明翻译；第Ⅲ部分由蔺峰、翟征和张华勇翻译。译文由杨可审阅和校译。此外，此书出版还要感谢人民卫生出版社编辑部王晓虹和徐若霞老师的鼎力支持。

This translated edition is completed thanks to the work of a diligent team. Mr. Wang Chen and Yang Ke are responsible for the general chapter arrangement and edition of the whole book. The book consists of three parts, of which Part I is translated by Xu Hailing and Li Ya, Part II is translated by Zhang Yang, Zhang Meng, Lü Chunchen, Xiao Jian and Ma Ming, and Part III is translated by Lin Feng, Zhai Zheng, and Zhang Huayong. The final English version is examined and reviewed by Yang Ke. In addition, heartfelt thanks shall also be extended to editor Wang Xiaohong and Xu Ruoxia from the editorial department of PMPH, without whose strong support, this book would not come out.

译者
2018 年 12 月
Translators
December 2018

Preface by the Author

阿勒泰地区的大部分耕地(80%以上)是在干旱瘠薄土地上开发出来的,但是,由于缺乏荒漠瘠薄土地的作物需水量及灌溉制度方面的试验研究成果,加之该区灌溉技术及灌溉管理落后,广大农牧民误认为在干旱荒漠地上只要有水就浇,多浇水就会多增产,结果造成在水多的地方灌水量偏大而闹水害,水少的地方则因缺水而闹水旱。

Most of the arable land (more than 80%) in Altay Prefecture, Xinjiang wasreclaimed fromarid and barrenland. Because oflimited study of crop water requirement (CWR) and irrigation scheduling for arid and barren land, and due to the fact that the irrigation technology and management lagged behind in this region, the farmers and herdsmen here used to have a mistaken belief that the arid and barren land should be irrigated with as much water as it is possibly available, which would help increase crop yield. The consequence was water-logging happened where there was plenty of water for irrigation, while droughts occurred where there was limited water for irrigation.

为了合理地、充分地利用水资源,在1989—1992年,结合当地生产条件和地域环境特点,通过4年期间对各种作物的对比试验,收集和获得第一手作物灌溉资料,揭示了土壤水分与作物生长发育及产量的关系,确定了经济合理的灌溉制度,为荒漠瘠薄土地作物的合理灌溉用水提供了科学依据,并于1996年出版了《干旱瘠薄土地作物需水量与灌溉制度研究(以新疆阿勒泰地区为例)》中文版,供科技人员和地方农业工作者参考。这本书在全国发行,至今已过去22年,各方面反应较好,尤其对于阿勒泰地区的农牧民提供了指导性的灌溉依据,比较实用,我个人也感到十分欣慰。

To provide a guidance for the people in agriculture in this area and to help them make better use of water resources, the Chinese edition of *A Case Study of*

Crop Water Requirement and Irrigation Scheduling in Arid and Barren Land (Altay Prefecture, Xinjiang) was published in 1996. This book was written on the basis of a lot of first-hand data on the crop irrigation, gathered from a number of comparison experimentsdone on different crops in 4 years (1989–1992). These experiments gave adequate consideration to local conditions and regional environment characteristics. The data gathered from these experiments revealed the relationship between soil moisture content and the growth and yield of crops, helped establish an economicaland reasonable irrigation scheduling, providing a scientific basis for determining the appropriate irrigation water for crops growing in the arid and barren land. This book was warmly welcomed by readers since its publication 22 years ago. I am very delighted that it provides practical guidance to farmers and herdsmen in the Altay Prefecture as for how to do irrigation.

今年初,当中水北方勘测设计研究有限责任公司的几位同志表达了要翻译此书的想法时,我感到十分荣幸,因为此书可以继续发挥它的余热和价值。但一方面由于本书专业性相对较强,翻译起来会耗费极大的心力;另一方面受众较少,影响面窄,所以我认为形成译本的意义不大,而且我觉得他们作为单位的骨干和中坚力量,应该把这部分时间放到更重要的事情上。不过他们认为把这本书翻译成英文,不仅便于提高本身的英文水平,更有利于持书与国外同行直接交流沟通,这种精神使得我十分感动,积极支持他们。

When severalcolleagues of mine from China Water Resources Beifang Investigation, Design and Research Co. Ltd. (BIDR) at the beginning of this year expressed their intention to have this book translated, I felt deeply honored to know that itkept making contribution and creating value. But initially I doubted whether it was necessary to translate this book. For one thing, this book is highly professionally, which makes its translation a challenging job; for the other, being a highly professional book, it has limited readers. I insisted they, as the backbone of BIDR, should devote time to doing more important things rather than translating this book. However, they believed that translating this book into English would not only help improve their proficiency in English, but also help facilitate direct communicationwith their foreign counterparts. At last, I was convinced by their enthusiasm for this job and agreed.

世界范围内干旱和极端干旱地区占总面积的约 18.8%,干旱地区的农业

原作者序言
Preface by the Author

灌溉生产直接影响着区域内水资源配置,希望本书能够给国外同行提供些参考。最后,此书出版之后如能起到一定作用,除感谢译者外,还应感谢出版社和其他所有支持他出版的人。我期望本书能比原著画上更令人满意的句号。

  Arid and extremely arid regions account for about 18.8% of the total land area in the world. Agricultural irrigation and production in arid areas has a direct bearing on regional allocation of water resources. Iwould be very glad if this book would be of any use for our foreign counterparts. Finally, I would like to extend my sincere thanks to the publishing house, to those who help its publication happen, as well as to its translators. I do hope this translated version will bring about a bigger success than the Chinese one.

Foreword

不同地区由于自然环境、水热条件、土壤耕作性能的差异,作物的需水量状况是不同的。而对作物生育期内不同阶段的需水量进行详尽试验研究,提出合理灌溉制度,不仅有利于作物增产,而且还可以为灌区规划、设计提供科学依据,避免水利工程设计偏大而造成浪费。同时由于适时适量的灌水,一方面起到了节约用水的目的,再则还有助于防止土地盐渍化和肥土流失。因此开展作物需水量及灌溉制度研究具有直接的生产实际意义。

Besides different environment, heat and water condition, soil ploughabikoty and so on, there is also different in water requirement of crops in different regions. By means of making experiments for detailedly studying on the water requirement in every growing stages of crops, it will be able to work out a rational irrigation system which is advantageous to not only increasing production of crops, but also providing scientific basis for reasonably planning irrigation area and reducing waste of water to a certain degree, through carring out advanced irrigation system can economize on water resources for one things and be favourable to preventing salinization and soil erosion for another things Therefore, it is essential for agricultural production to research on water requirement of crops and its irrigation system.

新疆北部阿勒泰地区的农业耕作区主要在荒漠瘠薄土地上,该区土地类型与一般土地在土质结构上有很大区别,土壤肥力差,有机质含量低,仅为 0.24%~0.49%,地表下 0~40cm 基本是沙壤土,40cm 以下由砂子、卵石等组成,且下伏地层为第三系不透水层泥岩,耕作层薄,土壤渗漏量大,田间持水率比一般沙壤土要低很多,因此,一方面其灌溉制度与灌水方法具有自身特点,另外极易造成土地盐渍化。因而,有必要在结合一系列工程、生物措施进行中低产田改造、盐碱地改良、沙漠化防治过程中,开展干旱瘠薄土地作物需水量及优化灌溉制度的试验研究,它不仅将对阿勒泰地区农牧业生产发展有着重要的意义,而且还可丰富在灌溉制度与灌水方法研究领域的理论和方法,填补干旱、瘠薄土地作物需水量研究的空白。为此,在自治区人民政府、自治区水利厅及阿勒泰地区各级领导的关怀和支持下,由中国科学院新疆地理所、新疆水文水资源局、新疆水利管理总站、阿勒泰地区水利处等有关单位专家和技术人员,结合阿勒泰地区生产条件和地域特点,重点对阿勒泰地区春小麦、玉米、苜蓿、甜菜等主要农作物需水量及灌溉制度进行研究。课题组成员不畏艰辛,经过4年的种植试验对比研究,收集和获得了大量第一手资料,《干旱瘠薄土地作物需水量及灌溉制度研究》一书便是本项目理论研究的最终总结。

In the north of Altay region of Xinjiang the crops are mainly cultivated on poor soil. This type of land, differing from others in land structure, is characterized as unproductive soil and poor organic matter (only 0.24%-0.49%). Surfale layer is composed of sand soil from zero down to 40cm, and almost sand and gravel fillet below 40cm. Under it there is the Tertiary unpermeable mudstone stratum. This soil has thin plouging layer, good permeability and poor field water-contenting rate. If the irrigating methods are irreasonable, it is easy to lead to salinization. Therefore, in the process of developing a series of engineering and biological measures for improving low-yield farmland transforming salint-alkaliland and preventing desertification, it is indispensable to engage in experimental research on water requirement of crops and its irrigation system in this region. This study not only will have practical sense for agricultural production of Altay region, but also enrich the theory of irrigation system and fill in the gaps in the fields of water requirement of crops on poor soil in aried land. For this reason, it is nec to study on water requirement and irrigation system of crops such as spring wheat, corn, lucerne, beet and so on. As a result of the comparative research on experimental plot for four years, the research team has overcome varions difficulties, and collected and obtained a lot of first hand data, the book, *Study on water requirement and irrigation system of crops on poor soil in arid land* is just the achievements in this theoretic research.

本书由3个部分组成,第一部分为综合研究,第二部分为作物需水量与需水规律,第三部分为作物灌溉制度研究。书中对阿勒泰地区春小麦、玉米、苜蓿及甜菜等作物的需水量、需水规律、生理需水量、潜在腾发量、作物需水量数学模型、作物需水量与产量的关系以及作物灌溉制度和灌水方法进行了详细

前 言
Foreword

地分析和总结,对于指导阿勒泰地区农业生产具有很大的实际意义。

This book detailedly describes and discusses on the experimental research on irrigation system of crops such as spring wheat, corn, lucerne, beet and so on, including the analysis on water requirement, water-requiring law, physiological water requirement, potential e-vapo-transpirative capacity mathematic model of water requirement, the relationship between water requirement and crops yields and the irrigating methods. Thus it can be seen that this study is provided with practical sense for agricultural production.

本书的第一、二部分由陈亚宁、程发林、巴新执笔,第三部分由程发林、哈飞章、欧阳军执笔完成,由程发林、陈亚宁统稿。在项目实施和研究过程中,丁新利、李锐、郑文新、杨翠璞、李遇超、张玉玲、阿米努、杨磊、王庆峰、赵银花以及阿勒泰地区灌溉试验站的全体职工,作为课题组成员做过大量实际工作。在成书过程中,还得到韩德麟研究员、樊晏清、李建邦、巴福禄、张焕德高级工程师的大力支持和诸多指导,谨此一并致谢。

Part I & Part II of this book is penned by Chen Yaning, Cheng Falin and Ba Xin; and Part III penned by Cheng Falin, Ha Feizhang and Ouyang Jun. The whole book is edited and finalized by Cheng Falin and Chen Yaning. Sincere thanks would go to Ding Xinli, Li Rui, Zheng Wenxin, Yang Cuipu, Li Yuchao, Zhang Yuling, Aminu, Yang Lei, Wang Qingfeng, Zhao Yinhua, and all the staff in the experimental irrigation stations in Atlay prefecture, who, as members of the research group, have done plenty of works during the study. Our special thanks would also go to researcher Han Delin, senior engineer Fan Yanqing, senior engineer Li Jianbang, senior engineer Ba Fulu, and senior engineer Zhang Huande for their great support and guidance which makes the publication of this book possible.

第一章　综合研究	(1)
1　Comprehensive Study	(1)
第一节　问题的提出及目的所在	(1)
1.1　Problem and Purpose	(1)
第二节　试验区环境特征	(3)
1.2　Environmental Characteristics of Experimental Area	(3)
第三节　田间基本参数测试	(13)
1.3　Test of Field Parameters	(13)
第四节　作物需水量与灌溉制度试验设计	(29)
1.4　Experimental of Water Requirement and Irrigation System of Crops	(29)
第五节　作物需水量与需水规律试验研究	(43)
1.5　Experimental Study on Water Requirement of Crops and Its Law	(43)
第六节　作物灌溉制度与灌水方法试验研究	(57)
1.6　Experimental Research of Irrigation System and Method	(57)
第七节　试验成果效益分析	(75)
1.7　Irrigation Benefit	(75)
第二章　作物需水量与需水规律	(83)
2　Water Requirement of Crops and Its Law	(83)
第一节　作物需水量分析	(83)
2.1　Water Requirement	(83)
第二节　作物需水规律研究	(121)
2.2　Study on Water-Requiring Law of Crops	(121)
第三节　作物生理需水量分析	(150)
2.3　Analysis on Physiological Water Requirement of Crops	(150)
第四节　作物潜在腾发量及作物系数	(183)
2.4　Potential Evapo-Transpirative Capacity of Crops and Its Coefficient	(183)
第五节　作物需水量的数学模型	(190)
2.5　Mathematical Model of Water Requirement of Crops	(190)
第六节　作物产量与需水量分析	(200)

2.6 **Crop Yields and Its Water Requirement** …………………………（200）
第三章 作物灌溉制度研究 ……………………………………………（232）
3 **Study on Irrigation System of Crops** …………………………………（232）
  第一节 春小麦灌溉制度与灌溉效益分摊系数研究 …………………（232）
  3.1 **Irrigation System of Spring Wheat and Its Irrigation Benefit** ………（232）
  第二节 玉米灌溉制度与灌溉效益分摊系数研究 ……………………（270）
  3.2 **Irrigation System of Corn and Its Irrigation Benefit** ………………（270）
  第三节 苜蓿灌溉制度研究 ……………………………………………（299）
  3.3 **Irrigation System of Lucerne** …………………………………………（299）
附表 ………………………………………………………………………………（329）
Postform ………………………………………………………………………（329）
参考文献 …………………………………………………………………………（333）
Reference ………………………………………………………………………（333）

第一章 综合研究
1 Comprehensive Study

第一节 问题的提出及目的所在
1.1 Problem and Purpose

不同地区由于自然环境、水热条件、土壤耕作性能的差异,作物的需水量状况是不同的。而对作物生育期内不同阶段需水量进行详尽试验研究,提出合理灌溉制度,不仅有利于作物的增产,而且还可以为灌区规划、设计提供科学依据,避免水利工程设计偏大而造成的浪费。同时,由于适时适量的灌水,一方面起到了节约用水的目的,再则还有助于防止土地盐碱化和肥土流失。因此开展作物需水量及灌溉制度研究具有直接的生产实际意义。

Due to different environment, heat and water condition, soil ploughability and so on, water requirement of crops in different regions is different. By means of making experiments for detailedly studying on the water requirement in every growing stages of crops, it will be able to work out a rational irrigation system which is advantageous to not only increasing production of crops, but also providing scientific basis for reasonably planning irrigation area and reducing waste of water to a certain degree. Meanwhile, advanced irrigation system can economize on water resources for one thing, and be favorable to preventing salination and soil erosion for another thing. Therefore, it is essential for agricultural production to research on water requirements of crops and its irrigation system.

新疆北部阿勒泰地区的农业耕作区主要在荒漠瘠薄土地上,荒漠瘠薄土地的灌溉面积占全地区总灌溉面积的 80%。该土地类型与一般土地在土质结构上有很大区别,土壤肥力差、有机质含量低,仅为 0.24%~0.49%,地表下 0~40 cm 基本是沙壤土,40 cm 以下由砂子、卵石等组成,且下伏地层为第三系不透水层泥岩,耕作层薄,土壤渗漏量大,田间持水率比一般沙壤土要低很多,因此其灌溉制度与灌水方法具有自身的特点,它与一般地区有着很大差别。

In Altay region of northern Xinjiang the crops are mainly cultivated on poor soil in desert land. 80% of irrigation area in the region is on barren land. This type of land, differing from others in land structure, is characterized as unproductive soil and poor organic matter (only 0.24%–0.49%). Surface layer is composed of sand soil from zero down to 40 cm, and almost sand and gravel fillet below 40 cm. Under it there is the Tertiary impermeable mudstone stratum. This soil has thin ploughing layer, good permeability and poor field capacity. Therefore, the irrigation system and methods in the region have distinct features and great difference with general land.

阿勒泰地区的大部分耕地(80%以上)是在干旱瘠薄土地上开发出来的,但是,由于缺乏荒漠瘠薄土地的作物需水量及灌溉制度方面的试验研究成果,加之该区灌溉技术及灌溉管理落后,广大农牧民误认为在干旱荒漠地上只要有水就浇,多浇水就会多增产,农作物灌溉定额大都在 1 000 m³/亩以上,有些地方的最大灌溉定额超过 2 000 m³/亩。结果造成在水多的地方灌水量偏大而闹水害,水少的地方则因缺水而闹水旱,不能合理地、充分地利用水资源。据统计,新中国成立以来,阿勒泰地区有 60 万亩耕地因盐渍化而弃耕,约占阿勒泰地区灌溉面积的 23%;土地沙化面积为 45 万亩,约占阿勒泰地区灌溉面积的 18%。每年还有 20 万~30 万亩耕地因缺水干旱而作物减产。土地盐渍化、沙漠化问题的不断加重,不仅给当地农业生产建设造成损失,使大量的水利资源浪费,而且因土地荒芜还迫使农牧民进行搬迁,给广大农牧民也带来很大损失。

Most of the arable land (over 80%) in Atlay prefecture is developed in arid land. However, due to lack of experimental study on water requirement and irrigation system of crops in arid land, together with the backward irrigation technology and management, farmers and herdsmen have a common misconception that more watering would bring higher output. So the irrigation quota of crops was generally more than 1 000 m³/mu and the maximum irrigation quota even exceeded 2 000 m³/mu. Because water resources cannot be utilized in a rational and full way, water disaster often occurred in area rich in water due to more irrigation, while drought happened in area scarce of water due to less irrigation. According to statistics, since the founding of the People's Republic of China, 600 000 mu of farmland has been abandoned due to salinization and 450 000 mu of that has been deserted due to desertification, which accounted for about 23% and 18% of total irrigation area in Atlay prefecture respectively. Every year, another 200 000 to 300 000 mu of farmland is faced with a reduction in crop yields due to drought. The increasingly serious soil salinization and desertification cause losses not only to the local agricultural production and construction, with a large amount of water resources wasted, but also to the farmers and herdsmen for they had to relocate due to land desolation.

阿勒泰地区土地盐渍化、沙漠化问题的不断加重,已严重影响到当地经济发展,并且在今后的土地开发中基本上全部属荒漠戈壁地类型,土地瘠薄,因此有必要在采取一系列工程、生物措施进行中低产田改造、盐碱地改良、沙漠化防治的基础上,开展干旱瘠薄土地作物需水量及优化灌溉制度的试验研究,它将不仅对阿勒泰地区的农牧业生产发展有着重要的意义,而且还可丰富在灌溉制度与灌溉方法研究领域的理论和方法,填补在干旱瘠薄土地作物需水量研究的空白。为此,在自治区人民政府、自治区水利厅及阿勒泰地区各级政府的关怀和支持下,由新疆水利管理总站、阿勒泰地区水利处等组织有关单位专家

第一章 综合研究

1 Comprehensive Study

和技术人员论证立项,开展新疆阿勒泰地区荒漠瘠薄土地作物需水量及灌溉制度的试验研究项目。本项目试验观测期为1989—1992年,主要结合当地生产条件和地域环境特点,重点对阿勒泰地区的春水麦、苜蓿、玉米、甜菜等主要农作物进行灌溉试验研究,旨在通过4年(1989—1992年)期间对各种作物的对比试验,收集和获得第一手作物灌溉资料,从而揭示土壤水分与作物生长发育及产量的关系,探求经济合理的灌溉制度,为我区荒漠瘠薄土地作物的合理灌溉用水提供科学依据。

The increasingly severe soil salinization and desertification in Altay have badly affected the development of its local economy. Worsely, most of its land to be developed is barren Gobi desert. Therefore, in the process of developing a series of engineering and biological measures for improving middle-low-yield farmland, transforming saline-alkali land and preventing desertification, it is indispensable to engage in experimental research on water requirement of crops on poor soil and its irrigation system in this region. This study not only will have practical sense for agricultural production of Atlay prefecture, but also enrich the theory of irrigation system and fill in the gaps in the fields of water requirement of crops on poor soil in arid land. For this reason, it is necessary to carry out experimental study on water requirement and irrigation system of crops on poor soil in arid land in Atlay prefecture. The period of experimental observation is 1989 – 1992. Combined with local production conditions and geographical environment characteristics, experimental study on irrigation of major crops such as spring wheat, lucerne, corn, beet and so on is paid great attention, aiming to collect and obtain first hand data from the comparative research on experimental plot for four years (1989–1992), reveal the relationship between soil moisture and the growth and output of crops, work out a cost-effective and rational irrigation system and provide a scientific basis for rational irrigation of crops in arid land of this region.

第二节 试验区环境特征
1.2 Environmental Characteristics of Experimental Area

试验区位于新疆北部乌伦古河流域的福海县境内,北部为额尔齐斯河,南依乌伦古河,西壤乌伦古湖,地势东北高,而西南低,由东向西倾斜。区内气候温凉,无霜期短,属典型的大陆性温带气候区。

Located in Fuhai County in Ulungur River basin in Northern Xinjiang, with the Irtysh River in the north, the Ulungur River in the south and the Ulungur Lake in the west, the experimental plots ("the Plots") enjoy a high terrain in the northeast and low in the southwest, sloping from east to west. It is a typical continental temperate climate zone with cold climate and short frost-free period.

一、地形地貌
1.2.1 Topography and Landform

试验区地处额尔齐斯河和乌伦古河两河河间平原区,北部的额尔齐斯河和南部的乌伦古河近于平行均由东向西流向,前者经富蕴、北屯、布尔津、哈巴河等地流入国外,后者在我国境内经青河、富蕴、福海等地最后注入乌伦古湖。两河河间平原区地貌类型简单,主要是由第三纪地层和古生代基岩构成的基底平原。东部为古生代变质岩系构成的剥蚀平原,地形起伏,地表覆盖层主要为残坡积物。西部地势较平坦,下伏为第三纪地层,上覆第四纪堆积厚度不大,仅30~50 cm,地表组成物质较粗,主要为砂质和砂壤质物质,下伏有砾石层,地下水位在1~3 m。

The experimental plots are located in the plain between the Irtysh River in the north and the Ulungur River in the south, both flowing from east to west nearly in a parallel way. The former passes through counties like Fuyun, Beitun, Burqin, Habahe into foreign countries while the latter flows through counties like Qinghe, Fuyun, Fuhai before finally pouring into Ulungur Lake. With simple landform, the river plain is mainly composed of tertiary strata and Paleozoic bedrock. Its eastern part is the erosion plain composed of Paleozoic metamorphic rock series, with rolling topography and residual slope deposits as the surface layer of landform, whilst the western part enjoys a relatively flat terrain with tertiary strata as its underlying layer and only 30−50 cm thick of deposits as its overlying strata. The components of the surface layer are coarser, mainly sandy and sandy soil materials with gravel as the underlying layer. The underground water level is found at the depth of 1−3 m.

区内地表径流很少,地形切割微弱,基本属无径流区。

With little surface runoff and weak topographic cutting, the Plots almost belong to a zero runoff region.

二、光热条件
1.2.2 Solar-thermal Conditions

本区光热条件对农业生产利弊兼有。依据福海县气象站多年观测资料,现就试验区光热条件分述如下。

The solar-thermal conditions of the Prefecture have both advantages and disadvantages to agricultural production. According to the observation data of Fuhai Meteorological Station, this book describes the solar-thermal conditions in the Plots in the following parts.

（一）太阳辐射能
1.2.2.1 Solar Radiation Energy

太阳辐射是地面获得的最根本的能量来源。太阳总辐射量主要受地理位置（纬度）和天气现象（云量、降雨量、沙暴浮尘等）等的影响，并且季节分配亦不均匀。本试验区由于纬度偏北，加之区内降水少，下垫面植被覆盖率低，易出现浮尘天气，因此年太阳总辐射量仅为 546.7 kJ/cm²（见表1），其中年内作物生长期（4~9月）的太阳辐射量为 390.1 kJ/cm²，占年太阳总辐射量的 71.4%。

Solar radiation is the most fundamental source of energy for the ground. The total solar radiation is mainly affected by geographical location (latitude) and weather phenomena (cloud, rainfall, sandstorm, floating dust, etc.), and varies from season to season. Due to its northward latitude, little rainfall and low vegetation coverage on the underlying layer, the Plots tend to be blanketed by floating dust and only has a total annual solar radiation of 546.7 kJ/cm² (see Table 1). Besides, the solar radiation during the crop growth period (from April to September) is 390.1 kJ/cm², which accounts for 71.4% of the total solar radiation.

表1　试验区太阳年总辐射量统计表（据福海县气象站）　　单位：kJ/cm²

月份	1	2	3	4	5	6	7	8	9	10	11	12	合计
辐射量	19.2	26.4	43.5	57.3	71.2	75.3	74.1	64.9	47.3	33.1	19.7	15.1	546.7

Table 1　Statistics of Total Annual Solar Radiation in the Plots (based on Fuhai Meteorological Station)　Unit: kJ/cm²

Month	1	2	3	4	5	6	7	8	9	10	11	12	Total
Radiation	19.2	26.4	43.5	57.3	71.2	75.3	74.1	64.9	47.3	33.1	19.7	15.1	546.7

（二）日照
1.2.2.2 Sunshine

日照时数的长短对作物成熟和作物产量有直接的影响。本试验区的日照时数较长（见表2），日照百分率在65%以上，其中作物生长期（4~9月）的日照时数占年日照总时数的62.94%。

The sunshine duration has a direct effect on crop ripening and crop yield. With longer sunshine duration (see Table 2), the Plots enjoy over 65% sunshine percentage and its sunshine duration of crop growth period (from April to September) accounts for 62.94% of the total annual sunshine duration.

表 2 试验区日照时数统计表（据福海气象站） 单位:h

项目	1	2	3	4	5	6	7	8	9	10	11	12	年平均
日照时数	162.7	181.4	232.5	267.3	309.8	321.1	323.0	312.3	279.8	214.9	148.4	128.0	2 881.2

Table 2　Statistics of Sunshine Duration in the Plots
(based on Fuhai Meteorological Station)　　Unit: hour(s)

Item	1	2	3	4	5	6	7	8	9	10	11	12	Annual Average
Sunshine Duration	162.7	181.4	232.5	267.3	309.8	321.1	323.0	312.3	279.8	214.9	148.4	128.0	2 881.2

（三）热量状况

1.2.2.3 Thermal Conditions

作物养分的溶解、水肥的吸收运转以及水分蒸腾及同化作用等植物生理现象均与热量条件有关。

Plant physiological phenomena such as the dissolution of crop nutrients, the absorption and operation of water and fertilizer, water transpiration and assimilation are related to the thermal conditions.

1. 气温

I　Temperature

本区多年平均气温为 3.4 ℃，极端最高气温为 39.6 ℃，最低气温为 -42.7 ℃，夏季短暂，仅 50 d 左右；冬季漫长，达 155 d（见表 3），并常有寒潮和低温天气。

The Prefecture enjoys a multi-year average temperature of 3.4 ℃ with extreme temperatures of 39.6℃ (highest) and -42.7℃ (lowest). Summer is short here, only about 50 days; winter is so long that could be 155 days (See Table 3) with frequent cold waves and cold weather.

表 3　试验区多年气温特征值统计表　　单位:℃

类别	平均气温	极端最高气温	极端最低气温
福海	3.4	39.6	-42.7

Table 3　Statistics of Multi-year Temperature Characteristics in the Plots　　Unit: ℃

Category	Average Temperature	Extreme Highest Temperature	Extreme Lowest Temperature
Fuhai	3.4	39.6	-42.7

2.积温

II Accumulated temperature

本试验区≥10℃积温多年平均为 2 904.9℃,其中 80%的积温保证率为 2 680℃,年平均气温为 3.4℃。由于其西侧乌伦古湖的"冷湖"效应,气温变化表现为春季升温慢,秋季降温快(见表 4)。

The multi-year average accumulated temperature in the Plots ≥10 ℃ is 2 904.9 ℃, and 80% guarantee rate for the accumulated temperature is 2 680 ℃, the annual average temperature is 3.4 ℃. Due to the "cold-lake" effect caused by Ulungur Lake, the Plots enjoy a slow temperature rise in spring but a fast temperature fall in autumn (see Table 4).

表 4   试验区日平均温度≥10℃积温统计表   单位:℃

类别	≥10℃积温		
	平均	年际变化	80%保证率
福海	2 904.9	783.6	268.00

Table 4   Statistics of Daily Average Accumulated Temperature ≥10℃ in the Plots   Unit: ℃

Category	Accumulated Temperature ≥10℃		
	Average	Interannual Variation	80% Guarantee Rate
Fuhai	2 904.9	783.6	268.00

3.无霜期

III Frost-free periods

本试验区无霜期较短,多年平均为 147 d,其中最长为 185 d,最短仅为 129 d(见表 5)。

Frost-free period of the Plots are shorter, with multi-year average of 147 days, of which the longest are 185 days while the shortest are only 129 days (see Table 5).

表 5   试验区多年无霜期统计表

类别	无霜期(d)			霜期初、终日	
	平均天数	最长	最短	初日	终日
福海	147	185	129	9.23	4.28

Table 5 Statistics of Multi-year Frost-free Periods in the Plots

Category	Frost-free Periods (days)			First/Last Day of Frost Periods	
	Average Days	Longest	Shortest	First Day	Last Day
Fuhai	147	185	129	9.23	4.28

三、水分条件
1.2.3 Water Conditions

(一) 大气降水
1.2.3.1 Rainfall

本试验区为荒漠戈壁地,气候干旱,雨水较少,年蒸发量为 1 830 mm,农业作物生长主要靠灌溉。统计多年平均降水量为 112.7 mm。在年内分配上,降水量主要集中在夏季 6~8 月 3 个月,约占年降水总量的 41%(见表 6)。

The Plots are composed of desert Gobi, with dry climate and little rainfall and an annual evaporation of 1 830 mm. Agricultural crops mainly depend on irrigation for growth. Statistics show that the multi-year average rainfall is 112.7 mm. In terms of yearly distribution, rainfall is mainly in three months (from June to August) of the summer, which accounts for 41% of the annual total rainfall (see Table 6).

表 6 试验区降水量统计表

类别	降水量 (mm)	降水年内分配(%)			
		3~5 月	6~8 月	9~11 月	12~次年 2 月
福海	112.7	20	41	27	12

Table 6 Statistics of Rainfall in the Plots

Cate	Rainfall (mm)	Yearly Distribution (%)			
		March to May	June to August	September to November	December to the next year February
Fuhai	112.7	20	41	27	12

(二) 积雪情势
1.2.3.2 Accumulated Snow

平原区积雪消融供给土壤水分,对土壤保墒、增墒都有作用。由表 6 可见,本区冬季降水较少,仅占年降水总量的 12%,积雪厚度一般变化在 12~15 cm,积雪初日在 10 月底

第一章 综合研究
1 Comprehensive Study

至11月初,终日在3月底至4月上旬。

The melting snow cover in the plain area provides soil moisture, which will help preserve and increase the moisture in soil. From Table 6, it is easy to see that the Prefecture has less rainfall in winter, accounting for 12% of the total annual rainfall; and the thickness of snow varies from 12 to 15 cm and the first snowing day occurs in late October to early November while the last snowing day in the end of March to the first 10 days (hereinafter referred to as "early") of April.

四、农业气象灾害
1.2.4 Agro-meteorological Disaster

(一)寒潮、低温
1.2.4.1 Cold Wave and Low Temperature

由于本区地处中纬大陆性温带寒冷气候区,受西伯利亚寒冷气流影响,常因寒潮入侵,气温急剧下降。寒潮主要出现在冬季,强冷空气主要出现在春、秋两季。

Located in the middle latitude continental temperate cold climate zone, the Prefecture is affected by the Siberian cold air flow and often blanketed by cold wave and sharp temperature drop. In most cases, cold wave occurs in winter while severe cold air appears in spring and autumn.

(二)大风天气
1.2.4.2 Gale weather

本区位于额尔齐斯河和乌伦古河两河河间平原区,由于地形作用,区内大风天数较多,其中8级以上大风达60 d(见表7),多出现在春季,有时大风加寒潮,危害较大。

Located in the river plain between the Irtysh River and the Ulungur River, together with the topographic effect, the Prefecture has many windy days, up to 60 days of that are ≥8 gales weather (see Table 7). Mostly, gale weather occurs in spring, sometimes with cold wave days, which causes greater hazards.

表7　　　　试验区大风特征值统计表

类别	平均风速（m/s）	≥8级大风(d)		
		平均	最多	最少
福海	3.0	43.5	60	15

Table 7 Statistics of Gale Characteristics Value in the Plots

Category	Average Wind Speed(m/s)	≥8 Gales (days)		
		Average	Maximum	Minimum
Fuhai	3.0	43.5	60	15

（三）雷暴、冰暴
1.2.4.3 Thunderstorms and Hail

本试验区的雷暴天数较多,年平均在 22 d 左右;冰雹较之为少。雷暴和冰暴天气主要出现在夏季 6~8 月,常对区域农牧业生产造成危害。

Thunderstorms occur frequently in the Plots, with an annual average of 22 days; while hail is relatively rare. The two types of weather mostly occur in summer (from June to August) and often do harm to the agricultural and pastoral production.

五、地表水
1.2.5 Surface Water

本区地表水资源丰富,北部有额尔齐斯河,南部有乌伦古河。试验区灌溉取水主要引用乌伦古河地表水。

Rich in surface water resources, the Prefecture enjoys the Irtysh River in the north and the Ulungur River in the south. The surface water from the Ulungur River is mainly used for irrigation in the Plots.

（一）河川径流
1.2.5.1 River Flow

1.河流梗概
Ⅰ Summary of rivers

乌伦古河发源于蒙古境内的阿尔泰山,在我国境内以大青河为干流,全长 573 km,流域面积约 3.6 万 km²,主要由大青河、小青河、查干河和布尔根河组成。布尔根河与青格里河汇合口以下被称为乌伦古河。二台以上阿勒泰山区为径流形成区,二台至喀拉布勒根段是径流运转区,喀拉布勒根至乌伦古湖为下游径流散失区。这里河床开阔,流速缓慢,水量沿程渗漏损失大。

Originated from the Altai Mountains in Mongolia, the Ulungur River within the borders of China mainly consists of the Daqinghe River (its main stream, with a total length of 573 km and a watershed area of 36 000 km², the Xiaoqinghe River, the Chagan River and the Burgen River. The Ulungur River starts where the Burgen River meets the Qinggeli River. The Altay Mountains above Saertuohai forms the river flow, the section of Saertuohai-Kalabulegen where the river runs and the Kalabulegen-Ulungur Lake is the disappearing area of the lower reaches. With open river bed and at slow flow rate, the river suffers large losses of water leakage along the way.

2. 径流特征

II Characteristics of runoffs

乌伦古河主要由大气降水和季节性融雪水组成,多年平均年径流量为 9.92 亿 m³(见表 8),径流的年际变化较大,多年平均 C_v 值达 0.47,洪水集中在 5~6 月。

The water of Ulungur River is mainly composed of atmospheric rainfall and seasonal snowmelt water. The multi-year annual average runoff is 992 million m³(see Table 8), with a relatively large interannual variation and a multi-year average C_v value of 0.47. Floods often occur in May and June.

表 8 乌伦古河保证率流量统计表

类别	多年平均		P=25%		P=75%		灌溉期(4~10 月)	
	灌量 (m³/s)	C_v	灌量 (m³/s)	典型年	灌量 (m³/s)	典型年	水量 (亿 m³)	占全年 (%)
二台	31.4	0.47	39.9	1960	20.7	1963	9.32	94

Table 8 Statistics of Ulungur River Guarantee Rate Flow

Category	Multi-year Average		P=25%		P=75%		Irrigation Period (April to October)	
	Irrigation Amount (m³/s)	C_v	Irrigation Amount (m³/s)	Typical Year	Irrigation Amount (m³/s)	Typical Year	Water Amount (0.1 billion m³)	Annual Percentage (%)
Saertuohai	31.4	0.47	39.9	1960	20.7	1963	9.32	94

需要指出的是,由于试验占地面积较小,用水量少,并完全能得到保证,因而就地表水资源和灌溉用水问题这里不多叙述。

It should be noted that issues related to surface water resources and irrigation water will not be discussed in the book, for the Plots can be fully satisfied due to its small size and low water consumption.

(二) 水质状况
1.2.5.2 Water Quality

乌伦古河水质较好,矿化度低,河水硬度小,属优质的灌溉生活用水(见表9)。

Characterized with low mineralization and low water hardness, the water quality of Ulungur River belongs to high quality water for irrigation and living (see Table 9).

表9 乌伦古河水化学分析表

采样地点	采样时间	pH	矿化度 (g/L)	阴阳离子毫克当量百分数(%)				
				HCl_3^-	Cl^-	SO_4^{2-}	Ca^{2+}	Mg^{2+}	$K^+ + Na^+$
二台站	1990.8	7.5	0.17	53.01	16.50	30.54	48.59	11.33	40.08

Table 9   Statistics of Chemical Analysis of the Water of Ulungur River

Sampling Site	Sampling Date	pH	Mineralization (g/L)	Milligram Equivalent Percentage of Ions (%)					
				HCl_3^-	Cl^-	SO_4^{2-}	Ca^{2+}	Mg^{2+}	$K^+ + Na^+$
Saertuohai	1990.8	7.5	0.17	53.01	16.50	30.54	48.59	11.33	40.08

六、植被土壤
1.2.6 Vegetation and Soil

本区气候干燥,植被属草原化荒漠类型,以低等旱生的小蓬、假木贼、蒿类为主,伴有少量羽茅。土层较薄,0~40 cm基本为砂壤土、其下为砂子和砾石,土壤持土率低,渗漏性大,下伏为第三纪不透水层,极易造成土地盐渍化。土壤类型主要为淡棕钙土,有机质含量低,仅为0.24%~0.49%,土壤瘠薄,呈荒漠戈壁地自然景观。

With a dry climate, the Prefecture belongs to steppe desert as most of its vegetation are lower drought-born nanophyton erinaceum, anabasis salsa, artemisia argyi and little achnatherum sibiricum. The soil layer is relatively thin with sandy loam in 0–40 cm. Under it there is sand and gravel with low rate of soil conservation and good permeability. The underlying layer is the tertiary impermeable stratum, which can easily cause soil salinization. The soil is mainly composed of light brown calcium soil with low organic matter content (only 0.24% – 0.49%) and is barren. The soil presents the natural landscape of desert Gobi.

第三节 田间基本参数测试
1.3 Test of Field Parameters

为保证试验成果的准确性和可靠性,按规范要求,在进行灌溉试验研究之前,我们对与试验有关的田间基本参数进行了化验测试。其内容主要包括田间持水率、计划湿润层深度、土壤渗透系数、土壤物理化学性质(土壤容重、土壤比重、土壤有机质、土壤盐分、土壤酸碱度)、灌溉水质等,现将主要测试结果分述如下。

In order to ensure the accuracy and reliability of the experimental results, the basic field parameters related to the experiment were tested before conducting the irrigation experiment, according to the standards and requirements. Parameters include field capacity, depth of planned wetting, coefficient of soil permeability, physical and chemical properties of soil (bulk density, specific gravity, organic matter, salinity, pH value), quality of irrigation water, etc. Main testing results are shown as follows.

一、田间持水率
1.3.1 Field Capacity

田间持水率是指在田间土层内所能保持的最大含水量。田间持水率是制定灌溉定额的重要依据。当灌溉水量超过田间持水率时,只能加深土壤的湿润层深度,而不能增加土层中含水量的百分数。因此,它是土壤中对作物有效水的上限。

Field capacity means the maximum moisture content conserved in the field soil layers. It is an important basis for establishing irrigation quota. Only the depth of soil wetting, not the percentage of soil moisture content, can be deepened if the irrigation quota exceeds field capacity. Thus, it is the upper limit of available water for crops in the soil.

(一) 田间持水率测试计算
1.3.1.1 Calculation of Field Capacity Testing

要想研究戈壁地的节水灌溉技术,必须首先测定戈壁地的田间持水率。在测定田间持水率时我们采用了围墙法,视土壤的差异我们选择了2个测定点,测定深度为60cm。1989年7月3日首先在一点进行取土测定土壤含水率为4.7%(占干土重),据灌水公式:

In order to study the water-saving irrigation technology of Gobi, it is necessary to measure its field capacity first. By measuring the field capacity, we applied the enclosure method, and selected 2 testing pits at the depth of 60cm considering the soil diversity. In July 3rd, 1989, we measured one testing pit and found its soil moisture content was 4.7% (the percentage in water-free soil), according to the irrigation formula:

$$Q = (A - W \cdot S) \cdot F \cdot h \times 1.5$$

式中 A——孔隙度,%;
　　　S——容重,g/cm³;
　　　W——土壤含水率,%;
　　　F——计算地块面积,m²;
　　　h——测定深度,m;
　　　1.5——加大系数。

Where A = porosity, %
　　　　S = bulk density, g/cm³
　　　　W = soil moisture content, %
　　　　F = calculated plot area, m²
　　　　h = measured depth, m
　　　　1.5 = increased coefficient.

计算灌水量为 0.301 m³,7 月 4 日 8:20 开始灌水(一切操作按有关规定进行),直到 23:20 将 0.301 m³ 水全部加入围墙中,7 月 5 日 3:17 灌入围墙的水全部渗完,然后用塑料布、毡子及大量的草(厚 50 cm)将其盖好。按常规要求砂土和壤土在灌水后 24 h 便取土样,我们观察发现,戈壁地土壤的渗透性非常大,为可靠起见我们在 5 日 10:20 开始取土测定,每天早晚各取一次,直取到 8 日。通过 4 d 的取土测定我们发现 6 日、7 日、8 日这 3 d 测得的各层含水率之差稳定在 1% 之内。其结果如下。

It is calculated that the irrigation amount is 0.301 m³. Under the relevant regulations, on July 4th, we began to irrigate the plots at 8:20 a.m. (operate as per relevant regulations) and until 11:20 p.m., we finished the amount of 0.301 m³ into the enclosure. Then at 3:17 a.m. of July 5th, we covered the pit with plastic cloth, felt blanket and lots of grass (with a thickness of 50 cm) after all the water permeated out of the enclosure. In accordance with the conventional requirements, soil samples were taken from sandy soil and loamy soil 24 hours after irrigation. It is observed that the soil permeability of Gobi was very large. For the sake of data reliability, at 10:20 a.m. on the 5th day, we began to take a soil measurement each morning and evening until the 8th day. During the 4-day soil measurement, it was found that the difference of moisture content in each layer was within 1% on the 6th, 7th and 8th day. The results were as follows.

第一层(0~20 cm)3 天的平均含水率为 12.837%;容重 1.478;

First layer (0–20 cm) 3-day average moisture content: 12.837%; bulk density: 1.478;

第二层(20~40 cm)3 天的平均含水率为 8.661%;容重 1.555;

Second layer (20–40 cm) 3-day average moisture content: 8.661%; bulk density:

第一章 综合研究
1 Comprehensive Study

1.555;

第三层(40~60 cm)3 天的平均含水率为4.429%,容重1.679。
Third layer (40–60 cm) 3-day average moisture content: 4.429%; bulk density: 1.679;

用加权平均计算0~60 cm的田间持水率:
The field capacity of 0–60 cm layer calculated by weighted average:

$$田间持水率(\%) = \frac{(12.837 \times 1.478 + 8.661 \times 1.555 + 4.429 \times 1.679 \times 20)}{(1.478 + 1.555 + 1.679) \times 20} = 8.463 (占干体重)$$

$$\text{Calculation of field capacity}(\%) = \frac{(12.837 \times 1.478 + 8.661 \times 1.555 + 4.429 \times 1.679 \times 20)}{(1.478 + 1.555 + 1.679) \times 20}$$
$$= 8.463 (\text{the percentage in water-free soil})$$

0~40 cm的田间持水率为10.696。
The field capacity of 0–40cm layer was calculated as 10.696.

另一点的试验结果如下:
The results of another pit were as follows:

0~20 cm:3 天平均含水率为10.969%,容重1.597;
0–20 cm: 3-day average moisture content: 10.969%; bulk density: 1.597;

20~40 cm:3 天平均含水率为12.686%,容重1.656;
20–40 cm: 3-day average moisture content: 12.686%; bulk density: 1.656;

40~60 cm:3 天平均含水率为10.07%,容重1.680。
40–60 cm: 3-day average moisture content: 10.07%; bulk density: 1.680;

用加权平均计算:
Calculated by weighted average:

0~60 cm 田间持水率为11.239%;
0–60 cm: 11.239%;

0~40 cm 田间持水率变化在8.7%~12.8%之间。
0–40 cm: varied from 8.7% to 12.8%.

试验小区戈壁地的土质结构较复杂,0~40 cm 为沙壤土,40~60 cm 为细砂,60~80 cm 基本都是砂卵石。而表层的沙壤土,由于土质结构太差,其田间持水率比一般的沙壤土要低的多。

The soil structure of Gobi in the experimental plots is relatively complicated, with sandy loam in 0-40 cm, fine sand in 40-60 cm, and sand and gravel in 60-80 cm. Due to poor soil structure, the field capacity of sandy loam in Gobi is much lower than that of the common sandy loam.

(二) 田间持水率测试结果分析
1.3.1.2 Analysis of Field Capacity Testing Results

根据实测结果表明,荒漠戈壁地土层在 0~40 cm 厚的田间持水率为 8.7%~12.8%,40~60 cm 的田间持水率为 4.4%~10.0%,60~80 cm 的田间持水率为 2%~4%(见表 10)。由此可见,荒漠戈壁地表层沙壤土的田间持水率偏低。

Results showed that the field capacity of desert Gobi is 8.7%-12.8% at the depth of the soil layer of 0-40 cm, 4.4%-10.0% in 40-60 cm, and around 2%-4% in 60-80cm (see Table 10). It can be seen that the field capacity of sandy loam in desert Gobi is relatively lower.

表 10　　　　　　　　试验小区田间持水率测试成果表

土层深度 (cm)	0~40	40~60	60~80
田间持水率 (%)	8.7~12.8	4.4~10.0	2~4

Table 10　　　　Testing Results of Experimental Plots' Field Capacity

Depth of Soil Layer (cm)	0-40	40-60	60-80
Field Capacity (%)	8.7-12.8	4.4-10.0	2-4

它较一般地区沙壤土的田间持水率(17%~30%)约低 20 个百分点(见表 11)。田间持水率越低,灌水定额应该越小,否则会产生深层渗漏。由于田间持水率的差别很大,造成戈壁地的灌溉与一般沙壤土的灌溉技术有很大差别。因此,就阿勒泰地区荒漠戈壁地的灌溉制度及灌水方法而言,应与一般地区的沙壤土有所不同,不能照办,应结合实际情况,因地而宜,制定相适应的灌溉制度和灌水方法。

The field capacity (17%-30%) of desert Gobi is 20% lower than that of sandy loam in other sites. The lower the field capacity is, the less the irrigation quota should be; or it will

第一章 综合研究
1 Comprehensive Study

lead to leakage in deeper layers. Because of the big difference in field capacity, irrigation technology differs greatly in Gobi from that in the common sandy loam. Therefore, in terms of applying irrigation system and methods in desert Gobi in Altay Prefecture, there should be difference comparing the common sandy loam and varied measures suitable to local conditions should be adopted. No uniform system or method should be followed.

表 11　　　　　　　　　　沙壤土田间持水率对比表

资料名称	田间持水率(%)
土壤与农作物	22~30
农田水利学	17~30
水工设计手册	22~30
阿勒泰地区戈壁土测定	8~12

Table 11　　　Comparison of Sandy Loam's Field Capacity

Source of Information	Field Capacity (%)
Soil and Crop	22-30
Irrigation and Drainage	17-30
Water Conservancy Project Design Handbook	22-30
Gobi Soil Measurement in Atlay Prefecture	8-12

二、作物根系分布与计划湿润层深度
1.3.2 Crop Root Distribution and Depth of Planned Wetting Layer

土壤计划湿润层深度是指在旱田进行灌溉时,计划调节控制土壤水分状况的土层深度。它随着作物根系活动层深度、土壤性质等因素而变。荒漠戈壁地由于土质特殊,土壤结构差,因而作物的计划湿润层深度不能按一般土壤的计划湿润层来决定。为了准确地掌握阿勒泰地区广大戈壁地作物的计划湿润层深度,我们对玉米、小麦等作物的根系分布情况进行了较为详尽的调查和测试。我们发现玉米的根系分布情况为:0~20 cm 土层深度中,根系重量占总重量的57%,20~60 cm 土层深度中,根系重量占总重量的36.3%,在60~80 cm 土层深度中,根系重量则仅占总重量的6.7%;小麦根系的分布情况为:0~20 cm 土层深度中,根系重量占总重量的9.3%,20~60 cm 深度,根系重量占总重量的86.3%;60~80 cm 深度土层中,根系重量只占总重量的4.4%(见表12)。测试结果表明,荒漠戈壁地作物根系约95%分布在0~60 cm 土层深度中,而60 cm 以下只占5%。它表明在本区戈壁地的土质状况下,作物计划湿润层不能采用通常一般地区的数值,即不能采用60~80 cm 和80~100 cm 深度值,而应视具体情况而定,采用40~60 cm 为宜。

The depth of planned wetting layer of soil refers to the depth in soil layer that is planned to adjust and control the moisture in soil when irrigating the arid field. It varies with factors such as the depth of root active zone, soil property, etc. Because of the uncommon soil property and poor soil structure in desert Gobi, the depth of crops' planned wetting layer cannot be decided by that of the common soil. In order to accurately know the depth of crops' planned wetting layer in Altay Prefecture, the book conducted a very comprehensive investigation and testing on the root distribution of crops like corn and wheat. It was found that the root distribution of the corn's root weight accounted for 57%, 36.3%, 6.7% of the total weight in the depth of 0-20 cm, 20-60 cm and 60-80cm respectively while that of the wheat's root weight accounted for 9.3%, 86.3%, 4.4% of the total weight in the depth of 0-20 cm, 20-60 cm, 60-80 cm respectively (see Table 12). Results showed that about 95% of the desert Gobi crop's root distributes in the depth of 0-60 cm while that number only takes up 5% below 60cm. Also, under the Gobi soil property in Altay Prefecture, the crops' planned wetting layer cannot be applied with the same value used in common areas; that is to say, the depth values of 60-80 cm and 80-100 cm are not applicable and that of 40-60 cm can be adopted accordingly.

表 12　　　　　　　　　　戈壁地主要作物根系分布情况表　　　　　　　　　　单位:%

作物	各层根重占 0~80 cm 土层内根重量的百分比		
	0~20 cm	20~60 cm	60~80 cm
玉米	57.00	36.30	6.75
春小麦	9.30	86.30	4.40

Table 12　　　　　　　**Root Distribution of Major Crops in Gobi**　　　　　　　Unit: %

Crop	Root Weight Percentage in Soil Layers (0-80 cm)		
	0-20 cm	20-60 cm	60-80 cm
Corn	57.00	36.30	6.75
Spring Wheat	9.30	86.30	4.40

三、土壤渗透系数
1.3.3　Coefficient of Soil Permeability

土壤渗透性与土壤质地、结构等土壤性状有关。土壤渗透性能的强弱与田间畦沟的规格和配水量的大小等密切联系着。

Soil permeability depends on soil properties such as soil texture and structure. It is also closely related to such factors as the size of furrow and the quantity of water distributed.

(一) 土壤渗吸速度(K_0)
1.3.3.1 Soil Infiltration Rate (K_0)

本区戈壁地不同深度的土壤渗吸速度有很大差别。根据我们实测结果戈壁土 0~40 cm 的砂壤土渗吸速度为 5.5 mm/min,40~60 cm 为砂砾层,渗吸速度为 6.5 mm/min,为计算方便将 0~60 cm 渗吸速度换算为一个统一值,经加权计算,4 种不同深度的计划湿润层 (0~40 cm、40~60 cm、60~70 cm 以及 70~80 cm)的土壤渗吸速度见表 13。

The soil infiltration rate at different depths varies greatly in the desert Gobi of the Prefecture. Results show that the soil infiltration rate of sandy loam at 0-40 cm is 5.5mm/min and that of the sandy gravel at 40-60 cm is 6.5 mm/min. For the convenience of calculating, the Book adopted a uniform value to calculate the soil infiltration rate at 0-60 cm and by weighted calculation, the soil infiltration rate of soil wetting layer at four different depths (0-40 cm, 40-60 cm, 60-70 cm and 70-80 cm) is shown as follows (see Table 13).

表 13　　　　　　　　　不同计划湿润层土壤渗吸速度表

土壤计划湿润层深(cm)	0~40	40~60	60~70	70~80
渗吸速度 K_0(mm/min)	5.5	6.5	6.1	6.2

Table 13　　　　Soil infiltration rate in Different Soil Wetting Layers

Depth of Soil Wetting Layer (cm)	0-40	40-60	60-70	70-80
Soil infiltration rate K_0(mm/min)	5.5	6.5	6.1	6.2

(二) 渗吸速度减小指数(α)
1.3.3.2 Soil Infiltration Rate Derease Index (α)

土壤渗吸速度减小指数一般为 0.3~0.8,并随土质黏重和灌溉前土壤含水率的减小而增大。试验区戈壁土 0~40 cm 为砂壤土,40 cm 以下为砂砾层,土壤结构较差,加之该地区是干旱区,一般在灌溉前土壤的含水率较低,经与有关专家讨论研究渗吸速度减小指数取 0.42(见表 14)。

Soil infiltration rate derease index is generally 0.3-0.8 and this value often increases with the decrease of the soil cohesion and the soil water content before irrigating. With poor structure, the Gobi soil in the Plots is composed of sandy loam at 0-40 cm and sandy gravel below 40 cm. In addition, the Prefecture belongs to arid areas with relatively low soil water content before irrigating. Considering the above reasons, experts discussed and agreed to use the value of 0.42 as the soil infiltration rate derease index (see Table 14).

表14　不同计划湿润层深度土壤渗吸速度减小指数

土壤计划湿润层深(cm)	0~40	40~60	60~70	70~80
渗吸速度减小指数 $K_0(\alpha)$	0.45	0.42	0.41	0.4

Table 14　Soil Infiltration Rate Derease Index in Different Soil Wetting Layers

Depth of Soil Wetting Layer (cm)	0-40	40-60	60-70	70-80
Soil infiltration rate decrease index $K_0(\alpha)$	0.45	0.42	0.41	0.4

四、土壤物理化学性质
1.3.4　Physical and Chemical Properties of Soil

(一) 土壤容重
1.3.4.1　Soil Bulk Density

土壤容重是指土壤在自然结构状态下单位体积的干土重量。通常情况下,如土壤容重小,表明土体坚实、孔隙小,土壤的透水性、通气性不良,保水性能差。土壤容重变动范围一般较大,通常与土壤有机质含量的多少有关。一般情况下,贫瘠紧实的土壤,其容重可达1.8 g/cm³,紧密未熟化的心土、底土,容重变化在1.3~1.5 g/cm³之间,而较肥沃的耕作层,土壤容重在1 g/cm³左右。本试验小区田间土壤容重的计算是按下式进行计算的:

Soil bulk density refers to the weight of water-free soil per unit volume in the state of natural structure. Generally, if soil bulk density is small, it indicates that the soil mass is solid with small pores, which leads to poor water permeability, poor ventilation and poor performance of water retention. With larger variation range, the soil bulk density usually depends on the soil organic matter content. Generally, the bulk density of poor but compact soil can be up to 1.8 g/cm³, and that of the compact yet unmellowed subsoil and substratum is 1.3–1.5 g/cm³, and that of the fertile arable layer is about 1 g/cm³. The soil bulk density in the field of the Plots is calculated according to the following formula:

$$S = \frac{100 M}{(100 + \bar{W})V}$$

式中　S——烘干土壤容重;
　　　V——环刀体积,100 cm³;
　　　M——环刀筒内湿土重,g;
　　　$\bar{W}$——自然含水率,%。

Where　S = the bulk density of dried soil
　　　　V = cutting ring volume, 100 cm³

M = the weight of wet soil in the cutting ring, g

$\overline{W}$ = natural moisture content, %.

由计算测试结果可知,田间土壤容重为 1.445~1.518 g/cm³(见表 15),表明试验小区的田间土壤为紧密未熟化的心土、底土类。

Results show that the soil bulk density of field is 1.445 - 1.518 g/cm³ (see Table 15), indicating that the field soil was compact yet unmellowed subsoil and substratum.

表 15   试验小区田间土壤容重测试成果表

剖面深度	0~60 cm	0~80 cm
试验区平均值	1.618	1.653

Table 15   Testing Results of the Experimental Plots' Field Soil Bulk Density

Section Depth	0-60 cm	0-80 cm
Average Value of the Plots	1.618	1.653

(二) 土壤比重
1.3.4.2 Soil Specific Gravity

土壤比重是指土壤颗粒重量(包括土壤中有机质、矿物质)与同体积水(4 ℃)重量的比值。测定土壤比重,有助于了解土壤的有机质含量、矿物组成以及成土母质特性。通常情况下,耕作土壤的平均比重在 2.4~2.7 之间。

Soil specific gravity refers to the ratio: the weight of soil particle (including organic matter and minerals in the soil) to the weight of water (4℃) with the same volume. The measurement of soil specific gravity will help us to understand the organic matter and minerals in the soil and the characteristics of soil parent material. Generally, the average specific gravity of cultivated soil is about 2.4-2.7.

试验小区田间土壤比重的测试,在经过样品处理后,按下式进行计算:

The measurement of the soil bulk density in the field of the experimental plots, after sampling treatment, is calculated as:

$$d_s = \frac{g}{g + g_1 + g_2} d_{wt}$$

式中 d_s——土粒比重;

d_{wt}——t℃时蒸馏比重;

g——烘干样品重;

g_1——t℃时比重瓶+水重;

g_2——t℃时比重瓶+水+样品重。

Where　d_s = the specific gravity of soil particles

　　　d_{wt} = the specific gravity of distillation

　　　g = the weight of dried soil samples

　　　g_1 = the weight of pycnometer plus water at t℃

　　　g_2 = the weight of pycnometer plus water plus samples at t℃.

计算结果表明,本试验水区田间土壤比重在 2.64~2.69 之间(见表 16),接近一般状况下的耕作土壤比重。

Results show that the soil bulk density in the field of the experimental plots is 2.64–2.69 (see Table 16), which is approximately to the soil bulk density of cultivated land under common situation.

表 16　　　　　　　　　试验小区田间土壤比重测试计算成果表

取样点	玉米小区			小麦小区			测坑		
深度(cm)	0~30	30~60	60 以下	0~30	30~60	60 以下	0~30	30~60	60 以下
比重	2.65	2.64	2.67	2.68	2.65	2.68	2.65	2.66	2.69

Table 16　Testing Results of the Experimental Plots' Field Soil Specific Gravity

Sampling Site	Corn Plot			Wheat Plot			Test Pit		
Depth (cm)	0-30	30-60	< 60	0-30	30-60	< 60	0-30	30-60	<60
Specific Gravity	2.65	2.64	2.67	2.68	2.65	2.68	2.65	2.66	2.69

(三) 土壤有机质

1.3.4.3　Soil Organic Matter

土壤有机质含量是土壤肥力高低的重要标志。它对土壤的保土、保肥、耕性以及土壤温度和通气状况有直接的影响。

The content of soil organic matter is one of the important indicators of soil fertility. It has a direct impact on soil conservation, fertilizer conservation, tillage, soil temperature and ventilation.

第一章　综合研究

1　Comprehensive Study

本试验小区田间土壤有机质的测定是在通过对样品处理后,采用下式进行计算的:

The measurement of the soil organic matter in the field of the Plots, after sampling treatment, is calculated as:

$$\text{有机质}(\%) = \frac{\frac{0.8 \times 5}{V_0} \times (V_0 - V) \times 0.003 \times 1.724 \times 1.1}{\text{样品重}} \times 100$$

$$\text{Organic matter}(\%) = \frac{\frac{0.8 \times 5}{V_0} \times (V_0 - V) \times 0.003 \times 1.724 \times 1.1}{\text{Weight}} \times 100$$

式中　V_0——5 mL 0.8 N 标准重铬酸钾空白滴定时用去的硫酸亚铁毫升数;

V——滴定待测液中过剩的 0.8 N 标准重铬酸钾所用去的硫酸亚铁毫升数;

0.003——1 毫克当量碳的克数;

1.724——由土壤有机碳换算成有机质的经验常数;

1.1——校正常数;

100——换算成百分数。

Where　V_0 = the amount of ferrous sulfate used for blank titration experiment of potassium dichromate (5 mL, 0.8 N standard)

V = the amount of ferrous sulfate of the superfluous potassium dichromate (0.8N standard) in the fluid under test

0.003 = the grammage of carbon per mEq

1.724 = the empirical constant of converting soil organic carbon into organic matter

1.1 = calibration constant

100 = percentage conversion.

我们在试验区进行测坑取样,同时还对玉米小区、小麦小区进行了 2 个不同深度(0~20 cm,20~40 cm)的取样分析,结果表明,本区田间有机质含量偏低,变化在 0.25%~0.49%之间,全氮为 0.013%~0.047%,碱解氮为 7~27 mg/kg,土壤肥力的详细分析结果见表 17。

We conducted test pit sampling in the experimental plots, and made sample analysis of the soil at different depths (0-20 cm, 20-40 cm) in the Corn Plot and the Wheat Plot. Results show that lower content of field organic matter was found in the Plots with a variation of 0.25%-0.49%, the content of total nitrogen is 0.013%-0.047%, and the content of available nitrogen is 7-27 mg/kg. Detailed analysis of the soil fertility will be elaborated in this book (see Table 17).

表17　试验区田间土壤肥力分析表

取样点	玉米小区		小麦小区		测坑	
深度(cm)	0~20	20~40	0~20	20~40	0~20	20~40
全氮(%)	0.030	0.026	0.026	0.047	0.029	0.013
碱解氮(mg/kg)	27	16	16	1.2	13	7
速效磷(mg/kg)	4	1	5	Trace	5	Trace
速效钾(mg/kg)	187	202	212	217	288	313
有机质(%)	0.46	0.34	0.38	0.25	0.49	0.34

Table 17　Analysis of the Experimental Plots' Field Soil Fertility

Sampling Site	Corn Plot		Wheat Plot		Test Pit	
Depth (cm)	0-20	20-40	0-20	20-40	0-20	20-40
Total Nitrogen (%)	0.030	0.026	0.026	0.047	0.029	0.013
Alkali Nitrogen (mg/kg)	27	16	16	1.2	13	7
Available Phosphorus (mg/kg)	4	1	5	Trace	5	Trace
Available Potassium (mg/kg)	187	202	212	217	288	313
Organic Matter (%)	0.46	0.34	0.38	0.25	0.49	0.34

(四) 土壤盐分
1.3.4.4　Soil Salinity

土壤盐分对作物生长的影响主要取决于土壤盐分的含量及组成。当土壤中所含可溶盐分达到一定数量以后,会对作物的发芽和生长产生直接影响。

The effect of soil salinity on crop growth mainly depends on the content and component of soil salinity. When the soil contains a certain amount of soluble salt, it will directly affect the sprouting and growth of crops.

本试验小区田间土壤盐分的测定是采用干渣法(重量法)进行测定的。样品处理后,所采用的计算公式如下:

Residue technique (gravimetric analysis) was adopted to measure the soil salinity in the field of the Experimental Plots. After sampling treatment, the formula is as follows:

$$\text{烘干残渣}(\%) \atop (\text{盐分总量}) = \frac{\text{称皿与残渣重} - \text{称皿重}}{W} \times 100$$

$$\text{Dried residual}(\%) \atop (\text{Total weight of salt}) = \frac{\text{he weight of weighing dish and residual} - \text{weight of weighing dish}}{W} \times 100$$

式中 W——吸取待测液体积相当于土壤样品重。

Where W–the volume of fluid under test is equivalent to the weight of soil samples.

计算结果表明,试验小区田间土壤全盐为 0.128~0.087 g/g 土(见表 18)。

Results show that the total salt in the field of the experimental plots is 0.128-0.087 g per 1 g soil (see Table 18).

表 18　　　　　　　试验小区田间土壤盐分测试成果表

取样点	玉米小区			小麦小区			测坑		
深度(cm)	0~30	30~60	>60	0~30	30~60	>60	0~30	30~60	>60
全盐(g%) 水土比(5∶11)	1.012	0.776	1.030	0.964	1.697	1.353	1.310	0.796	0.661

Table 18　　Testing Results of the Experimental Plots' Field Soil Salinity

Sampling Site	Corn Plot			Wheat Plot			Test Pit		
Depth (cm)	0-30	30-60	>60	0-30	30-60	>60	0-30	30-60	>60
Total Salt (g%) Water-Soil Ratio(5∶11)	1.012	0.776	1.030	0.964	1.697	1.353	1.310	0.796	0.661

(五)土壤酸碱度(pH)
1.3.4.5　Soil pH Value

本试验小区田间土壤酸碱度(pH)的测定是采用电位测定法。我们分别由测坑、玉米小区、小麦小区土壤进行了采样,采样深度分 3 个层次,0~30 cm,30~60 cm,60 cm 以下。结果表明,本试验田间土壤酸碱度(pH)的变化范围较大,pH 值在 6.86~9.23 之间,呈微碱性(见表 19)。

Potentiometry was adopted to measure the soil pH value in the field of the Plots. We conducted sampling test on the soil of the test pit, the Corn Plot, the Wheat Plot at three different depths: 0-30 cm, 30-60 cm, below 60 cm. Results show that the soil pH value in the field of the Plots varied greatly, with a pH value of 6.86-9.23, presenting slightly alkaline (see Table 19).

表19　　　　　　　　　试验区田间土壤pH值测试成果表

取样点	玉米小区			小麦小区			测坑		
深度(cm)	0~30	30~60	>60	0~30	30~60	>60	0~31	30~60	>60
pH值	8.50	8.30	7.82	8.99	7.60	6.86	9.23	8.40	9.20

Table 19　　Testing Results of Field Soil pH Values in the Experimental Plots

Sampling Site	Corn Plot			Wheat Plot			Test Pit		
Depth (cm)	0-30	30-60	>60	0-30	30-60	>60	0-31	30-60	>60
pH Values	8.50	8.30	7.82	8.99	7.60	6.86	9.23	8.40	9.20

五、灌溉水质
1.3.5 Irrigation Water Quality

田间灌溉水质与土壤的理化性质有直接关系。本试验小区田间灌溉用水主营源于乌伦古河地表水,水质较好,总盐为 0.341 mg/L,pH 为 8.2,呈微碱性。

The irrigation water quality of the field is directly related to the physical and chemical properties of soil. The main resource of irrigation water for the field in the Plots is the high-quality surface water from the Ulungur River, with a total salt of 0.341 mg/L and its pH value of 8.2, presenting slightly alkaline.

六、土壤化学成分
1.3.6 Soil Chemical Property

(一)土壤阴离子含量
1.3.6.1 Anions in Soil

试验区土壤化学成分中的阴离子含量变化较大,阴离子含量中,以 SO_4^{2-} 含量为主。其中 SO_4^{2-} 含量最低为 0.002 g/100 g 土,最高达 0.456 g/100 g 土,二者相差 200 余倍;Cl^- 含量的最高、最低值虽不及 SO_4^{2-} 含量,但变化也在 20 倍以上,而 CO_3^{2-} 和 CO_3^- 含量的最高、最低值相差不甚很大(见表 20)。

The content of anions in soil chemical composition in the Plots varies greatly, and anions are mainly composed of SO_4^{2-}. With its highest content of 0.002 g/100 g soil and the lowest content of 0.456 g/100 g soil, the difference is more than 200 times. Though the highest and lowest contents of Cl^- are less than those of SO_4^{2-}, the difference is more than 20 times. Besides, the difference of the highest and lowest content of CO_3^{2-} is small (see Table 20).

比较不同土壤深度的阴离子含量,无明显变化规律,各阴离子的含量斑驳无章。

By comparison, the content of soil anions at different depths shows no obvious change and presents disorderly.

表 20　　　　　　　　试验区田间土壤阴离子化学成分分析表

取样点		玉米小区			小麦小区			测坑		
深度(cm)		0~30	30~60	>60	0~30	30~60	>60	0~30	30~60	>60
CO_3^{2-}	me/100 g 土	0.044			0.221			0.336	0.088	0.088
	g/100 g 土	0.001			0.007			0.020	0.003	0.003
HCO_3^-	me/100 g 土	0.265	0.265	0.190	0.221	0.177	0.110	0.243	0.177	0.146
	g/100 g 土	0.016	0.016	0.012	0.013	0.011	0.007	0.015	0.011	0.009
Cl^-	me/100 g 土	0.076	0.070	0.066	0.228	0.228	0.279	1.741	1.802	0.675
	g/100 g 土	0.003	0.003	0.002	0.008	0.008	0.010	0.062	0.064	0.024
SO_4^{2-}	me/100 g 土	0.244	0.041	0.634	0.634	9.004	9.461	3.248	9.501	3.243
	g/100 g 土	0.012	0.002	0.030	0.030	0.432	0.454	0.156	0.456	0.156

Table 20　Chemical Composition Analysis of Soil Anions in the Experimental Plots

Sampling Site		Corn Plot			Wheat Plot			Test Pit		
Depth (cm)		0-30	30-60	>60	0-30	30-60	>60	0-30	30-60	>60
CO_3^{2-}	me/100 g soil	0.044			0.221			0.336	0.088	0.088
	g/100 g soil	0.001			0.007			0.020	0.003	0.003
HCO_3^-	me/100 g soil	0.265	0.265	0.190	0.221	0.177	0.110	0.243	0.177	0.146
	g/100 g soil	0.016	0.016	0.012	0.013	0.011	0.007	0.015	0.011	0.009
Cl^-	me/100 g soil	0.076	0.070	0.066	0.228	0.228	0.279	1.741	1.802	0.675
	g/100 g soil	0.003	0.003	0.002	0.008	0.008	0.010	0.062	0.064	0.024
SO_4^{2-}	me/100 g soil	0.244	0.041	0.634	0.634	9.004	9.461	3.248	9.501	3.243
	g/100 g soil	0.012	0.002	0.030	0.030	0.432	0.454	0.156	0.456	0.156

(二) 土壤阴离子含量
1.3.6.2　Cations in Soil

试验区田间土壤中阳离子含量以 Ca^{2+} 含量比例略高,其余的 Mg^{2+} 及 $K^+ + Na^+$ 含量接近。Ca^{2+} 含量最高值为 0.172 g/100 g 土,出现在小麦小区 60 cm 深度以下的样品中,次高值出现在 30~60 cm 深度的土层中,最低值为 0.005 g/100 g 土;Mg^{2+} 含量变化在 0.001~0.012 g/100 g 土之间,不同深度的土壤阳离子含量无明显变化规律;$K^+ + Na^+$ 含量在不同深度的变化过程也无明显规律可循(见表 21)。

Results show that in the Plots, the contents of cations in soil are mainly Ca^{2+}, and the contents of Mg^{2+} and $K^+ + Na^+$ are equally close to each other. The highest content of Ca^{2+} is 0.172 g/100 g soil and is found in the soil sample of the Wheat Plot at the depth of below 60cm. The second highest content of that is found at the depth of 30-60 cm and the lowest content is 0.005 g/100 g soil. The content of Mg^{2+} varies from 0.001 g to 0.012 g/100 g soil and cations indicate no obvious change at different depths of soil. The content of $K^+ + Na^+$ also indicates such kind of law (see Table 21).

表 21 试验区田间土壤阳离子化学成分分析表

取样点		玉米小区			小麦小区			测坑		
深度(cm)		0~30	30~60	60 以下 <60	0~30	30~60	60 以下 <60	0~30	30~60	60 以下 <60
Ca^{2+}	me/100 g 土	0.320	0.228	0.634	0.228	7.602	8.577	0.431	5.395	2.122
	g/100 g 土	0.006	0.005	0.013	0.005	0.152	0.172	0.009	0.108	0.043
Mg^{2+}	me/100 g 土	0.203	0.091	0.142	0.076	0.974	0.472	0.152	0.670	0.178
	g/100 g 土	0.002	0.001	0.002	0.001	0.012	0.006	0.002	0.008	0.002
$K^+ + Na^+$	me/100 g 土	0.106	0.063	0.114	1.000	0.833	0.801	4.985	5.503	1.842
Na^+	me/100 g 土	0.002	0.001	0.003	0.023	0.019	0.018	0.115	0.127	0.042

Table 21 Chemical Composition Analysis of Soil Cations in the Experimental Plots

Sampling Site		Corn Plot			Wheat Plot			Test Pit		
Depth (cm)		0-30	30-60	60 以下 <60	0-30	30-60	60 以下 <60	0-30	30-60	60 以下 <60
Ca^{2+}	me/100 g soil	0.320	0.228	0.634	0.228	7.602	8.577	0.431	5.395	2.122
	g/100 g soil	0.006	0.005	0.013	0.005	0.152	0.172	0.009	0.108	0.043
Mg^{2+}	me/100 g soil	0.203	0.091	0.142	0.076	0.974	0.472	0.152	0.670	0.178
	g/100 g soil	0.002	0.001	0.002	0.001	0.012	0.006	0.002	0.008	0.002
$K^+ + Na^+$	me/100 g soil	0.106	0.063	0.114	1.000	0.833	0.801	4.985	5.503	1.842
Na^+	me/100 g soil	0.002	0.001	0.003	0.023	0.019	0.018	0.115	0.127	0.042

第一章　综合研究
1　Comprehensive Study

第四节　作物需水量与灌溉制度试验设计
1.4 Experiment of Water Requirement and Irrigation System of Crops

一、试验项目的意义
1.4.1 Significance of the Experimental Project

新疆北部阿勒泰地区的农业生产主要是在荒漠瘠薄土地上开发和发展起来的。荒漠瘠薄土地的灌溉面积占阿勒泰地区总灌溉面积的80%，因而，选择荒漠瘠薄土地进行灌溉节水及灌水方法的试验研究对阿勒泰地区的农牧业生产发展有着重要意义。

The agricultural production in the Prefecture is developed on the arid land. As 80% of the total irrigation area in the Prefecture is on the arid land, it is of great significance for the development of agriculture and animal husbandry in the region to conduct the experimental research on irrigation water saving and irrigation methods on the arid land.

(一) 节水增地，提高灌溉水平
1.4.1.1 Water Saving, Farmland Increase and Improving the Irrigation Quality

在新疆干旱区，水是最宝贵的自然资源。阿勒泰地区虽然说是丰水地区，但由于水资源时空分布不均匀以及不能合理利用水资源，水问题仍然是影响地区农牧业生产的重要因素。如阿勒泰地区，现有灌溉面积210万亩，由于灌溉技术落后，基本都是大水漫灌，灌溉定额高达1 500 m³/亩，个别地方甚至达到2 000 m³/亩，加之灌排不配套，造成阿勒泰全地区约60万亩土地盐渍化，占全区灌溉面积的23%。而开展本项目试验研究，旨在通过对本地区主要作物不同生长发育期耗水量的观测研究，探求经济合理的节水灌溉制度，使之既可满足作物生长期的植株蒸腾和株间蒸发，又不使田间产生深层渗漏，节水灌溉，提高灌溉水平。

Water is one of the most precious natural resources in the arid area of Xinjiang. Although Altay Prefecture is a region with abundant water, water is still an important factor affecting the regional production of agriculture and animal husbandry due to the uneven spatial and temporal distribution and unreasonable use of water resources. For example, limited by backward irrigation technology, people in Altay Prefecture have to use flood irrigation (with an irrigation quota of 1 500 m³/mu and some places even reach 2 000 m³/mu) to deal with the existing irrigation area of 2.1 million mu. Additionally, the unmatching irrigation and drainage facilities have lead to soil salinization of approximately 600 000 mu land in the Prefecture, accounting for 23% of the total irrigation area. Through the experimental research of this project, this Book aims to explore a reasonable and cost-effective irrigation system through observation of water consumption during different growth and development periods of main crops in the Prefecture, so as to make it not only meet the transpiration and inter-plant evaporation in the crop growth period, but also prevent it from deep leakage in the field, thus achieving the goals of water saving and quality improvement of irrigation.

(二)科学灌水,增产增收
1.4.1.2 Scientific Irrigation to Achieve the Increase in Yield and Income

科学灌溉用水一直是农田水利管理运用和科研工作的一个重要内容。中国是一个农业大国,党中央、国家领导人及各级人民政府从来对农业生产建设给予高度重视,自50年代中期起,全国各省陆续建起灌溉试验站,根据各地不同自然条件进行灌溉试验研究。至今灌溉试验工作已开展近40年,全国建成灌溉试验站(基地)500多处,进行了大量的作物需水量、灌溉节水等方面的试验研究工作,取得了丰硕的成果。新疆的灌溉试验从20世纪80年代开始的,并且在自治区水利系统各级领导、科技干部和职工们的共同努力下,取得了很多成果,为新疆农业生产发展起到了非常积极的作用。然而,有关荒漠戈壁地灌溉节水试验则尚未开展。阿勒泰地区的农业生产主要在荒漠戈壁地上开发利用的,荒漠戈壁地的灌溉面积占阿勒泰全地区灌溉面积的80%以上。这种戈壁地与一般土地在土质结构、性状等方面有着很大差别,土壤有机质含量、田间持水率等也不同于其他土质。因而,开展荒漠戈壁瘠薄土地灌溉节水试验研究既有其独特性,又有代表性,推广意义很大。

Scientific irrigation has always been an important topic for the management and application of irrigation and water conservancy and for scientific research as well. China is such a big agricultural nation that the CPC central committee, state leaders and people's governments at all levels have always attached great importance to agricultural production and construction. Since mid-1950s, irrigation experimental stations have been built across China, so as to conduct irrigation experiments according to different natural conditions of different provinces. Up to now, irrigation experiments have been carried out for nearly 40 years. Over 500 irrigation experimental stations (bases) have been built across China, and thereafter, a large quantity of experimental researches on crop water requirement, irrigation and water saving have been carried out to deliver many achivements. With joint efforts of leaders, scientific and technological researchers and workers from the water conservancy authorities at all levels in Xinjiang, the irrigation experiments have started in the 1980s and achieved fruitful results, thus playing a very positive role for the development of agricultural production in Xinjiang. However, experiments on saving irrigation water for the arid desert Gobi have not started yet. Agricultural production in the Prefecture is mainly developed in the desert Gobi whose irrigation area accounts for more than 80% of the total irrigation area in the Prefecture. Soil texture and properties of such type of Gobi land are very different from those of the common land. Also, the soil organic matter content and field capacity of Gobi are also different from other types of soil. Therefore, it is of great significance to carry out experiments on saving irrigation water in desert Gobi.

再者，由于缺乏作物需水量及灌溉技术方面的试验成果，当地农牧民在农业生产中，认为在干旱地区只要多浇水灌溉，就会多增产，灌溉定额很高，结果非但没有达到增产增收的目的，反而造成水资源的浪费，灌区地下水位上升，土地盐渍化现象加重。本试验研究的目的还在于通过对阿勒泰地区主要农作物需水量、需水时间、灌溉次数等方面的试验对比研究，揭示土壤水分与作物生长发育及产量关系，结合戈壁瘠薄土地土质结构和性状，提出最佳作物灌溉节水方法和灌溉制度。

Moreover, due to lack of test results in terms of crop water requirement and irrigation technology, local farmers and herdsmen have a common misconception that more water for irrigation, with a higher irrigation quota, in arid areas would certainly lead to the increase of crop yields in agricultural production. However, results show that they fail to improve the crop yields and their economic income. Instead, they have wasted water resources, which has caused the rise of ground water level in the irrigation area and worsened the land salinization. By comparing factors like water requirement, irrigation duration and frequency of main crops in Altay Prefecture, the experimental research aims to explore the relationship of soil water-crop growth and development-crop yield, and put forward the optimal method and scheduling for saving irrigation water for crops.

二、试验场地
1.4.2 Experimental Plots

试验场地位于乌伦古湖东侧约 15 km 处，地理坐标大致位于东经 87°35′、北纬 47°10′，田间地表高度为海拔 505 m，其基本特性如下。

Roughly located at 87°35′E, 47°10′N, the experimental plots are 15 km away from the east side of the Ulungur Lake, with a field surface height of 505 m. The following are its basic characteristics.

（一）场地选择
1.4.2.1 Selection of Plots

灌溉试验场地具有代表性，各自然要素（土壤种类、气候状况、地质地貌条件、水文地质状况等）及环境符合所在地区的自然条件和农牧业生产情况，具有代表性。再是交通方便，紧靠公路，便于与农牧业、水利、土壤、水文地质等部门加强协作，试验成果易于推广应用。

Natural elements (soil types, climatic conditions, geological and geomorphological conditions, hydrogeological conditions, etc.) and the environment of experimental plots conform to local natural conditions and production situation of agriculture and animal husbandry, which makes the plots very typical and representative for experiments. Moreover, the Plots, close to the highway, have a convenient transporation system which can strengthen cooperation with departments and units like agriculture and animal husbandry, water conservancy, soil, hydrogeology and easily popularize and apply the experimental results.

（二）场地状况
1.4.2.2 Introduction to the Plots

试验小区有专用土地 20 亩,土壤肥力及耕作均匀一致。场地内有良好的灌溉条件,独立的灌溉系统和充足的水源,能够确保和满足试验用水的排溉需要。
The experimental plots have left 20 mu of special land as its soil fertility and cultivation are uniform. Good irrigation conditions, independent irrigation system and adequate water supply in the Plots can ensure the demand of irrigation water and meet the requirements of drainage and irrigation.

（三）试验小区设备
1.4.2.3 Facilities in the Experimental Plots

试验小区内建有办公、生活、试验室用房 365 m² 以及气象观测场,灌溉试验场地边缘与各种建筑物的距离大于其高度 5 倍之多。灌溉试验用地 20 亩,测坑试区分有底、无底共 0.8 亩,混凝土防渗渠道 350 m,水塔、水池各一个、蓄水容量 59 m³,排水渠一条,小型农具一套、发电机一台,此外,还具备各种试验所需用的设备(见图 1)。
There are 365 m² land for offices, living rooms, labs and a meteorologic station in the Plots. The distance between the edge of the irrigation experimental plot and other buildings is more than 5 times of its height. There are 20 mu left for irrigation experiment, 0.8 mu for test pits (with bottom and bottomless), 350 m for the concrete anti-seepage channel, one water tower, one water pond with water capacity of 59 m³, one drainage ditch, one set of small farm tools and one generator. In addition, the equipment needed for various experiments are also available (see Figure 1).

三、试验设计
1.4.3 Experimental Design

结合新疆北部阿勒泰地区水热状况和环境条件和荒漠戈壁地的土质性质,根据试验小区具体情况,在进行灌溉节水试验研究过程中,灌溉制度采用田测法,需水量采用坑测

第一章 综合研究
1 Comprehensive Study

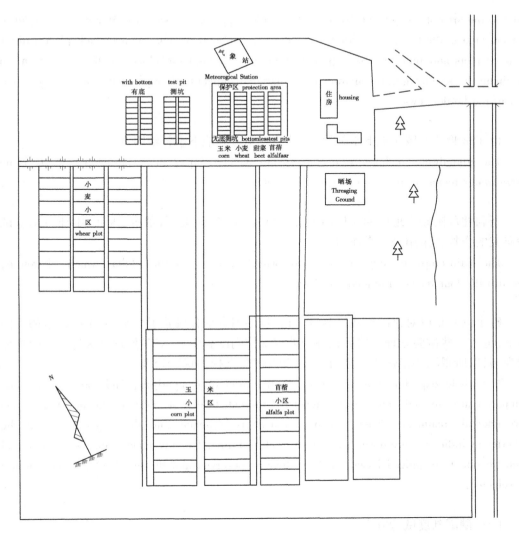

图 1 试验小区平面布置示意图
Figure 1 Floor Plan of the Experimental Plots

法。同时,为消减在观测试验中因土壤肥力差异造成的误差和土壤的差异,定每个试验小区面积为 0.12~0.15 亩,测坑面积为 0.01 亩,全部采用 3 次重复。并且,为防止旁渗,小区与小区之间、测坑与测坑之间都作了隔水处理(采用黑色塑料薄膜),还在各小区和测坑周边都设置了保护区,以消除最靠边小区与边行作物的特殊生育环境影响(边际影响)。

Considering the hydrothermal and environmental conditions and the soil properties of desert Gobi in Altay Prefecture as well as the specific conditions of the experimental plots, we decided to adopt field measurement method for measuring irrigation scheduling and soil pit method for measuring water requirement in the process of experimental research on saving for irrigation water. In addition, in order to reduce the errors and soil differences caused by soil fertility difference in the observational experiment, we determined the area of each plot as 0.12−0.15

mu and the area of test pit as 0.01 mu, all of which applied 3 repetitions. Moreover, in order to prevent side infiltration, we carried out water separation treatment (using black plastic film) between plots and pits, and set up protection zones around the plots and the pits, so as to eliminate the special growth environment (marginal impact) of crops in the most remote plot and in the border rows.

由于试验站地域无地下水补给,故试验时不考虑地下水补给因素。

The groundwater recharge is not taken into account during the experiment as there is no groundwater recharge in the area of the test station.

新疆北部阿勒泰地区的农作物主要为小麦、玉米、甜菜、苜蓿等,因而,本项目用以试验研究的作物亦采用该4类作物。

The main crops in Altay Prefecture are wheat, corn, beet and alfalfa, which, therefore, are also the four research objectives in this book.

整个试验是以对比的形式,把要进行比较的因素划分成若干个水平,不同因素的不同水平,组成一些试验处理。然后把各个处理安排到试验中去,对各处理试区进行试验指标的观测,以所观测的指标成果判断各处理的优劣,选择出优良处理。

The whole experiment is conducted in the form of comparison, and the factors to be compared are divided into several levels. Different levels of different factors form some experimental treatments. Then, each treatment is arranged into the experiment and the experiment indicators are observed in each treatment area. The advantages and disadvantages of each treatment are judged based on the indicators observed and the excellent treatment would be selected.

(一) 灌溉制度试验方案
1.4.3.1 Experimental Schemes of Irrigation Scheduling

1. 小麦试验方案设计

I Experimental scheme design for wheat

小麦试验方案设计的主导思想是,控制土壤含水率,通过正交试验,探索在荒漠戈壁地环境下,春小麦不同品种达到高产的灌溉模式。

Main idea of the experimental design for wheat is to control soil water content and explore the irrigation mode of different spring wheat varieties to achieve high yield in desert Gobi environment through orthogonal experiment.

试验因素和水平:设计在3个因素(品种、施肥量、土壤水分上下限)3个水平按正交进行对比试验。试验方案选用 $L_9(3^4)$ 正交表排列(见表22)。

Experimental factors and levels: three factors (variety, the amount of fertilizer, the upper and lower limits of soil water) are designed to be compared and tested at three orthogonal levels. $L_9(3^4)$ orthogonal table arrangement is chosen for the experimental scheme (see Table 22).

表22　　　　　　　　　春小麦灌溉制度试验方案

水平号	品种	施肥量 (kg/mu)	灌溉制度
1	1 阿春3号	1　2 000	1　50%
2	1 阿春3号	2　3 000	2　60%
3	1 阿春3号	3　4 000	3　70%
4	2 阿春2号	1　2 000	2　60%
5	2 阿春2号	2　3 000	3　70%
6	2 阿春2号	3　4 000	1　50%
7	3 阿春2号	1　2 000	3　70%
8	3 阿春2号	2　3 000	1　50%
9	3 阿春2号	3　4 000	2　60%

Table 22　Experimental Scheme of Irrigation Scheduling for Spring Wheat

Level	Varieties	Amount of Fertilizer (kg/mu)	Irrigation Scheduling
1	1 Achun #3	1　2 000	1　50%
2	1 Achun #3	2　3 000	2　60%
3	1 Achun #3	3　4 000	3　70%
4	2 Achun #2	1　2 000	2　60%
5	2 Achun #2	2　3 000	3　70%
6	2 Achun #2	3　4 000	1　50%
7	3 Achun #2	1　2 000	3　70%
8	3 Achun #2	2　3 000	1　50%
9	3 Achun #2	3　4 000	2　60%

试验方法：试验分9个处理，3个重复，共27个小区。每个小区为0.15亩，采用随机排列的方法，用量水堰量水秒表计时，采用畦灌形式，控制土壤含水率下限灌水。

Experimental methods: the experiment is composed of 9 treatments and 3 repetitions totaled 27 plots. Each plot is 0.15 mu and randomly arranged. Water stopwatch is used to measure the time on the measuring weir and the method of border irrigation is adopted to control the minimum irrigation of soil water content.

· 36 ·

2. 苜蓿试验方案设计
II Experimental scheme design for alfalfa

苜蓿灌溉制度的试验方案是采用单因子法,主要研究不同灌溉制度对苜蓿产量的影响。即在同等农业水平下,通过对不同的土壤水分上下限值的研究试验,探索出最优的灌水时间、灌水次数、灌水定额和灌溉定额。

The experimental scheme for alfalfa's irrigation scheduling adopts the method of single factor, aiming to study the effect of different irrigation scheduling on the yields of alfalfa. Under the same agricultural level, we get the results of the optimal irrigation time, irrigation frequency, irrigation water quota and irrigation quota through the study of different upper and lower limits of soil water.

试验因素和水平:设计在单因素下(土壤水分上下限),4个水平的对比试验(见表23)。

Experimental factors and levels: single factor (the upper and lower limits of soil water); contrast experiment at 4 levels (see Table 23).

表23　　　　　　　　　　　苜蓿灌溉制度试验方案

水平号	不同生育阶段土壤含水率下限(占田间持水率%)		
	幼苗期	蔓枝延长期	开花成熟期
1	65	70	65
2	45	70	65
3	65	45	45
4	65	70	40

Table 23　　　Experimental Schemes of Alfalfa Irrigation System

Level	Lower Limit of Soil Water Content at Different Growth Stages (accounting for ××% of field water-holding capacity)		
	Seedling Period	Branch Extension Period	Flowering Maturity Period
1	65	70	65
2	45	70	65
3	65	45	45
4	65	70	40

第一章　综合研究
1　Comprehensive Study

试验方法:苜蓿灌溉制度的试验分4个处理,3次重复,共12个小区,每个小区为0.12亩,形状为长方形,长19.0 m,宽4.2 m,采用顺次排列,灌溉方式采用畦灌,以控制土壤含水率。

Experimental methods: this experiment contains 4 treatments, 3 replicates and 12 rectangle experimental areas in total; 19.0 m long and 4.2 m wide, 0.12 mu of each area, distributed in sequence; border irrigation is used to control the soil water content.

3.玉米试验方案设计
Ⅲ　Corn experimental scheme design

玉米试验是采用"阿单一号"玉米品种,试验方案的设计思想是以灌溉水生产函数为主体,结合灌溉效益进行试验研究,通过一次试验,获得多项成果。

The corn of "Adan #1" is applied for this experiment. The experiment is conducted by following the irrigation water production function and combining the irrigation benefit to achieve multiple results by one time experiment.

试验因素:设计在不同的农业措施、土壤水分上下限2个因素下进行灌溉效益分摊系数试验(见表24)。

Experimental factors: design the experiment of share coefficient of irrigation benefit by taking varied agricultural measures and under the upper and lower limits of soil water content (see Table 24).

表24　玉米灌溉生产函数和分摊系数试验方案(土壤湿度)

农业措施	不同生育阶段土壤含水率下限(占田间持水率%)		
	吐丝灌浆期	拔节期	抽雄散粉期
农中1	60	65	60
农先2	60	65	60
农先3	40	65	60
农先4	60	45	60
农先5	60	65	40
农先6	40	45	40
农先7	40	45	40

Table 24　　　　Experimental Schemes of Corn Irrigation Production
Function and Sharing Coefficient (Soil Moisture)

Agricultural measures	Lower Limit of Soil Water Content at Different Growth Stages (accounting for ××% of field water-holding capacity)		
	Silking and Filling Period	Jointing Period	Tasseling and Pollinating Period
In the Middle of Agricultural Measures #1	60	65	60
Before Agricultural Measures #2	60	65	60
Before Agricultural Measures #3	40	65	60
Before Agricultural Measures #4	60	45	60
Before Agricultural Measures #5	60	65	40
Before Agricultural Measures #6	40	45	40
Before Agricultural Measures #7	40	45	40

试验方法:玉米试验区面积为 0.12 亩,种植行距为 60 cm,株距 25 cm,3 次重复,灌溉形式为畦灌。

Experimental methods: 0.12 mu of the corn experimental plot, with a row spacing of 60 cm and a plant spacing of 25 cm; repeat the experiment for 3 times; border irrigation.

(二)作物需水量试验方案

1.4.3.2　Experimental Schemes of Water Requirement of Crops

本项目在试验小区进行作物需水量试验的有小麦、苜蓿、玉米、甜菜 4 种作物。试验是在同一施肥水平和农业措施条件下,按作物生育阶段,设置 3 个土壤水分上下限,作对比试验。

4 kinds of crops, including wheat, alfalfa, corn and beet, are planted in the experimental plot for the crop water requirement experiment. As a contrast to each other, the experiments are conducted under the same fertilization levels and agricultural measures and by setting 3 upper and lower limits of soil water according to the crop growth stages.

1.小麦需水量方案设计

Ⅰ.　Experimental scheme design of wheat water requirement

小麦试验采用坑测法,共用无底测坑 9 个,测坑长 3.33 m,宽 2 m,种植面积 6.67 m²,坑深 1.7 m。春小麦的品种为新春二号,亩播种量为 22 kg,即每坑播 220 g,每米内下种 120 粒,用人工播种。

Soil pit method is used for the experiment: 9 bottomless pits in total, 3.33 m long, 2 m wide and 1.7 m deep; 6.67 m² of the plant area. The wheat "Xinchun No. 2" is used as the experimental object; a sowing rate of 22 kg per mu, namely 220 g of each pit and 120 seeds per meter; sowed by hand.

测坑试验是在同一施肥水平和农业措施条件下,按作物的生育阶段设置3个土壤水分下限作为对比试验。为减小误差,设置3个重复,共9个处理。设计方案见表25。

As a contrast to each other, the soil pit experiments are conducted under the same fertilization levels and agricultural measures and by setting 3 lower limits of soil water according to the crop growth stages. 3 replicates and 9 treatments are arranged so as to reduce error. See Table 25 for design scheme.

表25　　春小麦需水量试验设计方案

水平号	不同生育阶段土壤湿度下限(占田间持水率%)		
	分蘖—拔节期	孕穗—抽穗期	乳熟期
1	65	70	65
2	55	60	55
3	45	50	45

Table 25　Experimental Design Scheme of Spring Wheat Water Requirement

Level	Lower Limit of Soil Water Content in Different Growth Stages (accounting for xx% of the field water-holding capacity)		
	Tillering-Jointing Period	Booting-Heading Period	Milk Ripeness Period
1	65	70	65
2	55	60	55
3	45	50	45

土壤含水率的测定用取土烘干法,每5 d取土一次,取土深度1 m。次日根据土壤含水率灌水,灌水采用水表量水。灌水量和需水量均采用水量平衡法计算。

Soil drying method is used to determine the soil water content, collecting soil every 5-days from a depth of 1 m, according to the soil water content to conduct irrigation the next day under the monitor of water meter. The irrigation requirement and water requirement are both calculated by water balance method.

2.苜蓿需水量方案设计

Ⅱ　Scheme design of alfalfa water requirement

苜蓿的试验采用坑测法。共用无底测坑9个,测坑长3.33 m,宽2 m,坑深1.7 m。种植面积为 6.67 m²。苜蓿需水量试验方案见表 26。

Soil pit method is used for the experiment. 9 bottomless pits in total; 3.33 m long, 2 m wide and 1.7 m deep, 6.67 m² of the plant area. See Table 26 for the experimental scheme of alfalfa water requirement.

表 26　　　　　　　　　　　苜蓿需水量试验方案

水平号	不同生育阶段土壤湿度下限(占田间持水率%)		
	幼苗期	蔓枝延长期	开花成熟期
1	65	70	65
2	55	60	55
3	45	50	45

Table 26　　Experimental Scheme of Alfalfa Water Requirement

Leve	Lower Limit of Soil Water Content in Different Growth Stages (accounting for xx% of the field water-holding capacity)		
	Seedling Period	Branch Extension Period	Flowering Maturity Period
1	65	70	65
2	55	60	55
3	45	50	45

3. 玉米需水量方案设计

Ⅲ　Scheme design of corn water requirement

玉米需水量的试验采用坑测法,测坑南北走向,为无底测坑,长 3.33 m,宽 2.0 m,面积为 6.67 m²。并且采用防渗膜四周防渗。本项试验设 3 个处理,3 个重复,一个对照区,顺序排列。

Soil pit method is used for the experiment, distributed along the north-south direction; bottomless; 3.33 m long, 2.0 m wide and 6.67 m² of the area. Anti-infiltration membrane is applied to cover the corners to prevent infiltration. 3 treatments, 3 replicates and 1 contrast area are arranged in this experiment; distributed in sequence.

玉米试验品种采用阿单一号,播种量为 3 kg/亩,人播布种,播种间距为 0.25 m×0.60 m。一次性底肥,磷酸二铵($(NH_4)_2PO_4$)10 kg/亩,在生育期中耕施尿素 2 次,施肥量共计 20 kg/亩。

Corn "Adan #1" is employed for the experiment, with a sowing rate of 3 kg/mu and a sow spacing of 0.25 m×0.60 m; sowed by hand. Disposable diammonium phosphate ($(NH_4)_2PO_4$)

10 kg/mu is used for the base fertilizer; apply the urea (CH_4N_2O) two times while cultivating during growth and development period, with an application amount of 20 kg/mu.

灌水方法采用沟灌,用水表控制灌水量,以提高灌水的准确性。

Furrow irrigation is employed for irrigation and water meter is used to monitor the irrigation amount to improve the precision of irrigation.

玉米需水量试验方案设计见表27。

See Table 27 for experimental scheme of corn water requirement.

表27　　　　　　　　　玉米需水量试验方案

水平号	不同生育阶段土壤含水率下限(占田间持水率%)		
	拔节期	抽雄散粉期	吐丝灌浆期
1	60	65	60
2	50	55	50
3	40	45	40

Table 27　　Experimental Scheme of Corn Water Requirement

Level	Lower Limit of Soil Water Content in Different Growth Stages (accounting for xx% of the field water-holding capacity)		
	Jointing Period	Tasseling and Pollinating Period	Silking and Filling Period
1	60	65	60
2	50	55	50
3	40	45	40

4.甜菜需水量方案设计

Ⅳ Scheme design of beet water requirement

甜菜需水量试验拟采用坑测法。共用9个测坑。坑深1.7 m,测坑长3.3 m,宽2.0 m,面积6.60 m²。测坑四周用防渗薄膜防止水的旁渗。

Soil pit method is proposed to be used for beet water requirement experiment with 9 pits in total, 1.7 m deep, 3.3 m long, 2.0 m wide and an area of 6.60 m². Anti-infiltration membrane is applied around the pit to prevent side infiltration.

甜菜需水量试验处理设计,是在同一施肥水平和农业措施条件下,按作物生育阶段,设置3个土壤水分上下限,3个处理,3次重复,作对比试验。用土壤含水率控制处理水平,采用取土法测含水率控制灌水量。

As a contrast to each other, the beet water requirement experiments are conducted under the same fertilization levels and agricultural measures and by setting 3 upper and lower limits of soil water, 3 treatments and 3 replicates according to the crop growth stages. The treatment level is controlled based on the soil water content, and the irrigation amount is based on the water content tested by soil sampling。

甜菜品种选用新甜二号,5月25日播种,播种量为1 kg/亩,即每坑棵数66棵,每亩6 600棵。

The beet "Xintian #2" is used for the experiment; sowing starts on May 25; 1 kg/mu, namely 66 seeds of each pit and 6 600 of each mu.

采用磷酸二铵(($NH_4)_2PO_4$)播种,施肥量为20 kg/亩,中耕2次,施尿素1次,施量为20 kg/亩。

$(NH_4)_2PO_4$ is used for seeding, 20 kg/mu; cultivate 2 times; apply CH_4N_2O for one time, 20 kg/mu.

灌水方法采用洼灌。甜菜试验处理及方案设计见表28。

Hollow irrigation is used for irrigation. See Table 28 for beet experiment treatment and scheme design.

表28　甜菜需水量试验方案

水平号	不同生育阶段土壤湿度下限(占田间持水率%)		
	前期	中期	后期
1	60	70	65
2	50	60	55
3	40	50	45

Table 28　Experimental Scheme of Beet Water Requirement

Level	Lower Limit of Soil Water Content in Different Growth Stages (accounting for xx% of the field water-holding capacity)		
	Prophase	Metaphase	Anaphase
1	60	70	65
2	50	60	55
3	40	50	45

第五节 作物需水量与需水规律试验研究
1.5 Experimental Study on Water Requirement of Crops and Its Law

作物需水量的大小及需水变化规律与天气气候条件、土壤性质以及农业技术措施等有关。而这些因素对作物生育期内不同阶段的需水量变化的影响既错综复杂,又相互联系。通过1989~1991年3年对春小麦、苜蓿、玉米及甜菜等作物进行不同水平的试验研究,现将上述4类作物生育期内的需水量及不同阶段的需水变化规律归纳如下:

The water requirement of crops and its law are related to the weather, soil property and agricultural measures, etc. The influences of those factors on the water requirement at different stages during the growth and development period are complex and interconnected. According to the experimental analysis of spring wheat, alfalfa, corn and beet in different levels from 1989 to 1991, the water requirements over the growth and development period and the change law of the above crops at different stages are summarized as follows:

一、春小麦需水量及需水变化规律
1.5.1 Water Requirement and Its Changing Law of Spring Wheat

(一)春小麦各生育阶段的需水量变化
1.5.1.1 Water Requirement Changes of Spring Wheat at Each Growth Stage

春小麦生育期内各生育阶段的需水量变化有明显规律可循。由表29可见,春小麦在其生育期内的阶段需水量和日需水量变化很大,而在小麦的拔节—孕穗和抽穗—灌浆期,阶段需水量最大,分别高达77.1 mm 和71.3 mm,这两个阶段的需水量之和约占小麦生育期需水总量的34%左右,而这两个阶段的生育期仅历时19 d,占小麦全生育期的18%,其中抽穗—灌浆期的日需水量达14.3 mm。

There is distinct change law of water requirement to follow at each growth stage of spring wheat. As Table 29 shows, the stage water requirement and daily water requirement are very enormous over its growth and development period, peaking over jointing-booting period and heading-filling period, at 77.1 mm and 71.3 mm respectively. Total water requirements of the two periods account for about 34% of total water requirement over the growth and development period, while the two stages only last 19 days, 18% of the total period. The daily water requirement over heading-filling period reaches 14.3 mm.

通过3年(1989—1991年)的试验观测我们发现,在一般水文气象年份,6月下旬是

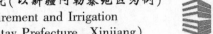

春小麦的抽穗—灌浆期,该阶段为小麦需水高峰期,也是需水的关键时期,对水分的敏感性特别强,可作为本地区小麦需水的临界期。该期的阶段需水量为 71.3 mm,日需水量达 14.3 mm。

According to the experimental observations from 1989 to 1991, conclusions are made that in general hydro-meteorological seasons, late-June is the heading-filling period that a large amount of water is required and that is a critical period for water, being highly sensitive to water. Given such circumstances, this period can be considered as the critical period of local wheat water requirement. The water requirement is 71.3 mm and the daily water requirement is 14.3 mm in this period.

表29 春小麦各生育阶段的需水量分析表

生育期	起止日期	天数(d)	阶段需水量(mm)	日需水量(mm/日)	阶段需水模系数(%)
播种—出苗	3/4~24/4	22	28.0	1.3	6.46
出苗—分蘖	25/4~12/5	18	62.1	3.5	14.38
分蘖—拔节	13/5~28/5	17	41.2	2.4	9.50
拔节—孕穗	30/5~12/6	14	77.1	5.5	17.78
孕穗—抽穗	13/6~19/6	7	52.0	7.4	12.0
抽穗—灌浆	20/6~24/6	5	71.3	14.3	16.43
灌浆—成熟	25/6~19/7	25	102.2	4.1	2.51
合计	3/4~19/7	108	433.7	4.0	100

Table 29 Analytical Table for Water Requirement of Spring Wheat at Each Growth Stage

Growth and Development Period	Start and End Date	Number of days (Days)	Stage Water Requirement (mm)	Daily Water Requirement (mm/day)	Water Requirement Modulus of Stage (%)
Sowing-Seedling	3/4~24/4	22	28.0	1.3	6.46
Seedling-Tillering	25/4~12/5	18	62.1	3.5	14.38
Tillering-Jointing	13/5~28/5	17	41.2	2.4	9.50
Jointing-Booting	30/5~12/6	14	77.1	5.5	17.78
Booting-Heading	13/6~19/6	7	52.0	7.4	12.0
Heading-Filling	20/6~24/6	5	71.3	14.3	16.43
Filing-Maturing	25/6~19/7	25	102.2	4.1	2.51
Total	3/4~19/7	108	433.7	4.0	100

第一章　综合研究

1　Comprehensive Study

由表 29 可见,春小麦生育期各阶段的需水量变化规律表现为:出苗 $\xrightarrow{\text{增加}}$ 分蘖 $\xrightarrow{\text{减少}}$ 拔节 $\xrightarrow{\text{增加}}$ 孕穗 $\xrightarrow{\text{减少}}$ 抽穗 $\xrightarrow{\text{增加}}$ 灌浆 $\xrightarrow{\text{减少}}$ 成熟。

As Table 29 shows, the change law of water requirement of spring wheat at each growth stage is manifested as: Seedling $\xrightarrow{\text{Increasing}}$ Tillering $\xrightarrow{\text{Decresing}}$ Jointing $\xrightarrow{\text{Increasing}}$ Booting $\xrightarrow{\text{Decresing}}$ Heading $\xrightarrow{\text{Increasing}}$ Filling $\xrightarrow{\text{Decresing}}$ Maturing.

(二) 小麦产量与需水量变化关系
1.5.1.2　Relation of Wheat Yield and Water Requirement Change

作物产量与需水量变化有着极为密切的关系。

The crop yield is greatly influenced by the water requirement change.

通过分析计算可得小麦需水量与产量之间有着较好的函数相关关系,二者函数曲线呈抛物线形态,即当春小麦需水量在小于 433.7 mm 时,产量随灌溉定额的增加而增加,当灌溉定额达 433.7 mm 时,小麦产量在不采取其他任何措施时已达顶峰,再增加灌溉定额,产量不再增加,如灌溉定额太大,产量反而会下降,从分析计算可得,小麦的需水系数为 1.70。春小麦的最优需水量为 433.7 mm(289.13 m³/亩)。

The analytical calculation shows that there is a positive function correlation between wheat water requirement and yield. The function curve of the two is in a parabola form, which means that when the irrigation quota is below 433.7 mm, the yield will increase with the increasing quota; the yield peaks when the irrigation quota reaches 433.7 mm, if no other measures put in place; after this point the yield may decrease if the quota is too large. According to the analytical calculation, the water requirement coefficient of wheat is 1.70, and the optimal water requirement is 433.7 mm (289.13 m^3/mu).

(三) 春小麦生理需水分析
1.5.1.3　Analysis of Physiological Water Requirement of Spring Wheat

作物在生长发育过程中,水分的消耗主要是植株蒸腾和棵间土壤蒸发。由试验分析计算得知,小麦一生中的植株蒸腾蒸发在 6 月下旬至 7 月上旬时最大,日蒸发量在 2.2~3.7 mm 之间,即灌浆期。植株蒸腾量随叶面积的增大而增大,需水量相应地也增大,试验表明,叶面积系数与日需水强度之间有良好的相关关系。

The water is mainly consumed by plant transpiration and soil evaporation between plants in the crop growth and development period. According to the experimental analysis and calculation, wheat has the maximum amount of transpiration and evaporation in filling stage from late-June to early-July, with a daily evaporation amount of 2.2 to 3.7 mm. The evaporation and water requirement increase with increasing leaf. The leaf area coefficient is positively related to the daily water requirement intensity according to the experiment.

小麦一生中棵间蒸发为 177.5 mm,占全部需水量的 40% 左右,植株蒸腾为 256 mm,占总需水量的 60% 左右。

The total evaporation amount of wheat is 177.5 mm, accounting for about 40% of the total water requirement, and the total transpiration amount is 256 mm, about 60%.

二、苜蓿需水量及需水变化规律
1.5.2 Water Requirement and Its Changing Law of Alfalfa

(一) 苜蓿生育期的需水量变化分析
1.5.2.1 Analysis of Water Requirement Change in the Growth and Development Period of Alfalfa

在阿勒泰地区,苜蓿在全年中可收获 3 次,即打三茬。三茬中第一茬的生长期最长,为 74 d,占全生育期的 43%;而第二茬的需水量最大,为 285.5 mm。在每一茬中各生育阶段的需水强度都有所不同,从表 30 中可以看出,苜蓿在整个生长期中有 3 个需水高峰:第一茬为 5 月底至 6 月上旬的开花成熟期,历时 14 d,阶段需水量为 103.1 mm,占该茬总需水量的 51%,然而需水强度高达 7.4 mm/d;第二茬时间为 6 月下旬至 7 月上旬的蔓枝伸长期,阶段需水量为 131.2 mm,占该茬需水总量的 60.07%,最大需水强度为 9.5 mm,阶段平均需水强度为 6.5 mm/d;第三茬时间为 8 月上旬至下旬阶段,需水量为 200 mm,占该茬总需水量的 70%,最大需水强度为 9 mm,平均需水强度为 6.5 mm/d,这 3 个时期可作为本地区苜蓿的需水临界期。

Alfalfa in Altay Prefecture can be harvested for 3 times throughout the year, namely 3 croppings. The 1st cropping has the longest growth period of 74 days, accounting for 73% of the total growth and development period; the second cropping has the maximum water requirement, namely 285.5 mm. The water requirement intensity varies with the growth stages of each cropping. Table 30 shows that there are 3 water requirement peaks in the total growth period: the period for the first cropping is from late-May to early-June, lasting 14 days, in which the water requirement is 103.1mm, accounting for 51% of the total, whereas the water requirement intensity reaches 7.4mm/d; the period for the 2nd cropping is the branch extension period, from late-June to early-July, with a water requirement of 131.2 mm, accounting for 60.07% of the total, in which period the maximum water requirement intensity is 9.5 mm and the average water requirement of a stage is 6.5 mm/d; the period for the 3rd cropping is from early-August to late-August, with a water requirement of 200 mm, accounting for 70% of the total, whereas the maximum water requirement intensity is 9 mm, and the average water requirement intensity is 6.5 mm/day. The above 3 periods can be considered as the critical period of water

1 Comprehensive Study

requirement of local alfalfa.

表30 苜蓿各生育阶段的需水量分析表

茬期	生育阶段	起止日期	天数(d)	阶段需水量(mm)	日需水量(mm/d)	模系数(%)
第一茬	苗期	27/3-20/4	25	15.2	0.61	1.88
	蔓枝伸长期	21/4-25/5	35	137.2	5.4	23.11
	开花成熟期	26/5-8/6	14	103.1	7.4	12.73
第二茬	苗期	8/6-21/6	13	42.5	3.3	5.25
	蔓枝伸长期	22/6-11/7	20	131.2	5.5	16.2
	开花成熟期	12/7-21/7	10	44.7	4.5	5.52
第三茬	苗期	22/7-31/7	10	39.9	4.0	4.92
	蔓枝伸长期	1/8-31/8	31	200.2	6.5	24.72
	开花成熟期	1/9-15/9	15	45	3.1	5.70
全生育期		27/3-15/9	173	310.1	4.6	100

Table 30 Analysis of Water Requirement of Alfalfa at Every Growth and Development Period

Cropping Period	Growth and Development Period	Start and End Date	Number of days (Days)	Stage Water Requirement (mm)	Daily Water Requirement (mm/day)	Modulus (%)
1st Cropping	Seedling Period	27/3-20/4	25	15.2	0.61	1.88
	Branch Extension Period	21/4-25/5	35	137.2	5.4	23.11
	Flowering & Maturing Period	26/5-8/6	14	103.1	7.4	12.73
2nd Cropping	Seedling Period	8/6-21/6	13	42.5	3.3	5.25
	Branch Extension Period	22/6-11/7	20	131.2	5.5	16.2
	Flowering & Maturing Period	12/7-21/7	10	44.7	4.5	5.52
3rd cropping	Seedling Period	22/7-31/7	10	39.9	4.0	4.92
	Branch Extension Period	1/8-31/8	31	200.2	6.5	24.72
	Flowering & Maturing Period	1/9-15/9	15	45	3.1	5.70
Whole Growth and Development Period		27/3-15/9	173	310.1	4.6	100

由表30可见,苜蓿打三茬,整个生长期为173 d,需水量为810.0 mm,其中第一茬生长期为74 d,需水量为305.4 mm;第二茬生长期为43 d,需水量为218.4 mm;第三茬生长期为56天,需水量为286.3 mm。

Table 30 shows that alfalfa has 3 croppings, the total growth period is 173 days, and the water requirement is 810.0 mm. The 1st cropping lasts 74 days, in which period the water requirement is 305.4 mm; the 2nd cropping lasts 43 days, 218.4 mm; the 3rd cropping lasts 56 days, 286.3 mm.

苜蓿在整个生育期中的阶段需水量变化规律表现为:苗期 $\xrightarrow{增加}$ 蔓枝伸长期 $\xrightarrow{减少}$ 开花成熟期。

The water requirement changing law of a stage in the whole growth and development period of alfalfa is as follows: Seedling Period $\xrightarrow{Increasing}$ Branch Extension Period $\xrightarrow{Decreasing}$ Flowering & Maturing Period。

日最大需水量为 11 mm/d。

The maximum daily water requirement is 11 mm/d.

(二)苜蓿产量与需水量变化关系
1.5.2.2 Relation between Yield and Water Requirement Change of Alfalfa

苜蓿的产量与需水量变化关系曲线亦呈抛物线关系。在灌水量少于 810.0 mm 时,苜蓿的产量随灌溉定额的增加而增加,且产量增加幅度大,一方水产 1.44 kg 苜蓿;当灌水量大于 810.0 mm 时,产量虽有增加,但是增加幅度很小,一方水只产 0.56 kg 苜蓿。当灌水量达 911.5 mm 时,产量增加量微小,甚至不再增加。当再加大灌溉定额时产量反而会下降。由此可见,在本区苜蓿的需水量定为 810 mm 是较经济的。

The relation curve between yield and water requirement change of alfalfa is also in a parabola form. When the irrigation amount is less than 810.0 mm, the alfalfa yield increases at a greater rate with the increasing irrigation quota, namely one cubic meter of water produces 1.44 kg alfalfa; when larger than 810.0 mm, the yield increases slightly, one cube meter of water producing 0.56 kg alfalfa only. As the irrigation amount reaches 911.5 mm, the yield increases to a very small extent or even stops increasing. The yield will decrease if the irrigation keeps increasing. This shows that 810 mm arranged as the water requirement of alfalfa in this plot is comparatively economical.

三、玉米需水量及需水变化规律
1.5.3 Corn Water Requirement and Its Changing Law

(一) 玉米各生育阶段的需水量变化
1.5.3.1 The Water Requirement Change of Corn at Every Growth and Development Period

由分析试验结果可见(见表31),玉米生育期内的日需水量和阶段需水量变化很大,最大阶段需水量为小麦抽雄—灌浆期,这个阶段的需水量约占总需水量的36%,但生育期天数只占17%,日需水量可达7.1 mm。在正常水文年份中,7月中旬至下旬是玉米的抽雄—灌浆期,也是玉米全生育期中需水高峰,对水分的敏感性特别强,可作为玉米的需水临界期,这个阶段的需水量可达148.7 mm。

The experimental analysis (see Table 31) shows that there is great variation of daily water requirement and water requirement of a stage in the growth and development period; the maximum water requirement of a stage is found in tasseling-filling period, accounting for about 36% of the total water requirement; with a daily water requirement of 7.1 mm, this period accounts for 17% only of the total growth period. During normal hydrological year, corn tasseling-filling period is from mid-July to late-June, which is also a peak period for water; this period, highly sensitive to water, can be considered as the critical period of corn water requirement, with a water requirement reaching 148.7 mm.

表31 玉米各生育阶段的需水量分析表　　　　　　　　单位:mm

生育阶段	起止日期	天数	阶段需水量	日需水量	阶段需水模系数(%)
播种—出苗	9/5~15/5	7	18.20	2.60	4.47
出苗—拔节	16/5~24/6	40	94.90	2.40	23.29
拔节—抽雄	25/6~8/7	14	47.40	3.40	11.64
抽雄—灌浆	9/7~29/7	21	148.70	7.10	36.48
灌浆—成熟	30/7~14/8	16	62.80	3.90	15.42
成熟—收割	15/8~7/9	23	35.40	1.50	8.70
全生育期	9/5~7/9	121	407.40	3.40	100

Table 31 Analysis of Corn Water Requirement at Every Growth and Development Period

Unit: mm

Growth and Development Period	Start and End Date	Days	Stage Water Requirement	Daily Water Requirement	Water Requirement Modulus of a Stage (%)
Seeding–Seedling	9/5–15/5	7	18.20	2.60	4.47
Seedling–Jointing	16/5–24/6	40	94.90	2.40	23.29
Jointing–Tasseling	25/6–8/7	14	47.40	3.40	11.64
Tasseling–Filling	9/7–29/7	21	148.70	7.10	36.48
Filling–Maturing	30/7–14/8	16	62.80	3.90	15.42
Maturing–Harvesting	15/8–7/9	23	35.40	1.50	8.70
Whole Growth and Development Period	9/5–7/9	121	407.40	3.40	100

玉米生育期的阶段需水量变化趋势为：出苗 $\xrightarrow{增加}$ 拔节 $\xrightarrow{减少}$ 抽雄 $\xrightarrow{增加}$ 灌浆 $\xrightarrow{减少}$ 成熟。

The change trend of stage water requirement in corn's growth and development period is: Seedling $\xrightarrow{Increasing}$ Jointing $\xrightarrow{Decreasing}$ Tasseling $\xrightarrow{Increasing}$ Filling $\xrightarrow{Decreasing}$ Maturing.

玉米生育期的阶段日最大需水量为7.1 mm，阶段最大需水模系数为36.48。

The maximum daily water requirement is 7.1 mm and the maximum water requirement modulus is 36.48 of a stage in corn's growth and development period.

(二) 玉米产量与需水量变化关系
1.5.3.2 Relation between Corn Yield and Water Requirement Change

玉米产量与需水量之间存在抛物线函数相关。灌水定额当小于409.3 mm时，玉米的产量随灌水定额的增加而增加，当灌水量达到409.3 mm时，产量达至最高峰，需水系数为0.98。在不采取其他农业措施时，再加大灌溉定额，产量不再增加，反而因灌水过量而产量下降。因而最优需水量为409.3 mm。

There is a parabola function relation between corn yield and water requirement. When the irrigation water quota is less than 409.3 mm, the corn yield increases with the increase of irrigation quota. When the irrigation water reaches 409.3 mm, the yield peaks, with a water requirement coefficient of 0.98. The yield will stop increasing or even decreases without taking other agricultural measures if the irrigation quota keeps increasing. Therefore, the optimal water requirement is 409.3 mm.

第一章 综合研究
1 Comprehensive Study

（三）玉米生理需水分析
1.5.3.3 Analysis of Corn Physiological Water Requirement

由试验分析得，玉米在整个生长期中的植株叶面蒸腾量占需水总量的67%左右，棵间土壤蒸发占33%左右，植株蒸腾量最大时间为7月下旬和8月中旬，即玉米的灌浆和成熟期，分别占全生育期需水量的75%和84%、蒸腾强度为5.35 mm/d 和 5.04 mm/d。

According to experimental analysis, water consumed by leaf transpiration accounts for about 67% of total water requirement in the whole growth period. Water consumed by soil evaporation between plants accounts for about 33%. The maximum transpiration is found in late-July and mid-August, the filling & maturing period, accounting for 75% and 84% respectively of total water requirement in the whole growth and development period, with the transpiration intensities of 5.35 mm/d and 5.04 mm/d respectively.

四、甜菜需水量及需水变化规律
1.5.4 Beet Water Requirement and Its Changing Law

（一）甜菜各生育阶段的需水量变化
1.5.4.1 Water Requirement Change of Beet at Every Growth and Development Period

由试验成果可知，甜菜前期需水量较小，中期需水量为最大，这个阶段的需水量占甜菜全生育期总需水量的49.98%，但阶段生育期天数只占全生育期的36%。该阶段平均日需水量为6.34 mm。甜菜后期的需水量最小，阶段需水量为133 mm，平均需水强度为3.2 mm/d。从几年的试验结果来看，甜菜的前期主要是生长叶面积，中期主要以生长根茎为主，这时对水要求极高，如这时缺水，对产量是有很大影响的，我们认为这个时期可作为甜菜的临界期，这个阶段的需水量为293.3 mm（见表32）。

The experimental result shows that beet has a small water requirement in the initial stage, and a maximum water requirement accounting for 49.98% of the total in the growth and development period in the middle stage, while the number of days of this stage only accounts for 36% of the whole growth and development period. The average daily water requirement in this stage is 6.34 mm. Beet has the minimum water requirement decreasing to 133 mm in the later stage, with an average water requirement intensity of 3.2 mm/d. According to experimental results from 1989 to 1991, the initial stage is mainly for leaf growth, and the middle stage mainly for rhizome growth, in which period a great amount of water is required. Lacking water in this period may make a big negative influence on yield, reasonably this period, with a water requirement of 293.3mm, is considered as the critical period of beet (see Table 32).

表 32　　甜菜各生育阶段的需水量分析表　　　　　　　单位：mm

生育阶段	日期	天数	阶段需水量	日需水量	模系数(%)
前期	25/5~7/7	42	160.5	2.8	27.4
中期	7/7~23/8	46	293.3	6.34	49.94
后期	23/8~4/10	41	133.0	3.2	22.55
合计	25/5~4/10	129	586.8	4.5	100

Table 32　　Analysis of Beet Water Requirement at Each Growth Stage　　Unit: mm

Growth and Development Period	Date	Days	Stage Water Requirement	Daily Water Requirement	Modulus (%)
Initial Stage	25/5–7/7	42	160.5	2.8	27.4
Middle Stage	7/7–23/8	46	293.3	6.34	49.94
Later Stage	23/8–4/10	41	133.0	3.2	22.55
Total	25/5–4/10	129	586.8	4.5	100

甜菜生育期内各阶段的需水量变化规律为：前期 $\xrightarrow{增加}$ 中期 $\xrightarrow{减少}$ 后期。

The changing law of beet water requirement at each stage in the growth and development period is:

Initial Stage $\xrightarrow{Increasing}$ Middle Stage $\xrightarrow{Decreasing}$ Later Stage.

(二) 甜菜产量与需水量变化关系

1.5.4.2　Relation of Beet Yield and Water Requirement Change

通过对试验资料的分析整理可知,甜菜产量与需水量呈良好的相关关系,当甜菜的灌溉定额小于 586.8 mm 时,产量随灌溉定额的增加而增加；而当灌溉定额达 586.8 mm 时,产量达高峰,这时需水系数为 0.22。在不采取其他农业措施的条件下,再增加灌水量时,甜菜产量不再增加,反而因灌溉定额太大,产量下降。由此得出,甜菜最优需水量为 586.6 mm。

According to the preparation and analysis of experiment data, there is a positive relation between beet yield and water requirement. When the irrigation quota is less than 586.8 mm, the yield increases with the increasing irrigation quota. When the irrigation quota reaches 586.8 mm, the yield peaks, with a water requirement coefficient of 0.22. After this point the yield, if no other measures are put in place, may decrease if the quota is too large. It can be concluded that the optimal water requirement of beet is 586.6 mm.

第一章 综合研究
1 Comprehensive Study

(三) 甜菜生理需水分析
1.5.4.3 Analysis of Physiological Water Requirement of Beet

甜菜在该区全生育期叶面蒸腾占需水总量的69%,棵间土壤蒸发占31%左右,叶面最大蒸腾量是在中期,即7月7日至8月23日,占阶段需水量的72%,蒸腾强度为3.92 mm/d。

Water consumed by leaf transpiration accounts for about 69% of total water requirement in the whole beet growth and development period in this plot, with the soil evaporation between plants accounting for about 31%; the maximum leaf transpiration is found in the middle stage, from July 7 to August 23, accounting for 72% of the water requirement in this stage, with a transpiration intensity of 3.92 mm/d.

五、作物的经济需水量
1.5.5 Economical Water Requirement of Crops

通过分析试验和计算比较,在新疆北部干旱瘠薄土地的春小麦的全生育期的最优需水量为289.14 m³/亩(433.71 mm);苜蓿的需水量为540 m³/亩(810 mm);玉米的需水量为272.87 m³/亩(409.3 mm);甜菜的需水量为391.20 m³/亩(586.8 mm),见表33。

According to the analysis and calculation, the optimal water requirement of spring wheat in the arid land in northern Xinjiang is 289.14 m³/mu (433.71 mm); the water requirement of alfalfa is 540 m³/mu (810 mm); that of corn is 272.87 m³/mu (409.3 mm); that of beet is 391.20 m³/mu (586.8 mm); see Table 33 for more information.

表33 作物全生育期需水量统计分析表

作物名称		春小麦	苜蓿	玉米	甜菜
需水量	mm	433.71	810	409.3	586.8
	mm³/亩	289.14	540	272.87	391.20
生育期	起止日期	4月3日~ 7月19日	3月27日~ 9月15日	5月9日~ 9月7日	5月25日~ 10月4日
	天数	108	173	121	129

Table 33 Statistics of Crop Water Requirement in the Whole Growth and Development Period

Crop Name		Spring Wheat	Alfalfa	Corn	Beet
Water Requirement	mm	433.71	810	409.3	586.8
	mm³/mu	289.14	540	272.87	391.20
Growth and Development Period	Start and End Date	April 3rd to July 19th	March 27th to September 15th	May 9th to September 7th	May 25th to October 4th
	Days	108	173	121	129

六、作物最优处理水平选择
1.5.6 Selection of the Optimal Crop Treatment Level

通过对不同水平的作物产量及灌水量分析,从而根据生产效益确定和选择最优处理水平。

The crop yield and irrigation amount at varied levels are analyzed to identify and select the optimal treatment level based on the production benefit.

(一)春小麦的最优处理水平
1.5.6.1 Optimal Treatment Level of Spring Wheat

比较3个不同水平的试验结果(见表34)可以看出,在3个不同处理中,以2号水平的生产效益最好,2号水平的需水量为289.14 m³/亩,产量为174.98 kg,需水系数为1.65 m³/kg,即每产1 kg 小麦,仅需水1.65 m³;而1号水平和3号水平产1 kg 小麦则需水量分别为2.48 m³和1.75 m³水。结合小麦产量与需水量函数关系及变化曲线分析,2号水平为最优选择水平。不同生育阶段土壤含水率占田间持水率的百分数为:分蘖—拔节期为55%,孕穗—抽穗期为60%,乳熟期为55%。

According to the comparison of experimental results of the 3 levels (see Table 34), the production benefit of Level #2 is the optimal; the yield is 174.98 kg, the water requirement is 289.14 m³/mu, which means 1 kg wheat only consumes 1.65 m³ water, whereas the water requirement of 1kg wheat of Level #1 and Level #3 is 2.48 m³ and 1.75 m³ respectively. According to combined analysis of function relation and change curve between wheat yield and water requirement, it is concluded that Level #2 is optimal. The percentages of soil water content and field capacity at different growth stages are: 55% in the tillering-jointing period, 60% in the booting-heading period and 55% in the milky maturity period.

表34 春小麦不同水平生产效益分析表 单位:m³/kg

水平	1	2	3
需水系数	2.48	1.67	1.75
水平选择	中	优	良

Table 34 Analysis of Production Benefit at Different Levels of Spring Wheat Unit: m³/kg

Level	1	2	3
Water Requirement Coefficient	2.48	1.67	1.75
Level Selection	Middle	Excellent	Good

（二）苜蓿的最优处理水平
1.5.6.2 Optimal Treatment Level of Alfalfa

从 3 个不同水平的比较试验结果（见表 35）可以看出,苜蓿以 2 号水平的效益较好,3 个水平产量分别为 716.74、669.92、616.10 kg/亩,灌溉需水量依次为 608.23、540.38、459.53 m³/亩,需水系数分别为 0.849、0.804、0.755 m³/kg,结合苜蓿需水量与产量变化曲线分析,2 水平具有增产、省水两向优势,是本试验的最优试验水平。其不同生育期土壤含水率的控制下限占土壤田间持水率的百分数分别是幼苗期 55%,蔓枝伸长期 60%,开花成熟期 55%。

According to the contrast experiment results at 3 levels (see Table 35), benefit of Level #2 is better than the rest; the yields of the 3 levels are 716.74 kg/mu, 669.92 kg/mu and 16.10 kg/mu respectively; the water requirement for irrigation is 608.23 m³/mu, 540.38 m³/mu and 459.53 m³/mu respectively; the water requirement coefficients are 0.849 m³/kg, 0.804 m³/kg and 0.755 m³/kg; The analysis by combining alfalfa water requirement and yield change curve shows that Level #2 has both advantages in higher yield and less water, which made it the optimal experiment level. Soil water content's lower limit accounts for 55% (seedling period), 60% (branch extension period) and 55% (flowering & maturing period) of the field capacity respectively.

表 35	苜蓿不同水平生产效益分析表		单位：m³/kg
水平	1	2	3
需水系数	0.849	0.804	0.755

Table 35	Analysis of Production Benefit at Different Levels of Alfalfa		Unit: m³/kg
Level	1	2	3
Water Requirement Coefficient	0.849	0.804	0.755

（三）玉米的最优处理水平
1.5.6.3 Optimal Treatment Level of Corn

不同水平的玉米需水量状况和产量变化亦是不同的。玉米生育期的需水量以 1 号水平最高(306.17 m³/亩),2 号水平次之(298.93 m³/亩),3 号水平最低(256.70 m³/亩),其产量依次为 276.53、206.23、153.73 kg/亩,由玉米需水量和产量的函数关系式及相关曲线分析,玉米生育期需水量为 273 m³/亩时,生产效益最好,需水系数为 0.98 m³/kg,而该值接近玉米试验方案的 2 号水平。

Corn water requirement and yield vary with the treatment level. Level #1 has the maximum water requirement in corn growth and development period, followed by Level #2 and #3, namely 306.17 m³/mu, 298.93 m³/mu and 256.70 m³/mu respectively. The yields of Level #1, #2 and #3 are 276.53 kg/mu, 206.23 kg/mu and 153.73 kg/mu respectively; according to function relation between corn water requirement and yield and pertinent curves, corn has the best production benefit when the water requirement is 273 m³/mu, with a water requirement coefficient of 0.98 m³/kg, close to the value of Level #2.

(四) 甜菜的最优处理水平
1.5.6.4 Optimal Treatment Level of Beet

比较甜菜不同水平的产量与需水量变化,1号水平的产量最高(1 757 kg/亩),但需水系数也最大(0.21 m³/kg),2号水平次之,3号水平的产量最低,需水系数也较小。然而由甜菜产量和需水量的函数关系曲线则可以看出,当需水量为391.17 m³/亩时,甜菜产量达最高值,需水系数为0.22 m³/kg,此值接近试验方案的1号水平。甜菜生育期的土壤含水率范围为干土重的5.36%,占田间持水率的51.09%。

According to the comparison of beet yield and water requirement change in different levels, Level #1 has the highest yield (1 757 kg/mu) and the biggest water requirement coefficient (0.21 m³/kg), followed by Level #2 and Level #3. But the function relation curve between beet yield and water requirement shows that when the water requirement is 391.17 m³/mu, the yield peaks, and now the water requirement coefficient is 0.22 m³/kg, close to the value of Level #2. The soil water content in beet growth and development period accounts for 5.36% of the weight of water-free soil, and 51.09% of field capacity.

上述分析表明,春小麦的最优试验方案为2号水平,作物需水系数为1.70 m³/kg;苜蓿的最优试验方案为2号水平,作物需水系数为0.804 m³/kg;玉米的最优试验方案接近2号水平,作物需水系数为0.98 m³/kg;甜菜的最优试验方案接近1号水平,作物需水系数为0.22 m³/kg(见表36)。

The above analysis manifests that Level #2 is the optimal experimental scheme for both spring wheat and alfalfa with the water requirement coefficients of 1.70 m³/kg and 0.804 m³/kg respectively. Level #2 is basically the optimal experimental scheme of corn, with a water requirement coefficient of 0.98 m³/kg. Level #1 is basically the optimal experimental scheme of beet, with the water requirement coefficient of 0.22 m³/kg (see Table 36).

表36　　　　　　　　　作物最优试验方案的水平选择分析表

项目	春小麦	苜蓿	玉米	甜菜
需水系数 (m³/kg)	1.70	0.804	0.98	0.22
最优水平	2	2	2	1

Table 36　Analysis of Crop Level Selection of the Optimal Experimental Scheme

Item	Spring Wheat	Alfalfa	Corn	Beet
Water Requirement Coefficient(m^3/kg)	1.70	0.804	0.98	0.22
Optimal Level	2	2	2	1

第六节　作物灌溉制度与灌水方法试验研究
1.6 Experimental Research of Irrigation System and Method

通过3年对春小麦、苜蓿、玉米及甜菜的种植试验,我们发现,作物在生长发育不同阶段,其蒸腾失水量及其对缺水的敏感程度是不同的。而在某一阶段的缺水,往往不仅影响本生育阶段,而且对下阶段的作物生长发育也产生影响,因此,无论哪个阶段的缺水都会导致作物不同程度的减产。如何正确地确定荒漠瘠薄土地作物需水量和最优灌溉制度,是我们在原始资料基础上进行理论分析计算的关键一步。首先我们对3年的原始资料进行了认真的查阅,对明显不符合实际情况和没有代表性的资料找出原因,实事求是地加以分析处理,以保证成果的可靠性。然后对基础资料用各种方法进行认真的分析计算,找出作物产量与灌溉制度的相关关系,从而确立荒漠戈壁地作物最优灌溉制度。

3-years experiments of spring wheat, alfalfa, corn and beet have helped us find that the transpiration amount and sensitivity to water shortage vary with the growth stages. The water shortage of a stage will hinder the growth both in this and next stages, as a result, the water shortage of any stage will decrease the yield to a certain degree. To accurately determine the crop water requirement and the optimal irrigation system in arid land is critical for our theoretical analysis and calculation based on the original data. We first carefully review the original data in the past 3 years to figure out why some data fails to be practical and unrepresentative, and then through faithful analysis and collection guarantee the reliability of result. At last we calculate and analyze the basic data in whatever way that fits to figure out the relation between crop yield and irrigation system and identify the optimal irrigation system in arid Gobi desert land.

一、春小麦灌溉制度试验研究
1.6.1 Research of Irrigation System Experiment of Spring Wheat

(一) 最优灌水定额分析
1.6.1.1 Analysis of the Optimal Irrigation Water Quota

作物灌水定额主要是根据田间土壤土质条件和作物根系分布状况来决定的。由前所述可知,在新疆北部的阿勒泰地区80%种植面积为戈壁地,戈壁地土质较差,0~40 cm 一般为沙壤土,40 cm 以下基本上是砂子和卵石,田间持水率非常低,土壤渗漏性大。春小麦的根系活动层主要在0~60 cm 之内,60~80 cm 内根系重量仅为总重量的4.4%。而0~60 cm 对土壤湿度要求一般为田间持水率的70%左右,60 cm 以下由于蒸发量和蒸腾量都非常小,两次取土(相隔10 d)含水率之差仅在1%左右,这表明每年要有一次灌水,计划湿润层按80 cm 来计算,灌水时间一般在第一水,但最晚不能超过抽穗期,其灌水定额为40 m³/亩左右。而生育期其他时间灌水计划湿润层均按40~60 cm 来考虑。其灌水定额为20~30 m³/亩。超过该定量的水量,田间将会产生深层渗漏。

The crop irrigation water quota is determined by soil conditions and crop root distribution. The above shows that 80% of plant area in Altay Prefecture in northern Xinjiang is poor Gobi desert where 0-40 cm soil is sandy loam, below 40 cm are basically sands and pebbles with a low field capacity and a high soil percolation. The zone at a depth from 0 to 60 cm is mainly the root active zone, and the root at a depth from 60 to 80 cm only accounts for 4.4% of total weight. Soil moisture at a depth of 0-60 cm is generally required to be about 70% of the field capacity. As the evapotranspiration below 60 cm is very little, water content difference of the two soil samples (10 days apart) is only about 1%, indicating that at least once irrigation is required each year. If the soil wetting layer is taken as 80 cm, the first irrigation shall be conducted no later than heading period, with an irrigation water quota of about 40 m³/mu. The 40-60 cm soil wetting layer is considered for irrigation in the growth and development period. The irrigation water quota is 20-30 m³/mu. Any excess will lead to deep infiltration in the field.

(二)小麦灌溉定额与产量相关分析
1.6.1.2 Relative Analysis of Wheat Irrigation Quota and Yield

作物灌溉定额与产量有着极为密切的关系,即在一定范围内作物的产量与灌溉定额呈稳定的正相关关系,产量随着灌溉定额的增加而增加。当灌溉定额达到一定值时,再增加灌溉定额则产量与灌溉定额呈负相关关系,即随着灌溉定额的增加,产量减少,其原因是水分过多,小麦根系就会因空气缺乏而呼吸不良,生理活动受抑制。小麦前期水分过多,会引起麦苗黄瘦,中、后期水分过多,也会抑制根系活动和穗粒发育。

The crop irrigation quota is closely connected to the yield, namely within a certain limit there is a stable positive correlation between yield and irrigation quota, and the yield will increase with the increasing irrigation quota. This positive correlation will become a negative one if the irrigation quota reaches a certain value, featuring that the yield will decrease with the increasing quota, because excessive water will lead to air deficiency, causing adverse respiratory effects and hindering the physiological activity of wheat root system. Excessive water in the initial stage will lead to yellow and thin wheat seedling, and that in the middle and later stages will hinder the root system activity and ear grain growth.

第一章 综合研究
1 Comprehensive Study

这里我们对3个不同处理水平(50%、60%和70%)进行了灌溉定额和产量的对比试验(见表37)。并结合对试验方案和极差分析,得出70%处理水平的灌溉制度模式比50%和60%处理水平的灌溉制度模式效益都好,最优试验方案组合为A2B2C3,灌溉定额为283.61 m³/亩,这与我们根据春小麦灌溉定额和产量函数相关曲线求出的289.14 m³/亩基本接近,灌水次数为10次。

Contrast experiments (see Table 37) are conducted about irrigation quota and yield at three different treatment levels (50%, 60% and 70%). In combination with the analysis of experimental schemes and range, the irrigation scheduling of 70% treatment level is better than the rest in terms of benefit, the optimal experimental scheme combinations are A2B2C3, the irrigation quota is 283.61 m³/mu, close to 289.14 m³/mu obtained by relevant curves of irrigation quota and yield function of spring wheat, and the times of irrigation is 10.

表37　　　　　　　　小麦不同水平灌溉定额与产量分析表

年份	1990			1991			平均		
处理水平	50%	60%	70%	50%	60%	70%	50%	60%	70%
产量(kg/亩)	139.87	129.11	155.41	217.45	250.90	268.17	178.66	190.01	212.01
灌溉定额(m³/亩)	222.01	213.14	265.14	212.01	214.62	302.07	217.01	213.88	283.61
灌水次数(次)	7	5	10	5	8	11	7	7	10
需水系数(m³/kg)	1.59	1.65	1.71	0.97	0.86	1.13	1.21	1.13	1.34

Table 37　　Analysis of Wheat Irrigation Amount and Yield at Different Levels

Year	1990			1991			Average		
Treatment Level	50%	60%	70%	50%	60%	70%	50%	60%	70%
Yield (kg/mu)	139.87	129.11	155.41	217.45	250.90	268.17	178.66	190.01	212.01
Irrigation Quota (m³/mu)	222.01	213.14	265.14	212.01	214.62	302.07	217.01	213.88	283.61
Irrigation Frequency (times)	7	5	10	5	8	11	7	7	10
Water Requirement Coefficient (m³/kg)	1.59	1.65	1.71	0.97	0.86	1.13	1.21	1.13	1.34

(三)小麦最优灌溉制度
1.6.1.3 Optimal Wheat Irrigation Scheduling

在小麦生育不同时期作物的生理需水量是不同,因此正确掌握小麦不同生育期的灌水量十分重要的。

Given that the physiological water requirement varies with the wheat growth stages, it's important to accurately control the irrigation amount at different wheat growth stages.

小麦全生育期分苗前、苗期、分蘖期、拔节期、孕穗期、抽穗期、灌浆期和成熟期。通过3年试验研究,确定采用"量少次多"灌溉制度,各不同生育期灌水量见表38。

The whole wheat growth and development period contains pre-seedling period, seedling period, tillering period, booting period, heading period, filing period and maturing period. An irrigation system featuring "small amount, many times" based on the experimental research in the past 3 years. See Table 38 for the irrigation amount at each growth stage.

表38　　　　　　　　　　　春小麦灌溉制度优化表

生育阶段	灌水时间	灌水定额（m³/亩）	灌水次数
苗前	4月12日~4月14日	36.60	1
苗期	4月22日~4月24日	30.31	1
分蘖期	5月10日~5月13日	36.60	1
拔节期	5月20日~5月22日	21.44	1
孕穗期	6月3日~6月5日	25.17	1
抽穗期	6月10日~6月12日	32.60	2
	6月15日~6月17日	22.60	
灌浆期	6月20日~8月22日	31.10	1
成熟期	6月26日~6月28日	23.00	2
	7月1日~7月3日	30.32	
合计		289.14	10

Table 38　　　　　　　Optimized Irrigation Scheduling of Spring Wheat

Growth and Development Period	Irrigation Time	Irrigation Water Quota (m³/mu)	Irrigation Frequency
Pre-seedling period	April 12th to April 14th	36.60	1
Seedling period	April 22nd to April 24th	30.31	1
Tillering period	May 10th to May 13th	36.60	1
Jointing period	May 20th to May 22nd	21.44	1
Booting period	June 3rd to June 5th	25.17	1
Heading period	June 10th to June 12th	32.60	2
	June 15th to June 17th	22.60	
Filling period	June 20th to August 22nd	31.10	1
Maturing period	June 26th to June 28th	23.00	2
	July 1st to July 3rd	30.32	
Total		289.14	10

第一章 综合研究
1 Comprehensive Study

春小麦全生育期的灌溉定额为 289.14 m³/亩,灌水次数为 10 次。

The irrigation quota is 289.14 m³/mu in the whole growth and development period of spring wheat and the irrigation frequency are 10.

二、苜蓿灌溉制度研究
1.6.2 Research of Alfalfa Irrigation System

(一)苜蓿最优灌水量分析
1.6.2.1 Analysis of the Optimal Irrigation Amount of Alfalfa

根据荒漠瘠薄戈壁地特点,计划湿润层厚度按 60cm 计算。第一茬苜蓿的头水灌水定额在 40 m³/亩左右,但第二、三茬的头水灌水量都较大,达 45 m³/亩左右,因第一、二茬苜蓿收割后,要等晒干拉走以后才能浇水,一般都是 10 d 左右,加之气温较高,棵间蒸发快,土壤中的含水率下降快,间隔时间越长,土壤含水率就越低,造成第二、三茬头水的灌水量就越大。

The soil wetting layer is taken as 60 cm for irrigation in accordance with features of arid Gobi desert land. The first irrigation amount in the 2nd cropping and 3rd cropping is 45 m³/mu, whereas the first irrigation quota of alfalfa in the 1st cropping is about 40 m³/mu, the reasons are: irrigation must be conducted when the alfalfa harvested in the 1st and 2nd cropping has been dried and removed, which usually lasts 10 days, in this process the soil water content may decrease rapidly by evaporation because of high temperature, the longer this process lasts, the more the water content decreases.

(二)灌溉定额与产量的关系
1.6.2.2 Relation between Irrigation Quota and Yield

由灌水量与产量的函数关系可知,当灌水量在小于 566 m³/亩时,产量随灌水量的增加而增加,当灌水量达到 566 m³/亩时,产量达最高峰,每亩干草为 897.91 kg(其他条件相同)。当再增加灌水量时,产量增加很少或不增加,如灌水量超过太大时,产量不但不增加,反而会减小。

According to the function relation between irrigation amount and yield, when the irrigation amount is less than 566 m³/mu, the yield increases with the increasing irrigation amount. When the irrigation amount reaches 566 m³/mu, the yield peaks, hay weight per mu reaching 897.91 kg (all things being equal). If the irrigation amount keeps increasing, the yield may increase slightly or stop increasing, and if the irrigation amount is too large, the yield will not increase but decrease.

结合苜蓿灌溉制度的研究,我们设计 4 个不同水平,从灌溉制度实施结果中可以看出,各水平产量依次为 881.67、883.38、897.91、869.41 kg/亩,即各水平间产量差异不大,各水平中以 3 号水平产量最高,4 号水平产量最低;4 个水平的灌水定额依次为 589.23、578.7、566.79、463.0 m³/亩,各水平中以 4 号水平灌水定额最小,3 号水平的灌水定额次之;耗水系数依次为 1.50、1.53、1.60、1.88,其中以 4 号水平耗水系数最大,但产量最小,3 号水平耗水系数次之,但产量最高,从高产的效果来看,3 号水平是取得高产的最优方案(见表 39)。

4 levels are designed based on the research of alfalfa irrigation system. The experimental results of irrigation system show that the yield of each level is 881.67 kg/mu, 883.38 kg/mu, 897.91 kg/mu and 869.41 kg/mu respectively, almost the same. Level #3 has the maximum yield, and Level #4 the minimum. The irrigation quota is 589.23 m³/mu, 578.7 m³/mu, 566.79 m³/mu and 463.0 m³/mu respectively. Level #4 has the maximum irrigation water quota, followed by Level #3. The water consumption coefficient is 1.50, 1.58, 1.60 and 1.88 respectively. Level #4 has the maximum coefficient with the minimum yield, and Level #3 has the 2nd maximum coefficient with the maximum yield; in terms of high yield, Level #3 is the optimal scheme (see Table 39).

表 39　　　　　　　　　苜蓿不同水平灌溉定额与产量分析表

水平	灌水次数	灌溉定额(m³/亩)	产量(kg/亩)	耗水系数(kg/m³)
1	18	589.23	881.67	1.50
2	16	578.7	883.38	1.58
3	15	566.79	897.91	1.60
4	15	463.00	869.41	1.88

Table 39　　　Statistics of Alfalfa Irrigation Quota and Yield at Different Levels

Level	Irrigation Frequency	Irrigation Quota (m³/mu)	Output (kg/mu)	Water Consumption Coefficient (kg/m³)
1	18	589.23	881.67	1.50
2	16	578.7	883.38	1.58
3	15	566.79	897.91	1.60
4	15	463.00	869.41	1.88

(三)苜蓿最优灌溉制度
1.6.2.3 Optimal Alfalfa Irrigation Scheduling

通过对苜蓿灌溉制度的试验研究,确定苜蓿打三茬的灌溉定额为 566.79 m³/亩,全生育期的灌水次数为 15 次(见表 40)。

According to the experimental research of alfalfa irrigation scheduling, 566.79 m³/mu is determined as the alfalfa irrigation quota in 3 cropping, and 15 times of irrigation are required across the total growth and development period (see Table 40).

表 40 苜蓿灌溉制度优化表

茬数	灌水日期(日/月)	灌水定额(m³/亩)	灌水次数
第一茬	28/4	40.67	4
	6/5	34.64	
	21/5	35.06	
	3/6	36.86	
第二茬	12/6	58.82	5
	17/6	28.21	
	20/6	28.21	
	26/6	37.81	
	6/7	42.03	
第三茬	20/7	45.97	6
	20/7	36.29	
	6/8	33.23	
	16/8	35.57	
	26/8	33.53	
	1/9	38.89	
合计		566.79	15

Table 40 Table for Optimized Alfalfa Irrigation Scheduling

Cropping Number	Irrigation Date (Day/ Month)	Irrigation water quota (m³/mu)	Irrigation Frequency
1st Cropping	28/4	40.67	4
	6/5	34.64	
	21/5	35.06	
	3/6	36.86	
2nd Cropping	12/6	58.82	5
	17/6	28.21	
	20/6	28.21	
	26/6	37.81	
	6/7	42.03	
3rd cropping	20/7	45.97	6
	20/7	36.29	
	6/8	33.23	
	16/8	35.57	
	26/8	33.53	
	1/9	38.89	
Total		566.79	15

三、玉米灌溉制度研究
1.6.3 Study on Corn Irrigation Scheduling

(一) 玉米的灌水量与产量关系
1.6.3.1 Relation between Corn Irrigation Amount and Yield

根据1900—1991年玉米灌溉试验的资料,分别计算出玉米的总产量,平均产量、边际产量和生产弹性(见表41),并绘制玉米物质生产函数曲线(见图2)。

The total yield, average yield, marginal yield and production flexibility are all obtained by calculation based on the corn irrigation experiment data from 1900 to 1991, and a production function curve is drawn (see Figure 2).

表41　　　　　　　　　　玉米灌溉试验成果表

编号	处理号	灌溉定额 (X)	总产量 (Y)	平均产量 (A)	水量增加 ($\triangle X$)	产量增加 ($\triangle Y$)	边际产量 (M)	生产弹性 (E_P)
1	7	201.36	289.03	1.43				
2	6	213.09	356.05	1.67	11.28	37.02	5.97	3.57
3	5	185.13	329.18	1.78	−27.96	−26.87	0.95	0.54
4	4	189.00	329.25	1.74	3.87	0.07	0.02	0.01
5	3	243.31	375.46	1.54	54.31	46.21	0.85	0.55
6	2	269.22	437.19	1.62	25.91	61.73	2.38	1.59
7	1	288.94	409.54	1.42	19.72	27.65	1.40	0.99

Table 41　　　　　　　　　　Irrigation Experimental Results of Corn

No.	Treatment No.	Irrigation Quota (X)	Total Yield (Y)	Average Yield (A)	Increased Water Amount ($\triangle X$)	Increased Yield ($\triangle Y$)	Marginal Yield (M)	Production Flexibility (E_P)
1	7	201.36	289.03	1.43				
2	6	213.09	356.05	1.67	11.28	37.02	5.97	3.57
3	5	185.13	329.18	1.78	−27.96	−26.87	0.95	0.54
4	4	189.00	329.25	1.74	3.87	0.07	0.02	0.01
5	3	243.31	375.46	1.54	54.31	46.21	0.85	0.55
6	2	269.22	437.19	1.62	25.91	61.73	2.38	1.59
7	1	288.94	409.54	1.42	19.72	27.65	1.40	0.99

由图中不难看出,总产量曲线随水量的增加而增加,总产量的增加由慢变快以后又逐渐减慢,当耗水量达270 m^3/亩时,总产量达到顶点427 kg/亩左右,这时再增加灌水量,产量不再增加,如灌水量超额太大,反而会引起产量下降。

第一章 综合研究
1 Comprehensive Study

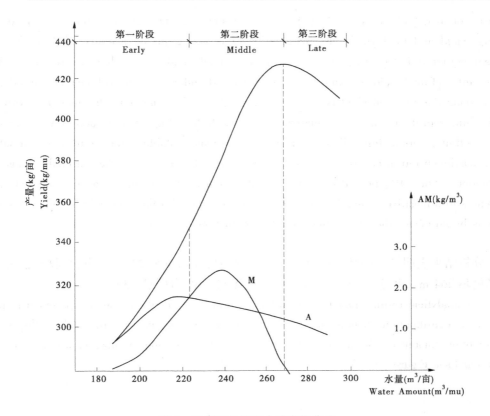

图 2 玉米产量与需水量变化曲线
Figure 2 Curves of Corn Yield and Water Requirement Change

It's not hard to find out from the figure that with the increasing water amount, the total yield increases from slowly to quickly, and then slowly; when the water consumption amount reaches 270 m^3/mu, the total yield peaks at about 427 kg/mu; the yield will stop increasing if the irrigation amount keeps increasing or it will decrease if the irrigation amount is too large.

通过分析我们把产量的增减划分为 3 个阶段,第一阶段特点是边际产量高于平均产量,因而引起平均产量逐渐提高,总产量的增加幅度大于资源增加幅度。第二阶段是平均产量最高点到总产量最高点之间,这一阶段边际产量低于平均产量,因而引起平均产量逐渐降低,总产量增加的幅度小于资源增加的幅度出现报酬递减。总产量曲线顶点是在一定技术条件下所能达到的一种界限,这时的边际产量为零。在这一阶段总产值按报酬递减的形式增长,可以获得较高的产量。在一定限度内,继续投入可获得利益。所以是资源投入的最适点,也是可以获得最大收益的点,因此称第二段为生产合理阶段。第三阶段是在总产量曲线的最高点以后,也就是边际产量转为负值以后的阶段。在这一阶段内越增加投入,经济效益越低,显然这是生产绝对不合理阶段。

Through analysis the yield change is divided into 3 stages. Stage I: the marginal yield is larger than the average yield which makes the average yield increase gradually, and the increase of total yield is more than that of the resources. Stage II: between the vertexes of average yield and total yield, the marginal yield is less than the average yield so that the average output will decrease, and the increase of total yield is less than the resources. The curve vertex of total yield is a limit that can be reached under a certain technical condition, and at this point the marginal yield is 0. In this stage, the total production value increases with the diminishing return, producing a relatively high yield. Inputting more resources will generate return within a certain limit. Reasonably this is the most suitable point for resources input in that a maximum return can be generated, hence stage II referred to "the reasonable stage for production". Stage III: behind the vertex of total yield curve; in this stage the marginal yield decreases to a negative value. The more the resources, the less the economic benefit, thus this stage is by no means the reasonable stage for production.

分析结果表明,玉米灌溉制度试验产量最高时,耗水量为 270 m³/亩,纯收益最大时,耗水量是 264 m³/亩,产量 425 kg/亩,最大效益 118 元/亩(见图 3)。

The analytical results show that when the yield of corn irrigation system experiment peaks, the water consumption amount is 270 m³/mu. When the net return peaks, the water consumption amount is 264 m³/mu, the yield is 425 kg/mu, and the maximum benefit is 118 yuan/mu (see Figure 3).

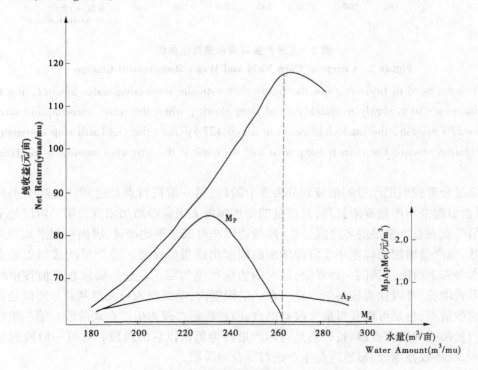

图 3 玉米产量与纯收益关系曲线
Figure 3　Relation Curves of Corn Yield and Net Return

(二) 玉米水分生产函数
1.6.3.2 Water Production Function of Corn

通过对玉米水分生产函数的分析计算,从而求出玉米各生育阶段水分敏感指数,确定玉米需水临界期。计算结果表明(见表42),在玉米生育期内,拔节期的水分敏感指数最大,达0.83,而苗期最小,仅为0.21,各生育阶段对水分敏感度反应大小的顺序为:拔节期>抽雄期>灌浆期>成熟期>苗期。

According to the water production function of corn, water sensitivity indexes of each growth stage are obtained and the critical period of corn water requirement is identified. The calculation (see Table 42) shows that in the corn growth and development period the jointing period has the largest water sensitivity index of 0.83, and the seedling period has the smallest, only 0.21; the order of water sensitivity indexes of each growth stage is: jointing period > tasseling period > filling period > maturing period > seedling period.

由表42可以看出,玉米在拔节期对水分的反应最敏感,是玉米需水临界期,必须保证水分的供应,苗期对水分的敏感度小,土壤水分只要满足其顺利出苗,度过蹲苗期即可。从水分敏感指数可以看出,玉米各阶段缺水率的高低对玉米产量有着不同程度的影响。

It can be seen from Table 42 that corn is most sensitive to water in jointing period, thus considered as the critical period of corn water requirement when a stable water supply must be guaranteed. Being less sensitive to water in seedling period indicates that the soil water content shall only meet the consumption amount in seedling period. According to the water sensitivity indexes, the influence on corn yield varies with water shortage at different stages.

表42　　　　　　　　玉米各生育阶段水分敏感指数分析表

生育阶段	苗期	拔节	抽雄	灌浆	成熟
水分敏感指数	0.21	0.83	0.62	0.48	0.26

Table 42　Analysis of Corn Water Sensitivity Index of Every Growth and Development Period

Growth and Development Period	Seedling Period	Jointing Period	Tasseling Period	Filling Period	Maturing Period
Water Sensitivity Index	0.21	0.83	0.62	0.48	0.26

(三) 玉米优化灌溉制度
1.6.3.3 Optimized Corn Irrigation Scheduling

在1989—1991年3月对玉米灌溉制度试验研究基础上,采用动态规划原理,建立玉米优化灌溉制度数学模型,将玉米分为5个阶段,采用逆向推法求出各种土壤初始含水量及各种总灌溉定额下的最优灌溉方案。即播前、苗期灌水2次(4月下旬~5月下旬);拔节期灌水1次(6月下旬);抽雄—吐丝期灌水3次(7月上旬~下旬);灌浆期灌水1次(7月下旬);乳熟期灌水1次(8月上旬),玉米全生育期灌溉定额为264 m³/亩,灌水次数8次(见表43)。

Based on experimental researches from 1989 to March 1991, a mathematical model of optimized corn irrigation scheduling is established with the principle of dynamic planning. In the model the growth and development period is divided into 5 periods, and the initial water contents of all soil and the optimal irrigation scheme under each total irrigation quota are obtained by using the method of backward inference. 2 times of irrigation in pre-seeding period and seedling period (late-April to late-May); 1 time in jointing period (late-June); 3 times in tasseling & silking period (early-July to late-July); 1 time in filling period (late-July); 1 time in milk ripeness period (early-August); the corn irrigation quota across the whole growth and development period is 264 m³/mu; 8 times of irrigation (see Table 43).

表43　　　　　　　　　　　　玉米灌溉制度优化表

生育期	灌水时间	灌水定额(m³/亩)	灌水次数
苗期	4月下旬 5月上旬	33 33	2
拔节期	6月下旬	33	1
抽雄—吐丝期	7月上旬 7月中旬 7月下旬	33 33 33	3
灌浆期	7月下旬	33	1
乳熟期	8月上旬	33	1

Table 43　　　　　　　　　Optimized Corn Irrigation Scheduling

Growth and Development Period	Irrigation Time	Irrigation Water Quota (m³/mu)	Irrigation Frequency
Seedling Period	Late-April Early-May	33 33	2
Jointing Period	Late-June	33	1
Tasseling-Silking Period	Early-July Middle-July Late-July	33 33 33	3
Filling Period	Late-July	33	1
Milk Ripeness Period	Early-August	33	1

(四) 玉米灌溉效益分摊系数
1.6.3.4 Sharing Coefficient of Corn Irrigation Benefit

研究结果表明,玉米灌溉效益分摊系数,随灌溉保证率升降而增减,在一般农业水平的情况下灌溉增长率为41.6%,在高水平农业技术情况下灌溉增长率为22.73%,效益分摊系数为:水占0.680 5,农业占0.319 5。

The research results show that the change of sharing coefficient of corn irrigation benefit is consistent with that of probability of irrigation; the irrigation growth rate is 41.6% under normal agricultural condition, and 22.73% under high-level agricultural technology condition; the sharing coefficient of benefit is: 0.680 5 for the water, and 0.319 5 for agriculture.

由上述对春小麦、苜蓿、玉米灌溉制度的试验研究结果表明,在荒漠戈壁瘠薄地土质条件下,为即可提高作物产量,节约用水,并可实现避免地下水位上升,防止土地盐碱化之目的,应采用"量少次多"的灌溉制度,根据作物不同生长阶段的需水状况灌水,选择最优灌溉定额(见表44)。

The above experimental results of spring wheat, alfalfa and corn irrigation scheduling show that an irrigation scheduling featuring "small amount, many times" shall be applied to improve yield immediately, save water, prevent the groundwater level rising and avoid land salinization in barren Gobi desert. Furthermore, an optimal irrigation quota shall be identified according to the unique water requirement at different growth stages (see Table 44).

表44 作物最优灌溉制度分析表

作物		春小麦	苜蓿	玉米
灌溉定额	m³/亩	289.14	566.79	264
	mm	433.71	850.19	396
灌水次数(次)		10	15	8
全生育期生长天数(d)		108	173	121

Table 44　Analysis Table for the Optimal Crop Irrigation Scheduling

Crop		Spring Wheat	Alfalfa	Corn
Irrigation Quota	m³/mu	289.14	566.79	264
	mm	433.71	850.19	396
Irrigation Frequency		10	15	8
Days of Whole Growth and Development Period (days)		108	173	121

四、适宜戈壁地的灌水方法
1.6.4 Irrigation Method Suitable for Gobi Desert

在作物生长发育和节水增地研究中,灌水方法是非常重要的。目前新疆北部阿勒泰地区的灌水技术尚比较落后,大水漫灌现象十分普遍,灌水新技术、新方法的应用较少,一些畦灌的地区,畦田也没有达到要求的标准,畦长都在150 m左右,造成畦尾的水还没灌到头,畦头已开始产生深层渗漏,造成毛灌溉定额很高,水资源浪费严重。

Irrigation method is critical for crop growth, water conservation and land increase. Currently in Altay Prefecture of northern Xinjiang, the irrigation technology is outdated, flood irrigation is commonly seen, as well as new technology and method of irrigation are rarely used. Close-end borders fail to meet the required standards in that the border with a length of 150 m will suffer deep infiltration in the head of border while the water is still the rear part of border, causing high gross irrigation quota and severe waste of water.

因此,必须改变长畦的灌水方法。荒漠戈壁地地形基本都是大平小不平,由于土层薄,不能进行大面积土地平整。针对荒漠戈壁地的实际情况,我们进行了畦田与地形坡降的灌水试验研究,根据不同的地面坡降条件,进行短畦范围内的土地平整;同时确定畦长、放水时间以及单宽流量。

Hence the irrigation method for long close-end border must be improved. The topography of Gobi desert land is basically even, but levelling on a large area shall be prohibited for its thin soil layer According to actual conditions of Gobi desert land, irrigation experiments based on close-end border and specific terrain gradient are conducted. Land levelling in short border area is implemented according to specific ground gradient. Border length, irrigation time and discharge per unit width are identified.

(一)放水时间、单宽流量及畦田长度的研究
1.6.4.1 Research of Irrigation Time, Discharge per Unit Width and Border Length

在试验研究过程中,我们仍是以春小麦、玉米、苜蓿及甜菜这4类主要作物为研究对象,根据这4类主要作物的灌水净定额和实灌定额进行放水时间、畦长及单宽流量的试验研究。

Spring wheat, corn, alfalfa and beet are still used as main research objects during the experiment process, and their irrigation quota and actual irrigation quota are used for research of irrigation time, border length and discharge per unit width.

根据非饱和土壤渗吸理论,畦灌灌水延续时间 t 内渗入的水量与计划的灌水定额相当,其计算公式应为:

第一章 综合研究
1 Comprehensive Study

According to the infiltration theory of unsaturated soil, the infiltrated water amount equals the irrigation amount in the duration (t) of border irrigation, and the computational formula shall be:

$$m = K_0 \cdot t^{1-\alpha}$$

式中 t——灌水延续时间,min;
       m——灌水定额,mm;
       K_0——土壤平均渗吸速度,mm/min;
       d——渗吸速度减小指数。

Where t = duration of irrigation, min
       m = irrigation quota, mm
       K_0 = the average soil infiltration rate, mm/min
       d = reduced index of infiltration rate.

根据上式可求得畦灌的延续时间 t:

Duration (t) of border irrigation can be obtained from above formula:

$$t = \left(\frac{m}{K_0}\right)^{\frac{1}{1-\alpha}} \tag{1}$$

进入畦田的灌水总量应与畦田长度 l 上达到灌水定额 m 所需的水量相等:

The total irrigation amount shall equal the amount of irrigation quota (m) in the close-end border with a length of l:

$$60 \cdot q \cdot t = m \cdot l \tag{2}$$

故单宽流量为:

Reasonably the discharge per unit width is:

$$q = m \cdot l / 60 \cdot t$$

式中 t——灌水延续时间,min;
       m——灌水定额,mm;
       l——畦田长度,m;
       q——进入畦田的单宽流量,L/(s·m)。

Where t = duration of irrigation, min
       m = irrigation quota, mm
       l = border length, m
       q = discharge per unit width (L/(s·m)) into the close-end border.

从上述的调查研究我们知道戈壁土地作物的根系主要分布在0~60 cm的土层中,我们按两种计划湿润层来研究,即0~40 cm、40~60 cm 灌溉前的土壤含水率按已达作物的调萎系数来计算,畦田的长度根据实践经验按30、40、50 m。计算结果见表45。

The above researches manifest that the crop root systems in Gobi desert are mainly distributed in the soil layer at a depth of 0 to 60 cm. Two kinds of soil wetting layers are used for research, namely the soil water contents from a depth of 0 to 40 cm and 40 to 60 cm are calculated based on the wilting coefficient, and 30m, 40 m and 50 m are considered as the border length based on practice experience. See Table 45 for calculation results.

表 45　　　　　　放水时间、单宽流时间灌量及畦田长度计算表

灌水定额(mm)	42			52.5			37.8			63		
实灌定额(mm)	55			62.5			68.8			75		
放水时间(min)	55			60			64			67		
畦长(m)	30	40	50	30	40	50	30	40	50	30	40	50
单宽流量(L/s·m)	0.45	0.61	0.76	0.52	0.69	0.87	0.54	0.72	0.90	0.56	0.75	0.93

Table 45　　　Calculation for Irrigation Time, Duration and Amount of Discharge per Unit Width and Border Length

Irrigation Quota (mm)	42			52.5			37.8			63		
Actual Irrigation Quota (mm)	55			62.5			68.8			75		
Water Discharge Time (min)	55			60			64			67		
Border Length (m)	30	40	50	30	40	50	30	40	50	30	40	50
Discharge per Unit Width (L/s·m)	0.45	0.61	0.76	0.52	0.69	0.87	0.54	0.72	0.90	0.56	0.75	0.93

对于每一个灌水定额,相应有灌水延续时间,当灌水延续时间已定时,如畦田的长度太长(毛渠间距),则要加大单宽流量,当单宽流量达到一定值时,则要冲刷土壤。

Each irrigation quota has corresponding irrigation duration. When the irrigation duration is defined, the discharge per unit width shall be increased if the close-end border is too long (field ditch spacing), and the soil will be eroded if the discharge per unit width reaches a certain value.

(二)畦首水深与地面坡降的关系
1.6.4.2　Relation between Headwater Depth and Ground Gradient

欲使确定的灌水定额在规定时间内以一定的单宽流量而实现,对于不同的地面坡降,只能以不同的水头即畦首水深来进行调节,否则,在规定的放水时间里,水流推进的长度

第一章　综合研究
1 Comprehensive Study

将相差很大。因为地面坡降越大,水流推进越快。对于较缓的地面,在密植的条件下,水流推进很慢,因而在一定的放水时间内,水流可能还未到达畦长的一半,要使水流达到畦尾,则要延长放水时间,这样,实际的灌水量将大大超过计划定额。为了使畦田上各点土壤湿润均匀,而又不致于产生深层渗漏,就必须研究在不同灌水定额、不同单宽流量的畦首水深与地面坡降的关系。根据水力学原理,水流推进速率为:

In order to reach the irrigation quota in a manner of discharge per unit width within the stipulated time, the unique ground gradient must be adjusted by different headwater depth, or the water advance distances may differ greatly. The larger the ground gradient gets, the quicker the water advances. For a relatively flat ground, the water advance process is very slowly under the high seeding density in that the water hasn't yet reached the middle of border length within the prescribed time. As a result the time must be extended for the water to reach the rear part of border. Consequently, the actual irrigation amount exceeds the planned amount substantially. In order to reach an even water distribution across the border and avoid deep infiltration, the relation between headwater depth and ground gradient under different irrigation quotas and discharges per unit width must be identified. According to the theory of hydraulics, the water advance velocity is:

$$V = C \cdot \sqrt{h \cdot i} \tag{3}$$

式中　V——水流推进速度,m/s;

C——流速系数,一般为 $15\sqrt{i} \sim 40\sqrt{i}$,其值由田面粗糙程度及作物密度决定,田面越粗糙,$C$ 越小,取 $C = 15\sqrt{i}$;

h——畦首水深,m;

i——地面坡降。

Where　V = water advance velocity, m/s

C = coefficient of velocity; $15\sqrt{i}$ to $40\sqrt{i}$ normally; determined by and consistent with the roughness of field (crop density), $C = 15\sqrt{i}$

h = headwater depth, m

i = ground gradient.

$$v = l/t' \tag{4}$$

式中　l——畦田长度,m;

t'——水流从畦首至畦尾的推进时间,s。

Where　l = border length, m

t' = time (s) of the water advance process from head to rear part.

根据灌溉要求,对于不同的地面坡降,要求有不同的封水长度,由于该区管理水平较低,仅以七成封口和八成封口进行分析。

The irrigation demands that different ground gradient be provided with unique water seal length. Due to the poor management in this area, 70 and 80 percent seal will only be considered.

由上面(3)、(4)式得：
It can be obtained from the above (3) and (4):

$$h = l^2/810\,000 t'^2 \times i^2$$

式中, t' 的单位为 min, h、l 的单位为 m。
Where, minute as the unit of t', m of h and l.

(四) 畦长、入畦流量与地面坡降的关系
1.6.4.3 Relation between Border Length, Flow into Border and Ground Gradient

在不同的地面坡降下，应采用不同的畦田长度，入畦流量也随之发生改变。

Different border length shall be applied according to unique ground gradient, and the flow into border will also vary.

针对新疆北部荒漠瘠薄戈壁地的土质条件和地面大平小不平的特点，在进行灌水方法试验中，还分析了畦长、入畦流量及地面坡降的相互关系，结果如下：

Considering the varied and uneven soil condition of barren Gobi desert land in Northern Xinjiang, the relation between border length, flow into border and ground gradient during the irrigation experiment is considered and the results are as follows:

当地面坡度为 1%~2% 时。采用封闭式畦灌(水到畦尾停灌)，畦长 20~30 m 为宜，入畦流量 4~5 L/s；地面坡度 3%~4% 时，畦长 30~40 m，采用封闭式畦灌，入畦流量 2~3 L/s；地面坡度 5%~8% 时，采用畦流通灌，畦长 40~60 m，入畦流量 1.7~2.2 L/s。

When the ground gradient is from 1% to 2%, Closed border irrigation (irrigation stops once the water reaches the rear part of border) shall be applied, with a suitable border length of 20 to 30 cm and a flow of 4 to 5 L/s; when the ground gradient is from 3% to 4% and the border length is from 30 to 40 cm, the closed border irrigation shall be applied with a flow of 2 to 3 L/s; when from 5% to 8%, then the open border irrigation shall be used, with a border length of 40 to 60 cm and a flow of 1.7 to 2.2 L/s.

由上述分析表明，结合荒漠瘠薄戈壁地的特点，灌水采用"量少次多，改长畦为短畦，缩短每畦灌水时间"的方法是比较符合实际情况的。

The above analysis proves the reasonability of irrigation method " small amount, many times", short border and the reduced duration of each irrigation process.

第一章 综合研究
1 Comprehensive Study

第七节 试验成果效益分析
1.7 Irrigation Benefit

新疆北部阿勒泰地区以农牧业生产为主,农业耕作区主要位于荒漠瘠薄戈壁地上,农牧业生产以灌溉形式为主,要想持续稳定地发展该区农牧业经济,就必须把水利作为基础产业来抓,坚持不懈地进行农田水利基本建设,提高农牧业灌溉技术,不断扩大种植面积和改善落后的管理方法。坚持增水与节水相结合、以节水为主的水利建设方针,在现有的水源基础上实行精量灌溉,提高水的利用率。本项目研究结合该区自然条件和环境特点,通过3年试验和对比分析,提出了适应该区荒漠戈壁地特点的主要作物需水量及灌溉制度等研究成果,为新疆北部阿勒泰地区的水利建设和水利管理提供了科学依据。其节水效益、工程效益、增产效益是十分显著的。

In Northern Xinjiang, the agricultural farmland is mainly located in the barren Gobi desert area, and the main industry in Altay Prefecture is agriculture and husbandry whose water resource is mainly from irrigation. As a result, the agriculture and husbandry industry will be promoted in a sustainable manner only by developing the irrigation works as a basic sector, sticking to the irrigation and water conservancy projects, improving irrigation technology, continuously expanding the plant area and innovating the outdated management. We must implement the precision irrigation method based on the existing water resources to improve the usage of water by staying firm to the irrigation works construction policy of combining water increase and conservation and focusing on water conservation. Based on experiments and comparative analysis in the past 3 years and specific natural conditions in this area, such achievements as water requirement and irrigation scheduling suitable for main crops in the barren Gobi desert area are obtained, providing a scientific basis for the irrigation works construction and management. Its benefits in water conservation, works and yield increase.

一、节水效益
1.7.1 Water Conservation Benefit

据统计在阿勒泰地区几乎每年都有20万~30万亩耕地受到干旱的威胁,使粮食作物减产。并且大多县灌溉定额都在1 000 m³/亩以上,最大的可达2 000 m³/亩(毛)。而通过试验表明,该区春小麦全生育期的最优(经济)需水量为289 m³/亩(净),主要用于作物蒸腾和棵间蒸发。考虑到田间土地不太平整,田间水利用率为0.9,渠系在没有防渗的情况下利用系数为0.36,则小麦的毛灌溉定额可按下式进行计算:

According to the statistics, there is almost 200 000 to 300 000 mu farmland suffering drought that reduces the yield. And the irrigation quotas in many prefectures are beyond 1 000 m³/mu, 2 000 m³/mu at most. The experiments show that 289 m³/mu is the optimal (economical) water requirement of spring wheat across the whole growth and development period, mainly consumed by plant transpiration and soil water evaporation. Considering the extremely uneven field, the irrigation water use efficiency is 0.9, or 0.36 without infiltration prevention treatment, and then the gross irrigation quota can be calculated by the following formula:

春小麦毛灌溉定额 = 净灌溉定额÷田间利用系数÷渠系利用系数
Gross irrigation quota of spring wheat = net irrigation quota ÷ water efficiency in field ÷ water efficiency of canal system

式中,田间利用系数为 0.9;渠系利用系数为 0.36。
Where, water efficiency in field is 0.9; water efficiency of canal system is 0.36.

故在该区春小麦的毛灌溉定额为:
Reasonably the gross irrigation quota of spring wheat in this area is:

$$289 \div 0.9 \div 0.36 = 892 \text{ m}^3/\text{亩}$$
$$289 \div 0.9 \div 0.36 = 892 \text{ m}^3/\text{mu}$$

通过上述计算我们可以发现,在现有的引水能力和灌溉管理水平上每浇 1 亩小麦,最少有 100 多方水被人为的浪费,最多约有 1 000 m³ 水浪费。现状毛灌溉定额我们按 1 200 m³/亩来计算,那么每亩就有 300 m³ 水被浪费。全地区小麦充足灌溉面积按 50 万亩计算,则一年光小麦灌溉就浪费 1.5 亿 m³ 水。再加上其他作物浪费的水近 2 亿 m³ 相当于一个大型水库的水量。如每亩灌溉定额按 892 m³ 进行配水,将节约下来的 2 亿 m³ 水,可灌溉受旱面积近 30 万亩。每亩按增产 50 kg 小麦计算,则全区可增加 1.5 万 t 小麦,折成人民币为 900 万元。

The above formula shows that 100 to 1 000 m³ of water are wasted manually per acre under the current water diversion ability and irrigation management. The gross irrigation quota is calculated by 1 200 m³/mu, then 300 cubic meters of water are wasted per acre. The total local irrigation area is calculated by 500 000 mu, and then 150 000 000 m³ of water are wasted each year, or 200 000 000 m³ of water plus the water wasted for other crops, equaling to the capacity of a large reservoir. For example, an irrigation quota of 892 m³/mu will save 200 000 000 m³ of water that are able to irrigate nearly 300 000 mu drought area. An increase of 50 kg wheat each mu means 15 000 t in the whole plot and RMB 9 000 000 yuan.

二、增产效益
1.7.2　Benefit of Yield Increase

农业增产与水、肥、土、种、管等措施有密切的联系。在降雨量较多的地区,灌溉只是

第一章 综合研究
1 Comprehensive Study

促进农业增产的措施之一。而在气候干旱的荒漠瘠薄土地,灌溉却是农业增产的重要措施。查阅有关资料和本次研究中所做的玉米灌溉效益分摊系数分析计算,在干旱荒漠瘠薄土地的灌溉效益分摊系数为 0.68,农业措施效益分摊系数只占 0.32。

Increase of agricultural production is closely connected to water, fertilizer, soil, plant condition and management, etc. Irrigation is just one of the measures to increase agricultural production in areas with more rainfall. While in arid and barren desert area, it plays a more significant role. According to the information from related data and the analysis of this experiment of sharing coefficient of corn irrigation benefit, the sharing coefficient of irrigation benefit in arid and barren desert area is 0.68, and the sharing coefficient of benefit of agricultural measures is only 0.32.

(一)适时适量的灌溉有利于作物的增产
1.7.2.1 Proper Irrigation can Help Increase Crop Yield

从小麦灌溉制度的试验研究分析可看出,在其他条件相同的情况下,平均每次灌水定额在 28 m^3 左右,灌水次数在 10 次左右,则产量都是较高的。而平均每次灌水定额在 36 m^3 左右,灌水次数在 6 次左右时,产量都较低,平均要低 30 kg 左右。通过分析计算可以发现在计划湿润层不变的情况下(60 cm),当平均每次灌水定额在 36 m^3,灌水次数在 6 次时,土壤中的含水率非常低,需要灌大量的水才能达到田间持水率。而由于浇水的间隔时间太长,土壤中的含水率变化很大,对作物的生长是极为不利的。特别是在小麦的需水高峰期(临界期),缺水使小麦产量大幅度下降。据我们试验结果和调查了解,在本区小麦灌水次数在 6~8 次时,产量一般在 200~250 kg,而灌水次数在 8~12 次时,产量一般都在 250~300 kg。因此,我们认为在该地区由于气候和土质条件的因素,对作物的灌溉应采取"量少、次多"的灌溉方法,即每次的灌水量不要太大,但灌水次数要适当增加,并且特别要注意作物临界期的灌溉。

From the wheat irrigation scheduling experiment, it can be seen that under the same conditions the yield will be comparatively high if each irrigation water quota is about 28 m^3 and the number of irrigation is about 10 tims. On average, the yield may decrease by about 30 kg if the average irrigation water quota is about 36 m^3 and the number of irrigation is about 6 tims. According to analysis and calculation, the soil water content will be extremely low and a great amount of water are required to maintain the field capacity if the average irrigation quota is 36m^3 and the number of irrigation is 6 tims under the same soil wetting layer (60 cm). The long interval of irrigation may influence soil water content, hindering the crop growth. This is so obvious in peak period for water (critical period) that water shortage leads to a sharply reduced wheat yield. It can be drawn from the experiment results and survey that the wheat yield will reach 200 to 250 kg if the number of irrigation in this plot is 6 to 8; or 250 to 300 kg if the number is 8 to 12. As a result, it is reasonable to believe that the irrigation method " small amount, many times" is suitable for crops in this area considering its climate and soil conditions, which means a relatively small irrigation amount and an increased number of irrigation with special attention to the irrigation in critical period.

(二)最优灌溉定额对作物的增产效益
1.7.2.2 Yield Increase Benefit Brought by the Optimal Irrigation Quota

在阿勒泰地区有不少的农牧民认为作物浇的水越多产量就越高,其实并非如此,在干旱地区,虽说灌溉对产量起着很重要的因素,但它是在一定范围内的,超出这个范围就会起到相反的作用。通过这几年的试验我们发现,不论是从小麦需水量与产量的关系曲线图,还是灌溉制度与产量关系曲线图中,均可明显地看到小麦的最优灌溉定额为 289 m³/亩左右,当超过此灌溉定额,小麦的产量就不再增产,如灌溉定额超过太大,产量反而会下降。据调查现状的小麦灌溉定额要比最优的灌溉定额最少多 100 m³/亩。即 389 m³/亩左右。从灌溉制度的产量与灌溉定额关系曲线中可看出,当灌溉定额为 289 m³/亩时,产量为 228.4 kg,当灌溉定额为 389 m³/亩时,产量只有 170 kg 左右,即每亩小麦因灌溉过量而减产 58 kg 左右。全地区小麦有效灌溉面积按 50 万亩计算,则每年因过量灌溉减产达 2.9 万 t,占全地区粮食总产的 10%。

There are many farmers and herdsmen in Altay Prefecture who hold the misconception that more water leads to more yield. On the contrary, irrigation, as a significant factor of yield in arid area, will only play a role in a certain limit. Years of experiments have helped identify that the optimal irrigation quota is about 289 m³/mu, which can be easily read from the wheat water requirement-yield curve and irrigation scheduling-yield curve. Any excess over this point will not do help to yield increase. According to the survey, the current wheat irrigation quota is larger than the optimal quota by 100 m³/mu at least, namely about 389 m³/mu. It can be seen from the yield-irrigation quota curve of irrigation scheduling that when the irrigation quota is 289 m³/mu, the yield is 228.4 kg, or only about 170 kg when the quota is 389 m³/mu, namely an reduction of about 58 kg per mu. If there are 500 000 mu of effective irrigation area in the whole prefecture, then 29 000 t will be reduced by excess irrigation, accounting for 10% of the total.

三、经济效益分析
1.7.3 Economic Benefit Analysis

我们把不同处理水平的试验结果和民间不同的灌溉管理方法所产生的经济效益列表进行分析(见表46)。

A table analysis of economic benefit is made between experimental results at different treatment levels and different civil irrigation management methods (see Table 46).

第一章 综合研究
1 Comprehensive Study

表46 春小麦平均每亩的投资和产值统计对比表

项目 处理水平		农业投资（元）	水量费		浇水人工费			春小麦		利润	
			净水量（m³/亩）	毛水量（m³/亩）	水费（元）	次数	人工费（元）	总投资（元）	产量（kg/亩）	产值（元）	纯收入（元）
试验	50%	85	217.61	670	2.7	7	3.5	91.2	179	98.45	7.25
	60%	85	214	661	2.6	7	3.5	91.1	190	104.5	13.4
	70%	85	284	877	3.5	10	5	93.5	212	116.6	23.1
民间	灌水次数少	85		1200	4.8	7	3.5	93.3	225	123.75	30.45
	灌水次数多	85		1200	4.8	10	5	94.8	275	151.2	56.45

Table 46 Comparison for Spring Wheat Investment and Output Value per Mu

Item Treatment Level		Agriculture Investment (yuan)	Water Cost		Manual Irrigation Cost			Spring Wheat		Profit	
			Net Water Amount (m³/mu)	Gross Water Amount (m³/mu)	Water Cost (yuan)	Number of Irrigation	Manpower Cost (yuan)	Total Investment (yuan)	Yield (kg/mu)	Output Value (yuan)	Net Profit (yuan)
Tests	50%	85	217.61	670	2.7	7	3.5	91.2	179	98.45	7.25
	60%	85	214	661	2.6	7	3.5	91.1	190	104.5	13.4
	70%	85	284	877	3.5	10	5	93.5	212	116.6	23.1
Civil	Few Times of Irrigation	85		1200	4.8	7	3.5	93.3	225	123.75	30.45
	Many Times of Irrigation	85		1200	4.8	10	5	94.8	275	151.2	56.45

从表中可以看出，在种植水平相同的情况下，灌水次数多的，其经济效益都高于灌水次数少的经济效益。

It can be seen from the table that economic benefits produced by the plots with many times of irrigation are larger than that of plots with few times of irrigation under the same plant conditions.

四、对灌区规划、设计及管理所产生的效益
1.7.4 Benefits Produced by Irrigation Area Programming, Design and Management

在进行灌区水利规划设计和灌溉管理及工程管理时,没有基础资料往往难以把握工程设计规模,因而常因将保险系数放大设计而造成浪费。而该研究试验成果则为该区水工建筑物的规划设计和管理提供了第一手科学资料。

It's difficult to define the project design scale without basic data in the design and programming of irrigation area and the management of irrigation and project. Therefore, the assurance coefficient is generally enlarged in design to cause waste. This experiment offers the first-hand scientific data for the programming, design and management of the hydraulic structure in this area.

(一) 节省水利工程量
1.7.4.1 Hydraulic Engineering Amount Saving

阿勒泰地区在以前进行灌区规划设计时,基本上是按 1 m³/s 浇灌 1 万亩地来控制的,有的甚至更大。其原因是没有第一手基础资料,不知灌溉定额和需水量采用多大为好。在设计时因担心水不够用,总是把渠道、闸门设计的大一些,结果造成工程量过大,产生许多浪费。有了作物需水量及灌溉定额方面的基础资料,在灌区规划设计时就可按不同作物的需水量和最优灌溉制度来设计建筑物的大小,从而尽可能地避免在工程设计时产生不必要的浪费。

Basically 1 m³/s was the minimum planned irrigation amount for each 10 thousand mu during the programming and design of irrigation area in Altay Prefecture. The reason for this planned irrigation amount is lacking the first-hand basic data of the precise irrigation quota and water requirement. The worry about water shortage leads to larger channels and gates during design, resulting in excessive engineering amount and enormous waste. The basic data of water requirement and irrigation quota help design a suitable structure based on the water requirement and the optimal irrigation scheduling, which may reduce unnecessary waste as much as possible during the engineering design.

(二) 提高管理水平
1.7.4.2 Management Improvement

该地区无论是灌溉管理或是工程管理都是较落后。特别是在灌溉管理方面更为落后,基本上是"有水灌个够,没水旱个够,"水资源未能得到合理调配和利用。其结果,在水资源并不缺乏的阿勒泰地区,每年都有几十万亩的小麦受旱,使该地区的粮食产量受到

第一章　综合研究
1　Comprehensive Study

很大影响。其重要的一个因素是各管理单位都没有作物的需水量和灌溉制度等方面的资料,也就无法作较科学的用水计划。没有用水计划也就谈不上配水计划和供水计划。水管单位是有水就放,农牧民是有水就灌,结果是有水时就产生大水漫灌,排水渠成了排地表水,造成排水渠冲毁,大量泥沙淤积在排水渠内,使排水渠起不到排水效果。另外,因没有配水计划和供水计划,农牧民白天抢着用水,一到晚上都不用水,把各自的进水闸都关死,结果造成渠道水无出路,只有从渠顶翻出,把渠道冲毁。就上述现象每年给灌排工程的维修就带来了大量的工作,消耗大量的人力物力。有了作物灌水量试验成果,各管理部门可在每年的放水前作出用水计划,再根据用水计划作出配水计划,做到定时定量的供水,这样即可避免大水漫灌和日浇夜退的现象,又可大量地减少工程维修量,使水利工程发挥其正常效益。

　　The irrigation management and the engineering management in this area are both outdated, especially in the irrigation management. Basically the main irrigation strategy is "irrigate water to an amount depending on the stored amount", lacking a reasonable allocation and use of water resource. Consequently, there are tens of thousands *mu* of wheat suffering from drought in Altay Prefecture where water resource is sufficient in this area, enormously reducing the crop yield. One of the critical reasons is that each management unit lacks data concerning crop water requirement and irrigation scheduling, etc., which is important for a scientific water use plan. If there is no scientific water use plan, then there is no water distribution plan and water supply plan. The water management unit and farmers and herdsmen usually have no water release or irrigation control plan in hand, leading to broad irrigation, furthermore, the surface water in the drainage channels may destroy the channels and leave a great amount of silt, as a result the channels may fail to play their role. What's worse, farmers and herdsmen, having no water distribution and supply plans in hand, rush to irrigate water in the daytime while they close the intake gates at night, resulting in water overflow from the channel top, which may destroy the channel. The aforementioned phenomenon has brought about enormous repair to the irrigation and drainage engineering, consuming a large amount of manpower and resources. With the crop irrigation amount experiments, the management units at all levels are able to work out a water use plan before release, based on which to work out a water distribution plan so as to achieve a regular and quantitative water supply, avoiding broad irrigation and irrigation only in the daytime and reducing repair to facilitate the hydraulic engineering to play its role properly.

五、防止土地盐碱化和肥土流失
1.7.5　Soil Salinization and Fertile Soil Loss Prevention

　　据有关资料统计,阿勒泰地区从新中国成立以来已有60万亩地因土地盐碱化而弃耕,造成大量的水利资金浪费,而且农牧民因土地废弃还得进行搬迁,给农牧民也带来很大损失,这都是因为大水漫灌、灌溉定额太大以及灌排措施不配套等原因形成的。此外,由于灌溉定额太大,使土壤中的肥力也随着深层渗漏或冲刷而流失,这对作物的生长也是

干旱瘠薄土地作物需水量与灌溉制度研究(以新疆阿勒泰地区为例)
A Case Study on Crop Water Requirement and Irrigation Scheduling in Arid and Barren Land (Altay Prefecture, Xinjiang)

十分不利的。有了这些基础资料,我们就可以控制灌水定额和灌溉定额,适时适量灌水,尽量不产生深层渗漏和少产生深层渗漏,控制地下水位的上升和肥土流失,使开耕的土地充分发挥其作用和效益。

According to relevant statistics, about 600 thousand mu of arable land have been wasted by soil salinization since the founding of China in Altay Prefecture, causing great damages to the water engineering and farmers and herdsmen for they have to find another place to live. All those are caused by such reasons as border irrigation, excessive irrigation quota and improper irrigation measures, etc. Besides, the excessive irrigation may cause the soil to lose its fertility by deep infiltration or erosion, hindering the crop growth. Those basic data will enable us to control the irrigation requirement and quota, achieve a regular and quantitative irrigation, prevent or reduce deep infiltration, control the groundwater table and fertile soil loss and facilitate the arable land to better play its role and generate more benefits.

第二章 作物需水量与需水规律
2 Water Requirement of Crops and Its Law

第一节 作物需水量分析
2.1 Water Requirement

作物需水量是指在作物生育期内的植株蒸腾量和棵间蒸发量之和。作物需水量是农业生产最主要的水分消耗部分,也是影响作物灌溉制度和灌水量的基本因素。它作为农田水利工程规划、设计管理和灌溉预报的基本参数,是农田水利基础理论研究的重要内容之一。

The crop water requirement is the sum of plant transpiration and evaporation in the growth and development period. As the main part consumed for agricultural production, the crop water requirement is also a basic factor for irrigation scheduling and irrigation amount. It is a basic parameter for the programming, design and management of farmland water conservancy project and irrigation forecast, as well as one of the key parts of basic research of farmland water conservancy.

根据本项目作物需水量试验方案,本章着重就春小麦、苜蓿、玉米、甜菜等本地区主要农作物的全生育期需水量、各旬需水量以及各旬需水强度等的试验内容及研究结果介绍如下:

According to experimental schemes of crop water requirement, this chapter puts emphasis on the water requirement in the whole growth period and of every 10 days and the water requirement intensity of every 10 days and so on of such main crops in this area as spring wheat, alfalfa, corn and beet, etc., and the experimental results are as follows:

一、作物全生育期需水量
2.1.1 Crop Water Requirement in the Whole Growth and Development Period

作物全生育期需水量测定是根据本课题试验处理及试验方案进行测试。观测期为

1989—1991 年 3 年时间,现将本次试验的主要农作物全生育期需水量试验结果分述如下:

The test of crop water requirement in the whole growth and development period is based on the experimental treatment and schemes in this project. The observation period is from 1989 to 1991, and the water requirement experimental results of crops in this experiment are as follows:

(一) 不同水平的小麦全生育期需水量
2.1.1.1 The Wheat Water Requirement in the Whole Growth and Development Period at Different Levels

小麦不同水平全生育期需水量测试是根据表 25 需水量试验方案进行测试。测试年份 3 年(1989—1991 年)。小麦全生育期的需水时段在本区为 4 月至 7 月中旬。测试结果表明,不同水平的春小麦需水量不同,1 号水平的 3 年平均需水量为 319.06 m³/亩;2 号水平的 3 年平均需水量为 289.14 m³/亩;3 号水平的需水量 216.10 m³/亩。3 个水平相比较,1 号水平的需水量最高,3 号水平的需水量最低(见表 47)。

The wheat water requirement experiment in the whole growth and development period at different levels is based on the experimental scheme in Table 25. Test period lasts 3 years (1989 to 1991). Period from April to middle-July is the requirement period in the total wheat growth and development period in this plot. The test results show that water requirement of spring wheat varies with the levels; the average water requirement of Level #1 is 319.06 m³/mu, the Level #2 is 289.14 m³/mu, and the Level #3 is 216.10 m³/mu. Level #1 has the highest water requirement; while Level #3 has the lowest (see Table 47).

表 47　小麦不同水平全生育期需水量统计表(1989—1991 年)　　单位:m³/亩

年份	1989	1991	1991	平均
1	355.21	278.14	324.13	319.06
2	312.39	266.71	315.30	289.14
3	267.31	214.74	166.38	216.14

Table 47　Statistics of Wheat Water Requirement in the Whole Growth and Development Period at Different Levels (1989 to 1991)　　Unit:m³/mu

Year	1989	1991	1991	Average
1	355.21	278.14	324.13	319.06
2	312.39	266.71	315.30	289.14
3	267.31	214.74	166.38	216.14

（二）不同水平的玉米全生育期需水量
2.1.1.2 The Corn Water Requirement in the Whole Growth and Development Period at Different Levels

玉米全生育期需水量是按表 27 玉米需水量试验设计方案测试。玉米全生育期的需水时段为 5 月中旬至 9 月上旬，较春小麦需水量迟 1 个半月。1989—1991 年 3 年测试结果表明，1 号水平的全生育期需水量平均值最高，为 314.66 m³/亩；2 号水平次之，为 282.18 m³/亩；3 号水平最低，为 249.06 m³/亩（见表 48）。

The corn water requirement experiment in the whole growth and development period is based on the experimental design scheme in Table 27. The period from mid-May to early-September is the water requirement period in the total corn growth and development period, which is one and a half months later than that of spring wheat. Experimental results show that Level #1 has the highest average water requirement of 314.66 m³/mu, followed by Level #2 of 282.18 m³/mu, and Level #3 has the lowest value of 249.06 m³/mu (see Table 48).

表 48 玉米不同水平全生育期需水量统计表（1989—1991 年）　　单位：m³/亩

年份	1989	1991	1991	平均
1	382.88	241.37	319.75	314.66
2	325.10	213.20	308.25	282.18
3	313.25	215.73	218.20	249.06

Table 48　Statistics of Corn Water Requirement in the Whole Growth and Development Period at Different Levels (1989 to 1991)　Unit: m³/mu

Year	1989	1991	1991	Average
1	382.88	241.37	319.75	314.66
2	325.10	213.20	308.25	282.18
3	313.25	215.73	218.20	249.06

（三）不同水平的苜蓿全生育期需水量
2.1.1.3 The Alfalfa Water Requirement in the Whole Growth and Development Period at Different Levels

不同水平的苜蓿全生育期需水量测试工作仅记录了 1990—1991 年 2 年的实测数据，在本区苜蓿全生育期的需水时段为 4 月中旬至 9 月中旬。根据表 26 苜蓿需水量试验方案，1990—1991 年 2 年实际测试结果表明，不同处理水平苜蓿的全生育期需水量也不相同。同样表现出 1 号水平全生育期的平均需水量最高，为 608.2 m³/亩；2 号水平的平均需水量次之，为540.4 m³/亩；3号水平的平均需水量最低，仅为459.5 m³/亩，较1号水平需水量低约150 m³/亩（见表49）。

Only the experimental data from 1990 to 1991 of alfalfa water requirement in the whole growth and development period at different levels are kept, and the period from mid-April to mid-September is the water requirement period in the whole growth and development period in this plot. Table 26 (Experimental Scheme of Alfalfa Water Requirement) and the experimental data from 1990 to 1991 show that the alfalfa water requirement varies with the treatment levels in the whole growth and development period. Besides, they also show that Level #1 has the highest average water requirement of 608.2 m³/mu, followed by Level #2 of 540.4 m³/mu, and Level #3 has the lowest of 459.5 m³/mu, about 150 m³/mu smaller than Level #1 (see Table 49).

表49　　　　　　　　　苜蓿全生育期需水量（1989—1991年）　　　　　　　　单位：m³/亩

年份	1991	1991	平均
1	566.63	649.83	608.23
2	494.86	585.9	540.38
3	422.66	496.59	459.53

Table 49　Alfalfa Water Requirement in the Whole Growth and Development Period (1989 to 1991)

Unit：m³/mu

Year	1991	1991	Average
1	566.63	649.83	608.23
2	494.86	585.9	540.38
3	422.66	496.59	459.53

（四）不同水平的甜菜全生育期需水量
2.1.1.4 The Beet Water Requirement in the Whole Growth and Development Period at Different Levels

甜菜需水量平均值按表28试验设计方案。甜菜全生育期需水时段为5～9月5个月。根据1989—1991年3年测试结果表明，因测坑类别处理水平不同，需水量不尽相同（见表50）。1号水平的3年（1989—1991年）需水量平均值为375.93 m³/亩；2号水平的3年需水量平均值为316.11 m³/亩；3号水平的平均值为273.22 m³/亩。表现为1号水平的需水量最高，3号水平的需水量最低，较1号水平的平均需水量低约100 m³/亩。

The experimental scheme of average beet water requirement is based on Table 28. Period from May to September is the water requirement period in the total beet growth and development period. The experimental results from 1989 to 1991 show that the water requirement varies with the treatment levels of test pits (see Table 50). The average water requirement (1989 to 1991) of Level #1 is 375.93 m³/mu, Level #2 is 316.11 m³/mu and Level #3 is 273.22 m³/mu. Level #1 has the highest water requirement, while Level #3 has the lowest, about 100 m³/mu less than Level #1.

表 50　　　　　　甜菜全生育期需水量（1989—1991 年）　　　　单位：m³/亩

年份	1989	1991	1991	平均
1	317.67	385.13	425.00	375.93
2	256.67	359.67	332.00	316.11
3	214.00	315.67	290.00	273.22

Table 50　Beet Water Requirement in the Whole Growth and Development Period (1989 to 1991)

Unit: m³/mu

Year	1989	1991	1991	Average
1	317.67	385.13	425.00	375.93
2	256.67	359.67	332.00	316.11
3	214.00	315.67	290.00	273.22

（五）不同水平的作物全生育期需水量研究小结
2.1.1.5　Research Summary of Crops Water Requirement at Different Levels

通过对本区主要农作物春小麦、苜蓿、玉米及甜菜的试验分析，各不同水平的作物需水量很不相同（见表 51），春小麦、苜蓿、玉米及甜菜这 4 类作物的需水量均以 1 号水平的需水量最高，3 号水平的需水量最低。

The water requirement experiments of spring wheat, alfalfa, corn and beet in this plot show a very varied result (see Table 51), and Level #1 has the highest water requirement, while Level #3 has the lowest.

表 51　　　　　　主要作物全生育期需水量分析表　　　　单位：m³/亩

年份	春小麦	苜蓿	玉米	甜菜
1	319.06	608.23	314.66	375.93
2	289.14	540.38	282.18	316.11
3	216.14	459.53	249.06	273.22
平均	274.78	536.05	282.00	321.75

Table 51　　Analysis of Water Requirement of Main Crops in the Whole Growth and Development Period

Unit: m³/mu

Year	Spring Wheat	Alfalfa	Corn	Beet
1	319.06	608.23	314.66	375.93
2	289.14	540.38	282.18	316.11
3	216.14	459.53	249.06	273.22
Average	274.78	536.05	282.00	321.75

二、作物生育期间各旬需水量

2.1.2　Crop Water Requirement of Every 10 Days in the Growth and Development Period

在作物生育期，不同时期作物的需水量是不同的。为了准确详尽了解作物生育期内不同时段的需水状况，我们对各旬作物需水量作了测试分析。结果表明，不同作物在全生育期内的各旬需水量是不同的，这对我们节水灌溉是非常有帮助的。

The water requirement varies with growth stages of crops. To gain a detailed understanding of water requirement at different growth stages, tests are made to analyze the water requirement of every 10 days. The results show that different crops have difference water requirements of every 10 days, which may help achieve water-saving irrigation.

（一）小麦生育期内各旬需水分析

2.1.2.1　Analysis of Wheat Water Requirement of Every 10 Days in the Growth and Development Period

新疆北部荒漠戈壁地的春小麦生育需水期为4~7月4个月，我们对这4个月各旬需水量及模系数进行了4年（1989—1991年）的测试分析，测试结果见表52。

It is the water requirement period (April to July) for spring wheat in barren Gobi desert land of Northern Xinjiang, and we had experimented with the water requirement and modulus of every 10 days for three years (1989 – 1991). Please see Table 52 for experimental results.

由表52可见，1号水平的需水量以6月最高，6月又以其下旬为高，达56.70 m³/亩；2号水平的需水量虽以6月最高，但6月内则以6月中旬为高，达59.41 m³/亩；3号水平的最大需水量也出现在6月中旬，由图4可见，3个水平的小麦需水量自4月至6月中下旬均呈多峰形式，需水量呈增长势态，达到顶峰后，曲线形态发生变化，需水量呈直线形式下降。

Table 52 shows that the water requirement of Level #1 is comparatively high in June, and

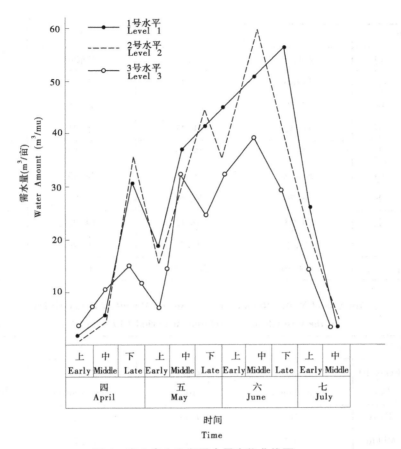

图 4 春小麦生育期需水量变化曲线图

Figure 4　Changing Curves of Spring Wheat's Water Requirement at Growth and Development Period

peaks in late-June at 56.70 m³/mu; the water requirement of Level #2 is also comparatively high in June, but peaks in mid-June at 59.41 m³/mu; the highest water requirement of Level #3 is in mid-June. We can see from Figure 4 that the water requirement curves of the 3 levels are distributing in a multimodal form from April to late-June, in which the curves show a rising tendency, and then descend in a straight line form after reaching the peak.

表 52　　　　春小麦生育期内各旬需水量分析表(1989—1991 年)　　　　单位:m³/亩

月份	水平 旬	1 需水量	2 需水量	3 需水量	平均需水量
	上	1.16	0.48	3.62	1.75
4	中	5.49	1.55	11.74	6.93
	下	31.39	26.59	14.85	27.34

续表 52

月份	水平 旬	1 需水量	2 需水量	3 需水量	平均需水量
5	上	18.86	14.97	7.62	13.82
	中	37.78	28.72	33.36	33.29
	下	41.42	44.25	25.21	36.96
6	上	45.89	35.89	32.57	37.85
	中	51.02	53.41	34.85	50.09
	下	56.70	40.29	29.82	42.27
7	上	25.93	20.92	14.90	20.58
	中	3.22	4.87	3.36	3.82
合计		319.86	289.14	216.10	274.77

Table 52　Analysis of Water Requirement of Spring Wheat of Every 10 Days in the Growth and Development Period (1989 to 1991)

Unit: m^3/mu

Month	Every 10 Days	1 Water Requirement	2 Water Requirement	3 Water Requirement	Average Water Requirement
4	Early	1.16	0.48	3.62	1.75
	Middle	5.49	1.55	11.74	6.93
	Late	31.39	26.59	14.85	27.34
5	Early	18.86	14.97	7.62	13.82
	Middle	37.78	28.72	33.36	33.29
	Late	41.42	44.25	25.21	36.96
6	Early	45.89	35.89	32.57	37.85
	Middle	51.02	53.41	34.85	50.09
	Late	56.70	40.29	29.82	42.27
7	Early	25.93	20.92	14.90	20.58
	Middle	3.22	4.87	3.36	3.82
Total		319.86	289.14	216.10	274.77

（二）苜蓿生育期内各旬需水量分析
2.1.2.2 Analysis of Alfalfa Water Requirement of Every 10 Days in the Growth and Development Period

苜蓿的全生育需水期为 4~9 月约 6 个月，在全生育期内的各旬需水量测试分析成果见表 53。由表 53 可见，1 号水平的需水量以 6 月下旬最高，达 70.2 m³/亩，次为 8 月上旬，为 69.86 m³/亩；2 号水平的需水量以 5 月下旬最高，达 61.32 m³/亩，次为 7 月上旬，为 51.06 m³；3 号水平的需水量以 6 月上旬最高，达 60.36 m³/亩，次为 8 月中旬，达 50.78 m³/亩。

The alfalfa water requirement period usually lasts 6 months (April to September) in the whole growth and development period. See Table 53 for the water requirement test results of every 10 days in the whole growth and development period. It can be seen from Table 53 that the water requirement of Level #1 peaks in late-June at 70.2 m³/mu, followed by 69.86 m³/mu in early-August. The water requirement of Level #2 peaks in late-May at 61.32 m³/mu, followed by 51.06 m³/mu in early-July. The water requirement of Level #3 peaks in early-June at 60.36 m³/mu, followed by 50.78 m³/mu in middle-August.

表 53　苜蓿全生育期各旬需水量分析表（1990—1991 年）　　单位：m³/亩

月份	旬	水平 1 需水量	2 需水量	3 需水量	平均需水量
4	中	11.2	10.14	5.7	9.01
	下	19.3	19.47	9.4	16.06
5	上	23.81	36.49	20.41	26.90
	中	34.11	41.03	29.02	34.72
	下	60.22	61.32	33.23	51.59
6	上	56.21	44.22	60.36	53.60
	中	11.24	15.85	17.12	14.74
	下	70.2	37.16	38.33	48.56
7	上	46.66	51.06	34.31	44.01
	中	43.99	30.26	27.93	34.06
	下	34.02	29.20	26.78	30

续表53

水平		1	2	3	平均需水量
月份	旬	需水量	需水量	需水量	
8	上	69.86	47.43	44.83	54.04
	中	48.16	46.74	50.78	48.56
	下	41.41	39.34	22.6	34.45
9	上	27.88	24.91	29.21	27.33
	中	9.96	5.82	9.49	8.42
合计		608.23	540.44	459.50	

Table 53　　Analysis of Alfalfa Water Requirement of Every 10 Days in the Growth and Development Period (1990 to 1991)　　Unit: m³/mu

Time		Level			Average Water Requirement
		1	2	3	
Month	Every 10 Days	Water Requirement	Water Requirement	Water Requirement	
4	Middle	11.2	10.14	5.7	9.01
	Late	19.3	19.47	9.4	16.06
5	Early	23.81	36.49	20.41	26.90
	Middle	34.11	41.03	29.02	34.72
	Late	60.22	61.32	33.23	51.59
6	Early	56.21	44.22	60.36	53.60
	Middle	11.24	15.85	17.12	14.74
	Late	70.2	37.16	38.33	48.56
7	Early	46.66	51.06	34.31	44.01
	Middle	43.99	30.26	27.93	34.06
	Late	34.02	29.20	26.78	30
8	Early	69.86	47.43	44.83	54.04
	Middle	48.16	46.74	50.78	48.56
	Late	41.41	39.34	22.6	34.45

Continue table 53

Time		Level			Average Water Requirement
Month	Every 10 Days	1 Water Requirement	2 Water Requirement	3 Water Requirement	
9	Early	27.88	24.91	29.21	27.33
	Middle	9.96	5.82	9.49	8.42
Total		608.23	540.44	459.50	

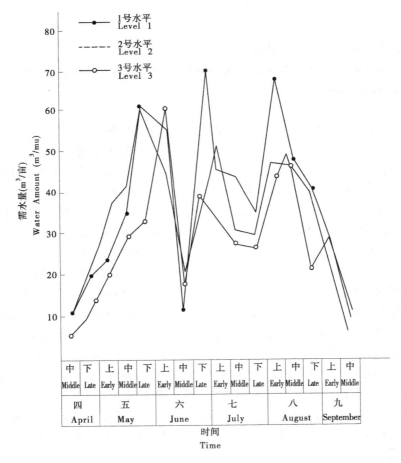

图 5 苜蓿全生育期需水量变化曲线
Figure 5　Changing Curves of Alfalfa's Water Requirement at Whole Growth Period

由图 5 可见,苜蓿生育期内的需水量变化幅度较大。3 个不同水平的苜蓿生长过程均变化出一致的特点,即在蔓枝伸长期需水量最大,苜蓿打三茬生育期内的需水量变化曲线呈三峰形式,峰谷相间,其中 1 号水平的最高峰和次高峰分别出现在首蓿第二茬和第三茬,即 6 月下旬和 8 月上旬,且需水量远较 2 号水平和 3 号水平高。2 号水平和 3 号水平的最高峰出现则在首蓿第一茬,需水量亦较 1 号水平少。

According to Figure 5, the water requirement in alfalfa growth and development period fluctuates greatly. One common feature is found in the three levels, namely the highest water requirement shown in branch extension period of alfalfa. The changing curves of alfalfa water requirement in 3 croppings are distributing with 3 peaks, valleys alternating with each other; the highest and the 2nd highest peaks of Level #1 are shown in the 2nd cropping and 3rd cropping, namely in late-June and early-August respectively, with a water requirement much larger than that of Level #2 and Level #3. The highest peaks of Level #2 and Level #3 are shown in the 1st cropping, with a smaller water requirement than Level #1.

(三)玉米生育期各旬需水量分析

2.1.2.3 Analysis of Corn Water Requirement of Every 10 Days in the Growth and Development Period

玉米全生育期的需水时段为 5 月中旬至 9 月上旬。1989—1991 年 3 年。实地测试结果表明(见表 54、图 6),1 号水平在作物生长初期(5 月中旬)的需水量较高,为 13.83 m³/亩,到 5 月下旬和 6 月上旬作物需水量降低,至 6 月中旬,1 号水平的需水量恢复到作物生长初期的 5 月中旬。到了 6 月下旬,1 号水平的需水量出现了第一次峰值,旬需水量为 47.78 m³/亩,进入 7 月上旬和中旬,需水量减少,而到了 7 月下旬和 8 月上旬,再度出现第二次需水高值区,这时 1 号水平的需水量达 50.02~51.02 m³/亩;2 号水平在作物生长初期虽需水量不多,5 月中旬至 6 月中旬的 40 d 内尚变化在 5.16~12.07 m³/亩之间,然而到了 6 月下旬,2 号水平的作物需水量骤然升高,由 6 月中旬的 9.45 m³/亩升至 6 月下旬的 37.05 m³/亩,并且进入 7 月上旬进一步增至 53.90 m³/亩,以后时段内需水量虽有增减,但再未达到 7 月上旬的需水量值;3 号水平的需水量除作物生长初期的 5 月中旬需水略高外(14.67 m³/亩),从 5 月下旬至 7 月上旬,需水量状况一直是个缓慢增加过程,由 5 月下旬的 9.66 m³/亩增至 7 月上旬的 32.18 m³/亩,出现第一次峰值。7 月中旬的需水量略有略低,而到了 7 月下旬,需水量再度增加,达 45.86 m³/亩。从 3 个不同水平全生育期各旬需水状况分析,玉米 1 号水平全生育期的需水量以 7 月下旬和 8 月上旬最高,次为 6 月下旬;2 号水的需水量以 7 月上旬最高,次为 7 月下旬;3 号水平的需水量以 7 月下旬最高,次为 7 月上旬。3 个不同水平的全生育期内各旬需水量对比分析,仍以 3 号水平的需水量最低。

The-corn water requirement period lasts from middle – May to early-September in the whole growth and development period. The field experiments (1989 – 1991) (see Table 54 and Figure 6) show that Level #1 has a comparatively high water requirement of 13.83 m³/mu in the initial crop growth stage (middle-May), and it decreases from late-May or early-June to middle-June and then rises to the level of middle-May. By late-June, the water requirement of

2 Water Requirement of Crops and Its Law

Level #1 peaks for the first time at 47.78 m³/mu, and then decreases in late and middle-July until peaks again by late-July and early-August, 50.02 to 51.02 m³/mu on an average; although the water requirement of Level #2 in the initial growth stage is comparatively low or ranges from 5.16 to 12.07 m³/mu, it increases sharply in late-June from 9.45 m³/mu in middle-June to 37.05 m³/mu in early-July and reaches 53.90 m³/mu, later on it fluctuates under this limit; in the initial growth stage of Level #3 the water requirement is comparatively high in middle-May; then it increases slowly from 9.66 m³/mu in late-May to the first peak of 32.18 m³/mu in early-July. The water requirement in middle-July decreases a little bit and then increases in late-July to 45.86 m³/mu. The analysis of the water requirement of every 10 days in the whole growth and development period of the 3 levels shows that Level #1 has the highest water requirement in late-July and early-August, and the 2nd highest in late-June; Level #2 has the highest in early-July and the 2nd in late-July; Level #3 has the highest in late-July and the 2nd in early-July. Level #3 has the minimum water requirement of every 10 days in the whole growth and development period.

表 54　　玉米全生育期各旬需水量分析表(1989—1991 年)　　单位:m³/亩

时段		5月		6月			7月			8月			9月
需水量		中旬	下旬	上旬	中旬	下旬	上旬	中旬	下旬	上旬	中旬	下旬	上旬
1989 年	1	22.32	—	5.15	28.55	67.43	25.58	22.01	52.81	76.84	34.38	32.48	17.71
	2	5.72	3.71	4.80	12.98	55.93	66.66	14.03	47.96	19.58	44.56	11.63	19.48
	3	31.76	2.21	—	25.23	35.01	41.52	21.92	66.27	21.75	36.98	15.96	16.42
1990 年	1	5.88	2.24	13.39	0.40	42.77	41.08	31.82	43.01	26.08	20.94	—	2.46
	2	—	3.31	15.06	—	29.04	62.25	12.43	37.37	7.71	20.94	—	2.46
	3	6.35	—	22.94	2.56	27.93	36.64	32.09	75.02	15.99	28.72	—	7.04
1991 年	1	13.27	23.14	4.82	9.95	33.15	42.60	43.35	57.23	48.18	40.93	—	31.3
	2	14.82	8.45	16.35	16.65	26.19	32.00	38.97	48.98	36.31	47.16	—	14.58
	3	5.90	6.98	8.44	18.45	17.40	18.39	25.28	39.23	26.25	29.12	—	14.12
3 年平均	1	13.83	8.46	7.79	12.97	47.78	36.35	32.40	51.02	50.02	35.17	10.83	17.71
	2	6.83	5.16	12.07	9.45	37.05	53.90	21.83	44.73	21.2	37.55	3.80	28.43
	3	14.67	9.66	10.46	15.41	26.78	32.10	26.43	45.86	21.43	31.61	5.32	12.53

Table 54 Analysis of Water Requirement of Every 10 Days in the Total Corn Growth and Development Period (1989 to 1991) Unit: m³/mu

Period Water Amount		May Middle	May Late	June Early	June Middle	June Late	July Early	July Middle	July Late	August Early	August Middle	August Late	September Early
1989	1	22.32	—	5.15	28.55	67.43	25.58	22.01	52.81	76.84	34.38	32.48	17.71
	2	5.72	3.71	4.80	12.98	55.93	66.66	14.03	47.96	19.58	44.56	11.63	19.48
	3	31.76	2.21	—	25.23	35.01	41.52	21.92	66.27	21.75	36.98	15.96	16.42
1990	1	5.88	2.24	13.39	0.40	42.77	41.08	31.82	43.01	26.08	20.94	—	2.46
	2	—	3.31	15.06	—	29.04	62.25	12.43	37.37	7.71	20.94	—	2.46
	3	6.35	—	22.94	2.56	27.93	36.64	32.09	75.02	15.99	28.72	—	7.04
1991	1	13.27	23.14	4.82	9.95	33.15	42.60	43.35	57.23	48.18	40.93	—	31.3
	2	14.82	8.45	16.35	16.65	26.19	32.00	38.97	48.98	36.31	47.16	—	14.58
	3	5.90	6.98	8.44	18.45	17.40	18.39	25.28	39.23	26.25	29.12	—	14.12
Average of the Three Years	1	13.83	8.46	7.79	12.97	47.78	36.35	32.40	51.02	50.02	35.17	10.83	17.71
	2	6.83	5.16	12.07	9.45	37.05	53.90	21.83	44.73	21.2	37.55	3.80	28.43
	3	14.67	9.66	10.46	15.41	26.78	32.10	26.43	45.86	21.43	31.61	5.32	12.53

由图 6 可见,玉米在生育期内有两次需水高峰期,第一次是 6 月下旬至 7 月上旬,第二次为 7 月下旬,因而呈两峰形式,而 7 月中旬玉米的需水量则较少,表现为谷值区。玉米需水量变化的另一特点是,自 5 月至 6 月底或 7 月上旬,需水量呈直线上增趋势,达到第一次峰值,而在此以后,需水量则呈波动变化,时多时少,峰谷相间。

It can be seen from Figure 6 that in the growth and development period the water requirement peaks twice in late-June to early-July and in late-July, and that middle-July has a comparatively low water requirement, hence called the valley area. The other feature of the corn water requirement change is that it rises in a straight line form from May, late-June or early-July until peaks for the first time, later on it fluctuates between the peak and valley.

(四)甜菜生育期各内旬需水量分析

2.1.2.4 Analysis of Beet Water Requirement of Every 10 Days in the Growth and Development Period

甜菜在本区的生育期内的需水时段为 5~9 月 5 个月。根据 1989—1991 年 3 年对 3 个不同水平的甜菜各旬需水量分析试测,我们发现甜菜的需水量状况同样由于处理方式不同而有差别(见表 55)。

The beet water requirement period lasts 5 month from May to September in this plot. The analysis of beet water requirement of every 10 days from 1989 to 1991 finds that the water

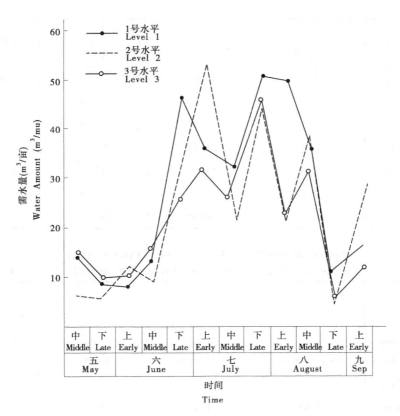

图 6 玉米全生育期内需水量变化曲线

Figure 6 Changing Curves of Corn's Water Requirement at Whole Growth and Development Period

requirement varies with the treatment mode

表 55 　　　　　甜菜生育期各旬需水量分析表(1989—1991 年)　　　　　单位:m³/亩

水平		1	2	3	平均需水量
月	旬	需水量	需水量	需水量	
5	上	—	—	—	
	中	8.30	11.35	9.04	9.56
	下	13.95	5.51	11.28	10.25
6	上	18.08	13.12	20.51	17.24
	中	19.20	17.1	11.40	15.90
	下	30.32	24.69	29.02	28.01
7	上	24.47	27.25	31.65	27.79
	中	37.95	20.23	17.09	25.09
	下	61.23	54.95	51.45	55.88

续表55

水平		1	2	3	平均需水量
月	旬	需水量	需水量	需水量	
8	上	38.29	29.95	25.49	31.24
	中	38.44	28.02	35.64	34.03
	下	45.07	55.50	41.99	47.52
9	上	28.33	21.14	16.79	22.09
	中	8.14	18.30	17.1	14.51
	下	19.40	21.74	12.26	17.80
合计		391.17	348.95	330.71	356.94

Table 55　Analysis of Beet Water Requirement of Every 10 Days in the Growth and Development Period (1989 to 1991)

Unit: m^3/mu

	Level	1	2	3	Average Water Requirement
Month	Every 10 Days	Water Requirement	Water Requirement	Water Requirement	
5	Early	—	—	—	
	Middle	8.30	11.35	9.04	9.56
	Late	13.95	5.51	11.28	10.25
6	Early	18.08	13.12	20.51	17.24
	Middle	19.20	17.1	11.40	15.90
	Late	30.32	24.69	29.02	28.01
7	Early	24.47	27.25	31.65	27.79
	Middle	37.95	20.23	17.09	25.09
	Late	61.23	54.95	51.45	55.88
8	Early	38.29	29.95	25.49	31.24
	Middle	38.44	28.02	35.64	34.03
	Late	45.07	55.50	41.99	47.52
9	Early	28.33	21.14	16.79	22.09
	Middle	8.14	18.30	17.1	14.51
	Late	19.40	21.74	12.26	17.80
Total		391.17	348.95	330.71	356.94

由表 55 可以看出,1 号水平的最大需水量时间为 7 月下旬,达 61.23 m³/亩,次为 8 月下旬;2 号水平的需水量变化曲线为典型双峰曲线(见图 7),在生育期 6～9 月内,各旬的需水量大多变化在 20～29 m³/亩之间,仅 8 月下旬和 7 月下旬出现两个时段的需水量高值,分别达 55.50 m³/亩和 54.95 m³/亩;3 号水平的需水量变化只有一个高峰,出现在 7 月下旬,达 51.45 m³/亩。其余时段波动起伏,峰谷相间,变化在 11～41.99 m³/亩之间。分析对比 3 个不同水平的需水量变化,除 6 月上旬 3 号水平的需水量相对 1 号、2 号水平的略高外,各旬均以 3 号水平的需水量最低。

It can be seen from Table 55 that the water requirement of Level #1 peaks in late-July at 61.23 m³/mu, then peaks again in late-August; the water requirement curve of Level #2 is a typical bimodal curve (see Figure 7), specifically speaking, the water requirement in the growth and development period ranges from 20 to 29 m³/mu of every 10 days in June to September, and only peaks in late-August and late-July at 55.50 m³/mu and 54.95 m³/mu respectively; the water requirement of Level #3 only peaks once in late-July at 51.45 m³/mu. In other period it fluctuates between 11 to 41.99 m³/mu. By contrast, we can find that the water requirement of Level #3 is lower than that of Level #1 and #2 of every 10 days except that the water requirement of Level #3 is slightly higher than that of Level #1 and #2 in late-June.

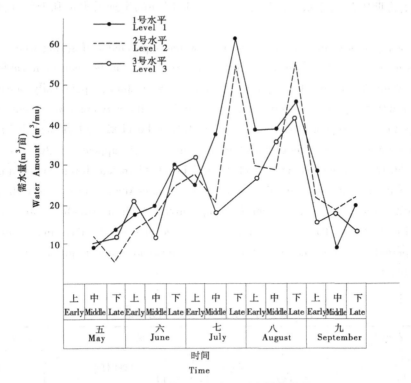

图 7 甜菜生育期内旬需水量曲线图

Figure 7 Curves of Beet's Water Requirement at Growth and Development Period

并且由图 7 还可看出,在对 3 个不同水平的甜菜需水量试验中,1 号水平和 2 号水平的生育期内需水量变化情势较接近,均呈双峰形式,最高峰和次高峰的出现时间亦同期,而3号水平的变化过程独特,仅有一次峰值,出现在8月下旬。

Furthermore, it is also found from Figure 7 that the beet water requirement change curves of Level #1 and #2 are similar to each other, featuring 2 peaks that appear in the same period, while the curve of Level #3 has only one peak in late-August.

(五)作物生育期内最大需水量分析
2.1.2.5 Analysis of the Maximum Crop Water Requirement in the Growth and Development Period

上述分析表明,春小麦、苜蓿、玉米、甜菜这4类作物的最大旬需水量出现时间是不同的。由表56可见,春小麦3个水平的最大需水量旬均出现在6月,其中1号水平出现在6月上旬,2号、3号水平的最大需水量出现在6月中旬;苜蓿则变化较大,1号、2号及3号水平的最大需水量旬分别出现在8月上旬、7月下旬和6月上旬;玉米3个水平的最大需水量旬均出现在7月,其中2号水平出现在7月上旬和3号水平出现在7月下旬;甜菜则7月下旬2次,8月下旬1次。从统计结果看,7月下旬需水量较大,次为7月上旬和6月上旬。作物生育期内各旬需水量的变化与作物不同时期的生理需水量和天气气候状况有密切的关系。

The above analysis shows that the maximum water requirement of spring wheat, alfalfa, corn and beet of every 10 days appears in different periods. It can be seen from Table 56 that the maximum water requirement of spring wheat appears in June, specifically speaking, in early-June of Level #1 and in mid-June of Level #2 and #3; the maximum water requirement of alfalfa appears in early-August of Level #1, in late-July of Level #2 and in early-June of Level #3; the maximum corn water requirement of the three levels appears in July, specifically speaking, Level #2 in early-July and Level #1 and Level #3 in late-July; the maximum beet water requirement appears in late-July twice, and in late-August once. According to the statistical results, the water requirement is comparatively high in late-July, followed by in early-July and early-June. The water requirement change of every 10 days in the growth and development period is closely connected to the physiological water requirement in different periods and climates.

表56 作物最大旬需水量及出现时间分析表

作物	春小麦		苜蓿		玉米		甜菜	
	需水量 (m^3/亩)	出现时间	需水量 (m^3/亩)	出现时间	需水量 (m^3/亩)	出现时间	需水量 (m^3/亩)	出现时间
1	56.70	6月下旬	69.86	8月上旬	51.02	7月下旬	61.23	7月下旬
2	59.41	6月中旬	51.06	7月上旬	53.90	7月上旬	55.50	8月下旬
3	39.85	6月中旬	60.36	6月上旬	45.86	7月下旬	51.45	7月下旬

Table 56　Analysis of the Maximum Water Requirement of Every 10 Days and Appearing Time

Crop	Spring Wheat		Alfalfa		Corn		Beet	
	Water Requirement (m^3/mu)	Appearing Time	Water Requirement (m^3/mu)	Appearing Time	Water Requirement (m^3/mu)	Appearing Time	Water Requirement (m^3/mu)	Appearing Time
1	56.70	Late-June	69.86	Early-August	51.02	Late-July	61.23	Late-July
2	59.41	Mid-June	51.06	Early-July	53.90	Early-July	55.50	Late-August
3	39.85	Mid-June	60.36	Early-June	45.86	Late-July	51.45	Late-July

三、作物各旬需水量占全生育期需水量百分比
2.1.3　Percentage of the Water Requirement of Every 10 Days against the Total in the Whole Growth and Development Period

作物在生长发育期,不同旬别的作物需水量占全生育需水量的百分比是不同的,而且不同作物类型也有很大差别。

The percentage of water requirement of every 10 days varies with the period; furthermore, it also varies with the categories of crops.

(一)春小麦各旬需水量占全生育期需水量比重
2.1.3.1　Percentage of Water Requirement of Every 10 Days against the Total in the Whole Growth and Development Period of Spring Wheat

春小麦生育期的需水量以6月中旬占百分比最大,3个水平的平均值为18.35%,次为6月下旬,为15.17%,二者合计约占全生育期春小麦需水总量的33.5%。由此看来,6月中、下旬是春小麦生长发育需水量最大时期(见表57)。比较3个不同水平的春小麦需水状况可见,1号水平不同时期需水量占需水总量百分比表现为,由4月上旬至6月下旬是一个需水比例不断增大的过程,6月下旬以后需水量骤然下降;2号水平和3号水平却表现为,由4月上旬至6月中旬需水量占总量百分比呈峰谷相间形式增长,与1号水平的需水量状况不尽一致。

The water requirement of spring wheat in mid-June accounts for a maximum percentage, with an average of 18.35% of the 3 levels, followed by 15.17% in late-June. As a result, the spring wheat has the maximum water requirement in mid and late-June (see Table 57). By contrast, we can find that the percentage of Level #1 continues to grow from early-April to late-June, and then decreases sharply; percentages of Level #2 and #3 grows in a peak-and-valley manner in the same period, slightly different with that of Level #1.

表 57　　春小麦各旬需水量占各生育期需水量百分比统计表　　单位:%

日	水平	1	2	3	平均
4	上	0.36	0.17	1.68	0.74
4	中	1.72	1.23	5.43	2.79
4	下	9.84	12.65	6.50	9.66
5	上	5.91	5.18	3.53	4.87
5	中	11.84	9.93	15.44	12.40
5	下	12.98	15.30	11.67	13.32
6	上	14.38	12.14	15.07	13.86
6	中	16.06	20.55	18.44	18.35
6	下	17.77	13.93	13.80	15.17
7	上	8.13	7.24	6.89	7.42
7	中	1.01	1.68	1.55	1.41
合计		100	100	100	100

Table 57　Statistics of Spring Wheat Water Requirement Percentage of Every 10 Days against the Total at Each Growth Stage　　Unit:%

Month	Level	1	2	3	Average
4	Early	0.36	0.17	1.68	0.74
4	Middle	1.72	1.23	5.43	2.79
4	Late	9.84	12.65	6.50	9.66
5	Early	5.91	5.18	3.53	4.87
5	Middle	11.84	9.93	15.44	12.40
5	Late	12.98	15.30	11.67	13.32
6	Early	14.38	12.14	15.07	13.86
6	Middle	16.06	20.55	18.44	18.35
6	Late	17.77	13.93	13.80	15.17
7	Early	8.13	7.24	6.89	7.42
7	Middle	1.01	1.68	1.55	1.41
Total		100	100	100	100

苜蓿在其生育期内的各旬需水量所占需水总量百分比均在13%以下,其中1号水平

第二章 作物需水量与需水规律
2 Water Requirement of Crops and Its Law

最大旬别的需水量百分比为 11.54%,出现在 6 月下旬,次为 11.49%,出现在 8 月上旬;2 号水平的需水量最大百分比为 11.35%,出现在 5 月下旬;3 号水平的需水量最大百分比为 13.14%,出现在 6 月上旬。比较 3 个不同水平的需水量百分比可见,它们均出现过 3 次所占百分比较高时段,第一次为 5 月下旬至 6 月上旬,需水量所占百分比变化在 9.24% ~ 13.14% 之间;第二次为 6 月下旬至 7 月上旬,需水量所占百分比变化在 8.34% ~ 11.54% 之间,较第一次峰值区减弱;第三次为 8 月上旬至 8 月中旬,变化在 8.78% ~ 11.49% 之间。需水量所占百分比最低值出现在 4 月上旬和 9 月中旬,3 不个同水平均表现出相似的特点(见表 58)。

The percentage of alfalfa is below 13% in the growth and development period, specifically speaking, the percentage of Level #1 peaks in late-June at 11.54%, followed by 11.49% in early-August; the percentage of Level #2 peaks in late-May at 11.35%, whereas that of Level #3 in early-June at 13.14%. By contrast, we can find that the 3 levels all have 3 periods with high percentages; the 1st period is from late-May to early-June, with a percentage ranging from 9.24% to 13.14%; the 2nd is from late-June to early-July, ranging from 8.34% to 11.54%, slightly lower than the 1st period; the 3nd period is from early to middle-August, ranging from 8.78% to 11.49%. The minimum percentages of the 3 levels are in early-April and in mid-September, similar to each other (see Table 58).

表 58　　苜蓿各旬需水量占各生育期需水量百分比统计表　　单位:%

月	水平	1	2	3	平均
4	中	1.84	1.88	1.24	1.65
	下	3.17	3.60	2.05	2.94
5	上	3.91	6.75	4.44	5.03
	中	5.61	7.59	6.32	6.51
	下	9.90	11.35	7.23	9.49
6	上	9.24	8.18	13.14	10.19
	中	1.85	2.93	3.72	2.83
	下	11.54	6.88	8.34	8.92
7	上	7.67	9.45	7.46	8.19
	中	7.23	5.60	6.08	6.30
	下	5.59	5.40	5.83	5.61
8	上	11.49	8.78	9.76	10.01
	中	7.92	8.65	11.05	9.21
	下	6.81	7.28	4.92	6.34
9	上	4.58	4.60	6.36	15.18
	中	1.64	1.08	2.01	1.59
合计		100	100	100	100

Table 58   Statistics of Alfalfa Water Requirement Percentage of Every 10 Days against the Total at Each Growth Stage

Unit: %

Month	Level	1	2	3	Average
4	Middle	1.84	1.88	1.24	1.65
	Late	3.17	3.60	2.05	2.94
5	Early	3.91	6.75	4.44	5.03
	Middle	5.61	7.59	6.32	6.51
	Late	9.90	11.35	7.23	9.49
6	Early	9.24	8.18	13.14	10.19
	Middle	1.85	2.93	3.72	2.83
	Late	11.54	6.88	8.34	8.92
7	Early	7.67	9.45	7.46	8.19
	Middle	7.23	5.60	6.08	6.30
	Late	5.59	5.40	5.83	5.61
8	Early	11.49	8.78	9.76	10.01
	Middle	7.92	8.65	11.05	9.21
	Late	6.81	7.28	4.92	6.34
9	Early	4.58	4.60	6.36	15.18
	Middle	1.64	1.08	2.01	1.59
Total		100	100	100	100

（四）玉米各旬需水量占全生育期需水量比重

2.1.3.3 Percentage of the Corn Water Requirement of Every 10 Days against the Total in the Whole Growth and Development Period

从玉米各旬需水量占全生育期需水量百分比分析(见表59)，3个不同水平的需水量状况表现基本相似特点，最大百分比均出现在7月下旬，所不同的是3号水平的7月下旬需水量所占比重较1号、2号水平所占各自比重大许多，为23.10%。在4个月(5月中旬~9月上旬)12个旬别测试中，仅此一个旬别的需水量占全生育期需水量超过20个百分点，其余均在12%以下，表明3号水平的需水量相对略为集中。而1号水平和2号水平的需水量所占百分比中，各自均有5个旬别的百分比在10%～17%之间，而无一次达到20个百分点。

According to the analysis of percentage of the corn water requirement of every 10 days against the total in the whole growth and development period (see Table 59), the water requirement of the 3 levels is basically similar to each other, featuring a peak in late-July; much larger than that of Level #1 and #2, the percentage of Level #3 in late-July reaches 23.10%. Among the experiments from mid-May to early-September, only 1% of water requirement of every 10 days exceeds 20% and the rest are all below 12%, indicating a relatively concentrated period for water. As for Level #1 and Level #2, each has 5 percentages varying from 10% to 17%, and none reaches 20%.

表 59　　　　　玉米各旬需水量占全生育期需水量百分数统计表　　　　单位:%

时间	旬	5月		6月			7月			8月			9月
		中旬	下旬	上旬	中旬	下旬	上旬	中旬	下旬	上旬	中旬	下旬	上旬
1989年	1	5.80	—	1.34	7.47	17.50	6.60	5.71	13.71	19.94	8.92	8.43	4.60
	2	1.86	1.21	1.56	4.22	18.22	4.57	21.71	15.62	6.38	14.51	3.79	6.34
	3	10.01	0.7	—	8.01	11.11	13.18	6.96	21.04	6.91	11.74	5.07	5.21
1990年	1	2.56	0.10	5.82	0.02	18.59	17.86	13.83	18.69	11.34	9.10	—	1.1
	2	—	1.74	7.90	—	15.23	32.66	6.55	19.60	4.0	10.99	—	1.29
	3	2.59	—	9.02	1.01	10.98	14.41	12.62	29.51	6.29	11.29	—	2.77
1991年	1	3.80	6.65	1.39	2.86	9.53	12.25	12.47	16.46	13.85	11.77	—	9.00
	2	4.78	2.73	5.27	5.05	0.45	10.58	12.57	15.77	11.71	15.21	—	7.93
	3	2.82	3.34	4.03	8.82	8.32	8.79	12.08	18.75	12.55	13.92	—	6.75
3年平均	1	4.06	2.25	2.85	3.43	15.21	12.24	10.67	16.28	15.04	9.07	2.81	4.9
	2	2.21	1.89	4.93	3.09	13.69	15.94	13.61	17.00	7.36	13.57	1.26	5.19
	3	5.11	1.14	4.35	5.75	10.14	12.13	10.55	23.1	8.58	12.11	1.69	4.91

Table 59　Statistics of Corn Water Requirement Percentage of Every 10 Days against the Total in the Whole Growth and Development Period　　Unit:%

Time	Ten days	May		June			July			August			September
		Middle	Late	Early	Middle	Late	Early	Middle	Late	Early	Middle	Late	Early
1989	1	5.80	—	1.34	7.47	17.50	6.60	5.71	13.71	19.94	8.92	8.43	4.60
	2	1.86	1.21	1.56	4.22	18.22	4.57	21.71	15.62	6.38	14.51	3.79	6.34
	3	10.01	0.7	—	8.01	11.11	13.18	6.96	21.04	6.91	11.74	5.07	5.21
1990	1	2.56	0.10	5.82	0.02	18.59	17.86	13.83	18.69	11.34	9.10	—	1.1
	2	—	1.74	7.90	—	15.23	32.66	6.55	19.60	4.0	10.99	—	1.29
	3	2.59	—	9.02	1.01	10.98	14.41	12.62	29.51	6.29	11.29	—	2.77
1991	1	3.80	6.65	1.39	2.86	9.53	12.25	12.47	16.46	13.85	11.77	—	9.00
	2	4.78	2.73	5.27	5.05	0.45	10.58	12.57	15.77	11.71	15.21	—	7.93
	3	2.82	3.34	4.03	8.82	8.32	8.79	12.08	18.75	12.55	13.92	—	6.75
Average of the Three Years	1	4.06	2.25	2.85	3.43	15.21	12.24	10.67	16.28	15.04	9.07	2.81	4.9
	2	2.21	1.89	4.93	3.09	13.69	15.94	13.61	17.00	7.36	13.57	1.26	5.19
	3	5.11	1.14	4.35	5.75	10.14	12.13	10.55	23.1	8.58	12.11	1.69	4.91

(四) 甜菜各旬需水量占全生育期需水量比重
2.1.3.4 Percentage of the Beet Water Requirement of Every 10 Days against the Total in the Whole Growth and Development Period

由表 60 可见,甜菜 1 号、2 号、3 号水平的各旬需水量占全生育期需水量的百分比均表明为双峰现象,有 2 次超过 10 个百分点,分别出现在 7 月下旬和 8 月下旬,3 个不同水平的需水量百分比变化不大,在 11.52% ~ 15.90% 之间,同时在 5 ~ 9 月 5 个月的生长期内,除 7 月下旬和 8 月下旬需水量所占百分比较大外,其余各旬的需水量所占百分比较均衡。

According to Table 60, there are two peaks of the percentages of the 3 levels, exceeding 10% twice in late-July and late-August and ranging from 11.52% to 15.90%; furthermore, the percentages from May to September (except late-July and late-August) are relatively even.

表 60　　　　甜菜各旬需水量占全生育期需水量百分比统计表　　　　单位:%

月	旬	1	2	3	平均
5	上	—	—	—	
	中	2.12	3.25	2.73	2.70
	下	3.56	1.61	3.41	2.86
6	上	4.62	3.76	6.20	4.86
	中	4.91	4.90	3.45	4.42
	下	7.25	7.08	8.78	7.70
7	上	6.25	7.81	9.57	7.88
	中	9.70	5.79	5.18	6.89
	下	15.65	15.75	15.56	15.65
8	上	9.78	8.58	7.71	8.69
	中	9.83	8.03	6.81	8.22
	下	11.52	15.90	12.69	13.37
9	上	7.24	6.06	5.08	6.17
	中	2.08	5.24	5.17	4.16
	下	4.95	6.23	3.71	4.96

Table 60　　Statistics of Beet Water Requirement Percentage of Every 10 Days against the Total in the Whole Growth and Development Period　　Unit:%

Month	Ten days	1	2	3	平均
5		—	—	—	
	Middle	2.12	3.25	2.73	2.70
	Late	3.56	1.61	3.41	2.86
6	Early	4.62	3.76	6.20	4.86
	Middle	4.91	4.90	3.45	4.42
	Late	7.25	7.08	8.78	7.70

Continue Table 60

Month	Ten days	1	2	3	平均
7	Early	6.25	7.81	9.57	7.88
	Middle	9.70	5.79	5.18	6.89
	Late	15.65	15.75	15.56	15.65
8	Early	9.78	8.58	7.71	8.69
	Middle	9.83	8.03	6.81	8.22
	Late	11.52	15.90	12.69	13.37
9	Early	7.24	6.06	5.08	6.17
	Middle	2.08	5.24	5.17	4.16
	Late	4.95	6.23	3.71	4.96

（五）作物各旬需水量占全生育期需水量百分比研究小结
2.1.3.5 Research Summary of Percentages of the Crop Water Requirement of Every 10 Days against the Total in the Whole Growth and Development Period

通过对作物各旬需水量占全生育需水量百分比的分析计算，现将主要结果归纳如下：

According to the analysis of percentages of the crop water requirement of every 10 days against the total in the whole growth and development period, the main summaries are as follows:

春小麦以 6 月中旬需水量占全生育期需水量比重最大，为 18.35%；苜蓿出现在 6 月上旬，为 10.19%；玉米出现在 7 月下旬，为 18.79%；甜菜出现在 7 月下旬，为 15.65%（见表 61），由此可见，6～7 月是作物需水量较大时期。

The percentage of spring wheat peaks in mid-June at 18.35%; that of alfalfa in early-June at 10.19%; that of corn in late-July at 18.79%; that of beet in late-July at 15.65% (see Table 61); therefore it's reasonable to conclude that the period from June to July has a large water requirement.

表 61　　作物各旬需水量占生育期需水量百分比分析表

作物	春小麦		苜蓿		玉米		甜米	
	最大值(%)	出现时间	最大值(%)	出现时间	最大值(%)	出现时间	最大值(%)	出现时间
1	16.06	6月中旬	11.54	6月下旬	16.28	7月下旬	15.65	7月下旬
2	20.55	6月中旬	11.35	5月下旬	17.00	7月下旬	15.90	8月下旬
3	18.44	6月中旬	11.05	8月中旬	23.10	7月下旬	15.56	7月下旬

Table 61 Analytic Table for Percentages of Crop Water Requirement of Every 10 Days against the Total in the Growth and Development Period

Crop	Spring Wheat		Alfalfa		Corn		Beet	
	Maximum (%)	Appearing Time	Maximum (%)	Appearing Time	Maximum (%)	Appearing Time	Maximum (%)	Appearing Time
1	16.06	Mid-June	11.54	Late-June	16.28	Late-July	15.65	Late-July
2	20.55	Mid-June	11.35	Late-May	17.00	Late-July	15.90	Late-August
3	18.44	Mid-June	11.05	Mid-August	23.10	Late-July	15.56	Late-July

春小麦最大需水量所占百分比值较高,呈单峰形式;苜蓿(打三茬)有3次需水量所占比例较高时期,呈三峰形式;玉米的需水量百分比的最高值和次高值接近,呈双峰形式;甜菜与玉米的需水量变化情势相似,亦呈双峰形式。

The percentage of spring wheat is comparatively high and only peaks once; the percentage of alfalfa (3 croppings) has 3 periods when it maintains at a high level and peaks 3 times; the percentage of corn peaks twice, each peak closing to each other; the percentage of beet is similar with that of corn.

四、作物生育期各旬需水强度分析
2.1.4 Analysis of Crop Water Requirement Intensity of Every 10 Days

作物需水强度是指生长发育期内阶段平均每天每亩所消耗的水量,是反映全生育期内作物需水量变化过程的重要组成部分,也是研究作物生理活动的重要因素之一。作物需水强度与作物的生理活动、气象条件有关。一般情况下,作物生长初期和后期的生理活动较弱,需水强度较小;而中期6~8月3个月正值作物生长发育的重要时期,生理活动旺盛,这时的需水强度一般较大。但也出现需水强度较小时段,因而详尽研究作物不同时段的需水强度,对指导作物供水量是十分有益的。

As an important indicator of the water requirement change in the whole growth and development period and a critical factor for the research of physiological activity, the crop water requirement intensity means the water consumed by crop every day per mu in the growth stages. The crop water requirement intensity is closely connected to the physiological activity and climate conditions. Generally, in the initial and later growth period, the crops have an inactive physiological activity with small water requirement intensity; the middle period (June to August) is critical for crop growth, requiring active physiological activity and large water requirement intensity. Period with small water requirement intensity is also found; therefore it is critical to comprehensively study the water requirement intensity in different stages for a precise water supply.

本区春小麦、苜蓿、玉米、甜菜这4类作物的需水强度大致有如下特点。

The water requirement intensities of spring wheat, alfalfa, corn and beet in this plot have the following features.

（一）春小麦生育期内需水强度分析
2.1.4.1 Analysis of Water Requirement Intensity in the Growth and Development Period of Spring Wheat

比较3个不同水平的小麦生育期内不同时段的需水强度可以发现，1号水平的最大需水强度出现在6月下旬，日需水量达5.67 m³/亩；2号水平和3号水平的最大需水强度均出现在6月中旬，日需水量分别为5.94 m³/亩和4.00 m³/亩，与1号和2号水平和最大需水强度相比，3号水平显得较小。

By contrast, it is found that the wheat water requirement intensity of Level #1 peaks in late-June at 5.67 m³/mu; that of Level #2 and #3 both peaks in mid-June at 5.94 m³/mu and 4.00 m³/mu respectively; unlike that of Level #1 and #2, the water requirement intensity of Level #3 is comparatively small.

在小麦生长初期的4月中、上旬和后期的7月中旬，作物需水强度最小，其中4月上旬的作物需水强度仅为0.05~0.12 m³/(亩·d)，以2号水平的日需水量最小，仅为0.05 m³/亩。

Wheat has the minimum water requirement intensity in early and middle-April (in the initial growth period) and in middle-July (in the later period), only 0.05 – 0.12 m³/(mu·day) in early-April, whereas Level #2 has the minimum daily water requirement, 0.05 m³/mu only.

由表62可见，小麦不同3个水平的需水强度的平均值以1号水平的最大，为2.87 m³/(亩·d)；2号水平次之，为2.60 m³/(亩·d)；3号水平最小，为1.95 m³/(亩·d)。

According to Table 62, Level #1 has the maximum average water requirement intensity, 2.87 m³/(mu·day); Level #2 follows, 2.60 m³/(mu·day); Level #3 is the least, 1.95 m³/(mu·day) only.

表62　　　　　春小麦不同水平各旬需水强度分析表　　　　单位:m³/(亩·d)

月份	旬	1		2		3		平均需水强度
		需水强度	模系数	需水强度	模系数	需水强度	模系数	
4	上	0.12	0.36	0.85	0.17	0.36	1.68	3.11
	中	0.55	1.72	0.36	1.23	1.17	5.43	0.69
	下	3.14	9.84	3.66	12.65	1.41	6.50	2.74
5	上	1.89	5.91	1.50	5.18	0.76	3.53	1.38
	中	3.78	11.84	2.07	7.17	3.34	15.44	3.06
	下	4.14	12.98	4.43	15.30	2.52	11.67	3.70

续表62

月份	旬	1		2		3		平均需水强度
		需水强度	模系数	需水强度	模系数	需水强度	模系数	
6	上	4.59	14.38	3.51	12.14	3.26	15.87	3.79
	中	5.13	16.08	5.94	20.55	4.00	18.44	5.02
	下	5.67	17.77	4.03	13.93	2.98	13.80	4.23
7	上	2.59	8.13	2.89	10.00	1.49	6.09	2.32
	中	0.32	1.01	0.49	1.68	0.34	1.55	0.38
平均		2.87	100	2.60	100	1.95	100	

Table 62　　Analysis of Water Requirement Intensity of Every 10 Days of Spring Wheat at Different Levels

Unit: $m^3/(mu \cdot day)$

Month	Every 10 Days	1		2		3		Average Water Requirement Intensity
		Water Requirement Intensity	Modulus	Water Requirement Intensity	Modulus	Water Requirement Intensity	Modulus	
4	Early	0.12	0.36	0.85	0.17	0.36	1.68	3.11
	Middle	0.55	1.72	0.36	1.23	1.17	5.43	0.69
	Late	3.14	9.84	3.66	12.65	1.41	6.50	2.74
5	Early	1.89	5.91	1.50	5.18	0.76	3.53	1.38
	Middle	3.78	11.84	2.07	7.17	3.34	15.44	3.06
	Late	4.14	12.98	4.43	15.30	2.52	11.67	3.70
6	Early	4.59	14.38	3.51	12.14	3.26	15.87	3.79
	Middle	5.13	16.08	5.94	20.55	4.00	18.44	5.02
	Late	5.67	17.77	4.03	13.93	2.98	13.80	4.23
7	Early	2.59	8.13	2.89	10.00	1.49	6.09	2.32
	Middle	0.32	1.01	0.49	1.68	0.34	1.55	0.38
Average		2.87	100	2.60	100	1.95	100	

小麦生育期内的需水强度由4月上至6月中、下旬是一个增大过程,由图8可见,3个水平的小麦需水强度变化均呈多峰(三峰)形式,以2号水平最为典型,表明小麦在生育期内需水强度变化是不均匀,在不同生育阶段的需水状况不尽一致。因此详尽研究其不同时期需水强度的变化过程是十分重要的。

The water requirement intensity of wheat continues to grow from early-April to middle and late-June. According to Figure 8, the wheat water requirement intensity change of the 3 levels are in a multimodal manner (3 peaks), of which the Level #2 is the most typical one, indicating that the water requirement intensity changes in an uneven manner in the growth and development period and the water requirement is unique in different growth stages. As a result, it is critical to comprehensively study the water requirement intensity change in different growth stages.

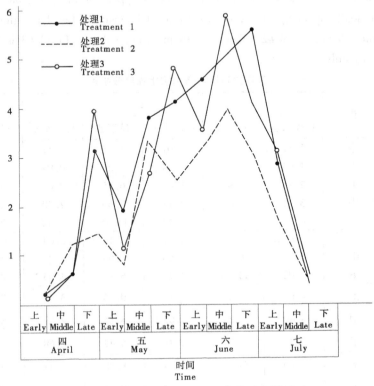

图 8 小麦各旬日需水强度变化过程线

Figure 8　Water Requirement Intensity Change Curves of Every 10 Days of Wheat

(二)苜蓿生育期内需水强度分析

2.1.4.2　**Analysis of Alfalfa Water Requirement Intensity in the Growth and Development Period**

苜蓿打三茬的全生育期为4月中旬至9月中旬,比较该期间3个不同水平的需水强度可以发现(见表63),1号水平的需水强度变化在1.12~7.02 m³/(亩·d)之间;2号水平0.58~6.13 m³/(亩·d)之间;3号水平0.57~6.04 m³/(亩·d)之间,而且各不同水平的苜蓿生长期内的需水强度均有3个高峰值。1号水平的出现在5月下旬、6月下旬和8月上旬,日需水量分比较苜蓿生育期内3个不同水平的平均需水强度可见,以1号水平的平均需水强度最大,为3.82 m³/(亩·d);2号水平次之,为3.38 m³/(亩·d);3号水平的最小,为2.94 m³/(亩·d)。

The total growth and development period of alfalfa with 3 croppings lasts from mid-April to mid-September. Through analysis of the water requirement intensities of the 3 levels, it is found that the water requirement intensity of Level #1 ranges from 1.12 to 7.02 m³/(mu·day); that of Level #2 from 0.58 to 6.13 m³/(mu · day); that of Level #3 from 0.57 to 6.04 m³/(mu · day); each level has 3 peaks. The peaks of Level #1 appear in late-May, late-June and early-August, specifically speaking, the water requirement intensity of Level 1 is the maximum one, 3.82 m³/(mu · day); Level #2 follows, 3.38 m³/(mu · day); Level #3 is the least, 2.94 m³/(mu · day) only.

表63　　　　　　　　苜蓿生育期内各旬需水强度分析表　　　　　　　单位:m³/(亩·d)

月份	旬	1		2		3		平均需水强度
		日需水量	模系数	日需水量	模系数	日需水量	模系数	
4	中	1.12	1.84	1.01	1.88	0.57	1.24	0.90
	下	1.93	3.17	1.94	3.59	0.94	2.05	1.60
5	上	2.38	3.91	3.65	6.75	2.04	4.44	2.69
	中	3.41	5.61	4.11	7.60	2.90	6.32	3.47
	下	6.02	9.9	6.13	11.35	3.32	7.23	5.16
6	上	5.62	9.24	4.41	8.17	6.04	13.14	5.39
	中	1.12	1.85	1.59	2.93	1.71	3.73	1.47
	下	7.02	11.54	3.70	6.85	4.93	8.34	5.22
7	上	4.67	7.67	5.11	9.45	3.43	7.47	4.40
	中	4.40	7.23	3.02	5.85	2.79	6.08	3.40
	下	3.40	5.59	3.0	5.42	2.68	5.83	3.03
8	上	6.99	11.49	4.74	8.78	4.48	9.76	5.40
	中	4.82	7.92	4.67	8.65	5.08	11.05	4.86
	下	4.14	6.81	3.93	7.28	2.26	4.92	3.44
9	上	2.80	4.60	2.49	4.61	2.92	6.36	2.74
	中	1.0	1.64	0.58	1.00	0.95	2.07	0.84
平均		3.82	100	3.38	100	2.94	100	

Table 63　Analysis of Alfalfa's Water Requirement Intensity of Every 10 Days in the Growth and Development Period　　Unit: m³/(mu · day)

Month	Every 10 Days	1		2		3		Average Water Requirement Intensity
		Daily Water Requirement	Modulus	Daily Water Requirement	Modulus	Daily Water Requirement	Modulus	
4	Middle	1.12	1.84	1.01	1.88	0.57	1.24	0.90
	Late	1.93	3.17	1.94	3.59	0.94	2.05	1.60
5	Early	2.38	3.91	3.65	6.75	2.04	4.44	2.69
	Middle	3.41	5.61	4.11	7.60	2.90	6.32	3.47
	Late	6.02	9.9	6.13	11.35	3.32	7.23	5.16

第二章 作物需水量与需水规律
2 Water Requirement of Crops and Its Law

Continue Table 63

Month	Every 10 Days	1		2		3		Average Water Requirement Intensity
		Daily Water Requirement	Modulus	Daily Water Requirement	Modulus	Daily Water Requirement	Modulus	
6	Early	5.62	9.24	4.41	8.17	6.04	13.14	5.39
	Middle	1.12	1.85	1.59	2.93	1.71	3.73	1.47
	Late	7.02	11.54	3.70	6.85	4.93	8.34	5.22
7	Early	4.67	7.67	5.11	9.45	3.43	7.47	4.40
	Middle	4.40	7.23	3.02	5.85	2.79	6.08	3.40
	Late	3.40	5.59	3.0	5.42	2.68	5.83	3.03
8	Early	6.99	11.49	4.74	8.78	4.48	9.76	5.40
	Middle	4.82	7.92	4.67	8.65	5.08	11.05	4.86
	Late	4.14	6.81	3.93	7.28	2.26	4.92	3.44
9	Early	2.80	4.60	2.49	4.61	2.92	6.36	2.74
	Middle	1.0	1.64	0.58	1.00	0.95	2.07	0.84
Average		3.82	100	3.38	100	2.94	100	

我们将苜蓿生育期内各旬需水强度变化过程绘制成图，由图9可见，苜蓿打三茬有3次需水强度高峰区，在需水强度过程线上3个不同水平的均呈三峰形式，尤以1号水平的典型。同时由图9还可见到，苜蓿打三茬生育期的需水强度变化呈峰、谷相间，规律性很强。

We plot the changing process of water requirement intensity in the alfalfa growth and development period (see Figure 9). It can be seen that alfalfa (3 croppings) has 3 periods with high water requirement intensity, appearing in a 3-peak manner in the process line, in which the Level #1 is the most typical one. What can also be seen is that the water requirement intensity of alfalfa (3 croppings) changes in a regular peak-valley manner.

（三）玉米生长期内需水强度分析
2.1.4.3 Analysis of Corn Water Requirement Intensity in the Growth and Development Period

玉米全生育需水期为5月中旬至9月上旬，在这个期间内，以7月的玉米需水强度最大，其中1号水平和2号水平和3号水平的最大需水强度出现在7月下旬，分别为5.10 m³/(亩·d)和6.02 m³/(亩·d)；2号水平的最大需水强度出现在7月上旬，为5.40 m³/(亩·d)，以3号水平的最大需水强度为高(见表64)。而在玉米生长初期(5月中旬~6月中旬)和后期8月下旬至9月上旬的需水强度较小，变化在0.53~1.54 m³/(亩·d)之间，3个不同水平平均表现出基本相似的特点。

The water requirement period of corn lasts from mid-May to early-September. In this

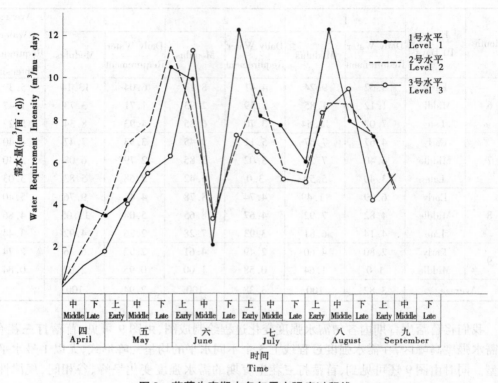

图 9 苜蓿生育期内各旬需水强度过程线

Figure 9 Alfalfa's Water Requirement Curves of Every 10 Days in the Growth and Development Period

period the corn has the maximum water requirement intensity in July, and the water requirement intensities of Level #1 and #3 peak in late-July at 5.10 m³/(mu · day) and 6.02 m³/(mu · day) respectively; the intensity of Level #2 peaks in early-July at 5.40 m³/(mu · day). Therefore, the water requirement intensity of Level #3 is the highest (see Table 64). The water requirement intensity of corn is comparatively low in the initial growth period (mid-May to mid-June) and the later period (late-August and early-September), ranging from 0.53 to 1.54 m³/(mu · day), a common feature of the 3 levels.

玉米生育期内的需水强度变化也不均匀, 由图 10 可见, 在 6 月下旬至 8 月下旬 2 个月内, 玉米需水强度出现 2 次高峰区和 2 次低值区, 峰、谷相间, 3 个不同水平均表现为相似的特点。由此说明, 玉米在生育期内的需水强度变化是有一定规律可循的。

The water requirement intensity of corn in the growth and development period changes in an uneven manner as well. From Figure 10 we can find 2 peaks and 2 valleys from late-June to late-August, a common feature of the 3 levels. It's fair to conclude that there is a distinct law of the water requirement intensity change to follow in the corn growth and development period.

第二章　作物需水量与需水规律
2　Water Requirement of Crops and Its Law

表 64　　　　　　　　　玉米各生育期需水强度分析表　　　　　单位:m³/(亩·d)

旬别		5月		6月			7月			8月			9月
		中旬	下旬	上旬	中旬	下旬	上旬	中旬	下旬	上旬	中旬	下旬	上旬
1989 年	1	2.35	—	8.52	2.86	6.70	2.56	2.20	5.27	1.86	3.44	3.25	1.77
	2	0.57	0.38	0.48	1.29	5.93	6.67	1.40	4.79	1.96	4.46	1.16	5.83
	3	3.18	2.21	—	2.52	3.52	4.15	2.19	6.63	2.18	3.70	1.68	1.62
1990 年	1	0.58	0.22	1.32	0.04	4.28	4.11	3.28	4.38	2.61	2.09	—	0.25
	2	—	0.33	1.51	—	2.90	1.23	1.25	3.74	0.77	2.09	—	0.25
	3	0.64	—	0.256	2.79	3.65	3.21	7.58	1.60	2.87	—	—	8.70
1991 年	1	1.33	2.31	0.48	0.99	3.31	4.26	4.34	5.72	4.82	4.09	—	3.13
	2	1.48	0.85	1.64	1.56	2.62	3.28	3.90	4.89	3.63	4.72	—	2.46
	3	0.59	0.70	0.86	1.85	1.74	1.84	2.53	3.90	2.66	2.91	—	1.41
3 年平均	1	1.42	0.84	0.77	1.30	4.76	3.64	3.27	5.10	5.04	3.20	1.08	1.71
	2	0.68	0.52	1.21	1.47	3.70	5.40	2.18	4.47	2.10	3.76	0.39	2.85
	3	1.47	0.97	1.85	1.54	2.68	3.21	2.64	6.02	2.15	3.16	0.53	1.24

Table 64　Analysis of Water Requirement Intensity at Each Growth Stage of Corn

Unit:m³/(mu·day)

Level		May		June			July			August			September
		Middle	Late	Early	Middle	Late	Early	Middle	Late	Early	Middle	Late	Early
1989	1	2.35	—	8.52	2.86	6.70	2.56	2.20	5.27	1.86	3.44	3.25	1.77
	2	0.57	0.38	0.48	1.29	5.93	6.67	1.40	4.79	1.96	4.46	1.16	5.83
	3	3.18	2.21	—	2.52	3.52	4.15	2.19	6.63	2.18	3.70	1.68	1.62
1990	1	0.58	0.22	1.32	0.04	4.28	4.11	3.28	4.38	2.61	2.09	—	0.25
	2	—	0.33	1.51	—	2.90	1.23	1.25	3.74	0.77	2.09	—	0.25
	3	0.64	—	0.256	2.79	3.65	3.21	7.58	1.60	2.87	—	—	8.70
1991	1	1.33	2.31	0.48	0.99	3.31	4.26	4.34	5.72	4.82	4.09	—	3.13
	2	1.48	0.85	1.64	1.56	2.62	3.28	3.90	4.89	3.63	4.72	—	2.46
	3	0.59	0.70	0.86	1.85	1.74	1.84	2.53	3.90	2.66	2.91	—	1.41
Average of the Three Years	1	1.42	0.84	0.77	1.30	4.76	3.64	3.27	5.10	5.04	3.20	1.08	1.71
	2	0.68	0.52	1.21	1.47	3.70	5.40	2.18	4.47	2.10	3.76	0.39	2.85
	3	1.47	0.97	1.85	1.54	2.68	3.21	2.64	6.02	2.15	3.16	0.53	1.24

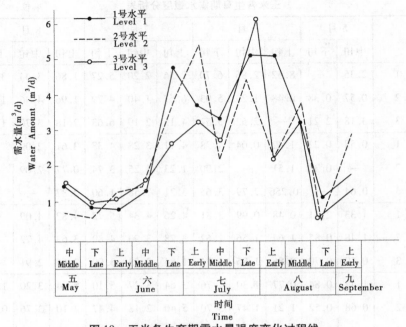

图 10 玉米各生育期需水量强度变化过程线

Figure 10 Corn's Water Requirement Curves of Every 10 Days in the Growth and Development Period

(四)甜菜生育期内需水强度分析

2.1.4.4 Analysis of Water Requirement Intensity in the Growth and Development Period of Beet

通过 3 年(1989—1991 年)对甜菜全生育期需水强度的分析测试,甜菜生育期的需水强度,最大值出现在 7 月和 8 月,正值甜菜生长中。7 月、8 月两个月的需水强度达 2.78~5.05 m³/(亩·d)之间。7 月则又以 7 月下旬需水强度最大,1 号、2 号和 3 号 3 个不同水平的需水强度分别为 5.57、5.00 m³/(亩·d)和 4.58 m³/(亩·d);8 月则以 8 月下旬的需水强度最大,1 号、2 号和 3 号 3 个不同水平的需水强度分别为 4.09 m³/(亩·d)和 3.81 m³/(亩·d)(见表 65)。在甜菜全生育期需水强度变化过程线上呈现两峰现象(见图 11)。由此可知,在甜菜生育期内,7~8 月 2 个月的用水是很重要的。

3-year-long analysis of water requirement intensity in the whole growth and development period of beet helps find that the water requirement intensities peak in July and August in the growth period. The intensities in this period range from 2.78 to 5.05 m³/(mu·day). In July, the intensities of the 3 levels peak in the last 10 days at 5.57 m³/(mu·day), 5.00 m³/(mu·day) and 4.58 m³/(mu·day) respectively; in August, the intensities peak in the last 10 days at 4.09 m³/(mu·day), 5.04 m³/(mu·day) and 3.81 m³/(mu·day) respectively (see Table 65). 2 peaks are seen in the changing process line of water requirement intensity in the whole growth and development period of beet (see Figure11). It's fair to conclude that water use is critical in period from July to August in the beet growth and development period.

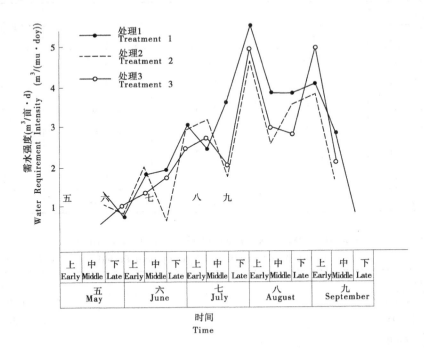

图 11 甜菜生育期内各旬需水强度变化过程线

Figure 11 Beet's Changing Curves of Water Requirement Intensity of Every Ten Days of the Growth and Development Period

表 65 甜菜生育期各旬需水强度分析表 单位:m³/(亩·d)

时数		水平			平均需水强度
月	旬	1	2	3	
5	上	—	—	—	
	中	1.39	0.56	1.13	1.03
	下	0.75	1.13	0.82	0.90
6	上	1.82	1.31	0.82	0.90
	中	1.92	1.71	1.14	1.59
	下	3.03	2.47	2.90	2.80
7	上	2.45	2.72	3.17	2.78
	中	3.79	2.02	3.21	3.01
	下	5.57	5.00	4.58	5.05

续表65

月	旬	1	2	3	平均需水强度
			水平		
		时数			
8	上	3.83	2.99	2.55	3.12
	中	3.84	2.80	3.56	3.07
	下	4.09	5.04	3.81	4.31
9	上	2.83	2.11	1.68	2.21
	中	0.81	1.83	1.71	1.45
	下	1.94	2.13	1.23	1.77
平均		2.53	2.25	2.14	

Table 65　Analysis of Water Requirement Intensity of Every 10 Days in the Beet Growth and Development Period

Unit: m³/(mu. day)

Period		Level			Average Water Requirement Intensity
Month	Every 10 Days	1	2	3	
5	Early	—	—	—	
	Middle	1.39	0.56	1.13	1.03
	Late	0.75	1.13	0.82	0.90
6	Early	1.82	1.31	0.82	0.90
	Middle	1.92	1.71	1.14	1.59
	Late	3.03	2.47	2.90	2.80
7	Early	2.45	2.72	3.17	2.78
	Middle	3.79	2.02	3.21	3.01
	Late	5.57	5.00	4.58	5.05
8	Early	3.83	2.99	2.55	3.12
	Middle	3.84	2.80	3.56	3.07
	Late	4.09	5.04	3.81	4.31
9	Early	2.83	2.11	1.68	2.21
	Middle	0.81	1.83	1.71	1.45
	Late	1.94	2.13	1.23	1.77
Average		2.53	2.25	2.14	

（五）作物生育期各旬需水强度研究小结
2.1.4.5 Research Summary of Water Requirement Intensity of Every 10 Days in the Crop Growth and Development Period

通过3年(1989—1991年)对春小麦、苜蓿、玉米和甜菜4类作物需水强度的试验研究，现归纳如下。

See the following section for summary of the experimental analysis of water requirement intensities of spring wheat, alfalfa, corn and beet from 1989 to 1991.

1. 春小麦生育期各旬需水强度

Ⅰ Water requirement intensity of every 10 days of spring wheat in the growth and development period

春小麦生育期内各旬的需水强度以6月中旬最高，3个不同水平的平均值为5.02 m³/(亩·d)，模系数约占全生育期的20%；次为6月下旬，3个不同水平的平均值为4.23 m³/(亩·d)。生育期内的需水强度变化过程呈三峰形式。

Spring wheat has the maximum water requirement intensity in mid-June, and the average intensity of the 3 levels is 5.02 m³/(mu·day), with a modulus accounting for 20% of the whole growth period; followed by late-June, 4.23 m³/(mu·day). The water requirement intensity in the growth and development period changes in a 3-peak form.

2. 苜蓿生育期各旬需水强度

Ⅱ Water requirement intensity of every 10 days in the alfalfa growth and development period

苜蓿打三茬的全生育期需水强度有3个高峰区，第一高峰区为5月下旬和6月上旬；第二峰值区在6月下旬和7月上旬；第三峰值区为8月上旬和中旬，日最大需水量为7.02 m³/亩(1号水平，第二茬)。生育期内的需水强度变化过程线呈三峰形式。

Alfalfa (3 croppings) has 3 peaks of water requirement intensity in the total growth and development period, the first in late-May and early-June, the second in late-June and early-July, and the third in early and mid-August, with a maximum daily water requirement of 7.02 m³/mu (Level #1, 2nd cropping). The changing process of water requirement intensity in the growth and development period has three peaks.

3. 玉米生育期各旬需水强度

Ⅲ Water requirement intensity of corn in every ten days of the growth and development period

玉米生育期内的需水强度以7月下旬和上旬最大，变化在5.10~6.02 m³/(亩·d)之间，总体表现为由前期5月中旬至中期7月需水强度不断增大，随后至玉米生长后期的9月上旬，需水强度不断减小。

The water requirement intensity of corn is the highest in the last 10 days (hereinafter " late") and the first 10 days (hereinafter " early") of July, varying from 5.10 to 6.02 m³/(mu · day). Generally, the intensity grows higher from the middle 10 days (hereinafter " middle" or "mid-") of May to mid-July, and gradually lowers till early September.

4. 甜菜生育期各旬需水强度
Ⅳ Water requirement intensity of beet every ten days of the growth and development period

甜菜生育期内的最大需水强度出现在 7 月下旬和 8 月下旬,需水强度分别为 4.58 ~ 5.57 m³/(亩·d)和 3.81 ~ 5.04 m³/(亩·d),全生育期需水强度的变化过程线呈双峰现象(见表 65)。

The water requirement intensity of beet is the highest in late July and late August, varying between 4.58 – 5.57 m³/(mu · day) and 3.81 – 5.04 m³/(mu · day) respectively. The changing process of water requirement intensity in the whole growth and development period has two peaks (see Table 65).

作物生育期最大需水强度对比分析见表 66。

The contrastive analysis of maximum water requirement intensity of crop in the growth and development period is shown in Table 66.

表 66　　作物生育期最大需水强度对比分析表　　单位:m³/(亩·d)

水平	春小麦		苜蓿		玉米		甜菜	
	需水强度	出现时段	需水强度	出现时段	需水强度	出现时段	需水强度	出现时段
1	5.67	6月下旬	7.02	6月下旬	5.10	7月下旬	5.57	7月下旬
2	5.94	6月中旬	6.13	5月下旬	5.40	7月上旬	5.04	8月下旬
3	4.00	6月中旬	6.04	6月上旬	6.02	7月下旬	4.58	7月下旬
最大平均	5.20	6月中旬	6.40	6月	5.51	7月下旬	5.06	7月下旬
全生育期平均	2.47		3.38		2.43		2.31	

Table 66 Contrastive Analysis of Maximum Crop Water Requirement Intensity
in the Growth and Development Period   Unit: m³/(mu · day)

Level	Spring Wheat		Alfalfa		Corn		Beet	
	Water Requirement Intensity	Emergence Period	Water Requirement Intensity	Emergence Period	Water Requirement Intensity	Emergence Period	Water Requirement Intensity	Emergence Period
1	5.67	Late-June	7.02	Late-June	5.10	Late-July	5.57	Late-July
2	5.94	Mid-June	6.13	Late-May	5.40	Early-July	5.04	Late-August
3	4.00	Mid-June	6.04	Early-June	6.02	Late-July	4.58	Late-July
Largest Average	5.20	Mid-June	6.40	June	5.51	Late-July	5.06	Late-July
Average of the Entire Period	2.47		3.38		2.43		2.31	

第二节 作物需水规律研究
2.2 Study on Water-Requiring Law of Crops

不同作物在其生长发育阶段有其自身的生理特点,因而作物需水量在全生长发育过程中的变化大体也有一定规律,表现为作物生育前、后期的生理活动较弱,需水量较少;而作物生育中期,生理活动加强,作物需水量增大,表现出大致有规律的变化过程。

Different crops have different physiological characteristics in respect to their growth and development stages. Therefore, the law of crop water requirement in the whole growth and development period can be concluded as weak crop physiological activity and low water requirement at the early and the late stages, and strong and high in the middle stage, which is a regular changing process on the whole.

详尽研究各类作物在各生长发育阶段的需水规律,对我们开展节水灌溉工作是非常重要的。

It is of vital importance to study the law of water requirement of crops at different stages for water-saving irrigation.

一、作物不同生长阶段的需水规律
2.2.1 Law of Crop Water Requirement at Different Growing Stages

作物不同生长阶段的需水状况是不同的,下面结合 3 年试验观测,就春小麦、苜蓿、玉米及甜菜 4 类作物在不同阶段的需水规律详尽述之:

The water requirement of crop differs from its growing stages. Here is the law of water requirement of spring wheat, alfalfa, corn and beet at different stages in accordance with a 3-year experiment and observation:

(一)春小麦不同生长阶段的需水规律
2.2.1.1 Law of Spring Wheat's Water Requirement at Different Growing Stages

春小麦全生长期要经过苗期、分蘖、拔节、孕穗、抽穗、灌浆、成熟 7 个阶段,而不同生长阶段,作物的需水量是不同的。从 3 个不同水平的试验结果看(见表 67),春小麦不同阶段的需水量有着基本相似的变化规律,均表现为在小麦灌浆—成熟阶段需水量最高,变化在 59.82~95.67 m³/(亩·d)之间;次为小麦抽穗—灌浆阶段;再次为拔节—孕穗阶段;而苗期的需水量在小麦整个生长发育阶段是需水量最少的,仅为 10.37~11.70 m³/亩(见图 12)。

The growth period of spring wheat includes 7 stages, i.e., in a chronological order, seedling, tillering, jointing, booting, heading, filling and maturing. Plant in different stages requires different amount of water. The results of three different levels (see Table 67) show that the water requirement of spring wheat changes in a similar way at different stages, that is, filling-maturing stage requiring most water, varying between 59.82 to 95.67 m³/(mu·day), followed by heading-filling, jointing-booting, and seedling requiring the least, only 10.37 – 11.70 m³/mu (see Figure 12).

表 67　　春小麦各生长阶段需水量分析表(1989—1991 年)　　单位:m³/亩

水平			1	2	3
生育期	起止日期	天数	需水量	需水量	需水量
苗期	3/4 ~ 24/4	22	10.37	18.67	11.70
分蘖	25/4 ~ 12/5	18	42.07	41.42	23.35
拔节	17/5 ~ 19/5	17	30.33	27.46	24.65
孕穗	30/5 ~ 12/6	14	41.43	51.40	29.79
抽穗	13/6 ~ 19/6	7	33.85	34.70	21.54
灌浆	20/6 ~ 24/6	5	65.34	47.52	45.25
成熟	25/6 ~ 19/7	25	95.67	67.97	59.82
合计	3/4 ~ 19/7	108	318.06	289.14	216.10

第二章　作物需水量与需水规律
2　Water Requirement of Crops and Its Law

Table 67　Analysis of Spring Wheat Water Requirement at Different Stages (1989 – 1991)

Unit: m³/mu

Period Growth and Development Period	Start and End Date	Days	1 Water Requirement	2 Water Requirement	3 Water Requirement
Seeding	3/4 ~ 24/4	22	10.37	18.67	11.70
Tillering	25/4 ~ 12/5	18	42.07	41.42	23.35
Jointing	17/5 ~ 19/5	17	30.33	27.46	24.65
Booting	30/5 ~ 12/6	14	41.43	51.40	29.79
Heading	13/6 ~ 19/6	7	33.85	34.70	21.54
Filling	20/6 ~ 24/6	5	65.34	47.52	45.25
Maturing	25/6 ~ 19/7	25	95.67	67.97	59.82
Total	3/4 ~ 19/7	108	318.06	289.14	216.10

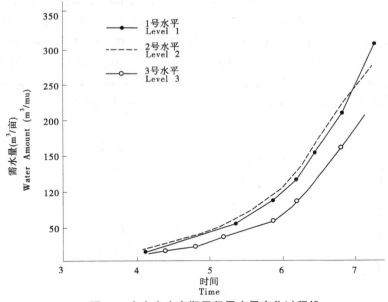

图 12　春小麦生育期累积需水量变化过程线
Figure 12　Changing Curves of Spring Wheat's Water Requirement Accumulated at Growth and Development Period

（二）苜蓿不同生长阶段的需水规律
2.2.1.2　Law of Alfalfa Water Requirement at Different Growing Stages

苜蓿的全生育期分苗期、蔓枝伸长阶段和开花成熟阶段 3 个生长阶段。从苜蓿全生育期不同生长阶段的需水状况观测分析,苜蓿的蔓枝伸长阶段,3个不同水平的需水量均表现为最大(见表68),并且远超过苜蓿开花成熟期和苗期的需水量,此阶段是苜蓿对水分的最敏感期,枝叶繁茂,叶面蒸腾量大,是苜蓿生长的需水临界期。从苜蓿发育期三茬的需水状况分析,第一茬生长期为74 d,需水量为146.04~203.79 m³/亩(305.4 mm);第二茬生长期43 d,需水量115.77~186.66 m³/亩(218.4 mm);第三茬生长期56 d,需水量181.29~228.57(286.3 mm)。3个不同水平的需水量均表现为第二茬需水量最少,历时最短,仅43 d(见表69),而且3个不同不平的苜蓿全生育期累积需水量变化过程也表现出基本相似的特点(见图13)。

The entire growth and development period of alfalfa includes 3 stages, that is, seedling, branch stretching and flowering & maturing. According to the observation of alfalfa water requirement at different stages, alfalfa requires most water at branch stretching stage regardless of the level (see Table 68) and far surpasses flowering and maturing stage and seedling stage. Alfalfa at the stretching stage, with luxuriant foliage which contributes to high transpiration, is most sensitive to water and enters a critical period for water requirement. According to the water requirement of three consecutive croppings of alfalfa in the growth and development period, the growing period of first cropping lasts 74 days, requiring 146.04 – 203.79 m³/mu (305.4mm), the second 43 days and 115.77 – 186.66 m³/mu (218.4 mm), and the third 56 days, requiring 181.29 – 228.57m³/mu (286.3mm). The results show that the second cropping requires least water and least growing duration (43 days, Table 69). Alfalfa of three different levels shows similar changing characteristics in terms of the accumulated water requirement in the entire growth and development period (see Figure 13).

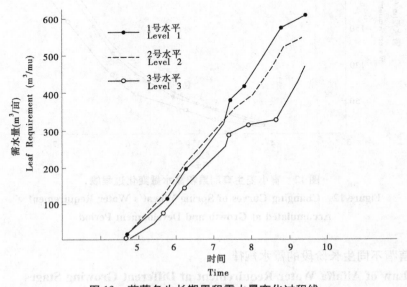

图13　苜蓿各生长期累积需水量变化过程线

Figure 13　Changing Curves of Alfalfa's Water Requirement Accumulated at Growth and Development Period

表68　　苜蓿各生长阶段需水量分析表(1989—1991年)　　单位：m³/亩

生育期	生育阶段	起止日期	天数	水平 1 需水量	2 需水量	3 需水量
第一茬	苗期	27/3~20/4	25	11.20	10.14	5.70
第一茬	蔓枝伸长期	21/4~25/5	35	104.59	124.91	73.93
第一茬	开花成熟期	26/5~8/6	14	77.81	68.74	66.41
第二茬	苗期	9/6~21/6	13	29.50	28.37	34.12
第二茬	蔓枝伸长期	22/6~11/7	20	114.88	87.48	70.51
第二茬	开花成熟期	12/7~21/7	10	42.68	29.89	27.57
第三茬	苗期	22/7~31/7	10	30.93	26.62	24.35
第三茬	蔓枝伸长期	1/8~31/8	31	159.43	133.51	118.72
第三茬	开花成熟期	1/9~15/9	15	37.95	31.82	38.70
全生育期		27/3~15/9	113	608.23	540.38	459.53

Table 68　Analysis of Alfalfa Water Requirement at Different Stages (1989 to 1991)

Unit: m³/mu

Growth and Development Period	Growth and Development Period	Start and End Date	Days	Level 1 Water Requirement	2 Water Requirement	3 Water Requirement
1st Cropping	Seedling	27/3-20/4	25	11.20	10.14	5.70
1st Cropping	Branch Extension	21/4-25/5	35	104.59	124.91	73.93
1st Cropping	Flowering & Maturing	26/5-8/6	14	77.81	68.74	66.41
2nd Cropping	Seedling	9/6-21/6	13	29.50	28.37	34.12
2nd Cropping	Branch Extension	22/6-11/7	20	114.88	87.48	70.51
2nd Cropping	Flowering & Maturing	12/7-21/7	10	42.68	29.89	27.57
3rd Cropping	Seedling	22/7-31/7	10	30.93	26.62	24.35
3rd Cropping	Branch Extension	1/8-31/8	31	159.43	133.51	118.72
3rd Cropping	Flowering & Maturing	1/9-15/9	15	37.95	31.82	38.70
Whole Growth and Development Period		27/3-15/9	113	608.23	540.38	459.53

表 69　　　　　　　　　　苜蓿三茬需水量分析表　　　　　　　　　　单位：m³/亩

水平	生育期			全生育期
	第一茬	第二茬	第三茬	
1	193.60	186.66	228.57	608.23
2	203.79	115.77	190.95	540.40
3	146.04	132.20	181.29	459.53
生育天数（d）	74	43	56	173

Table 69　Analysis of Alfalfa Water Requirement in Three Consecutive Croppings　Unit：m³/mu

Level	Growth and development period			Whole Growth and Development Period
	1st Cropping	2nd Cropping	3rd Cropping	
1	193.60	186.66	228.57	608.23
2	203.79	115.77	190.95	540.40
3	146.04	132.20	181.29	459.53
Growth and Development Duration (Days)	74	43	56	173

（三）玉米不同生长阶段的需水规律

2.1.1.3　Law of Corn Water Requirement at Different Growth Stages

玉米全生育期要经过播种—出苗、出苗—拔节、拔节—抽雄、抽雄—灌浆、灌浆—成熟及成熟—收割 6 个生长发育阶段，各阶段由于作物的生理特点和气象因素不同，作物的需水量变化极大。但在各生长发育阶段，3 个不同水平的需水量变化过程基本相似，均表现为抽雄—灌浆阶段需水量最大。由表 70 可见，在玉米整个生长发育阶段，需水规律表现为，生育前的需水量较小，然后逐渐增大，到抽雄—灌浆期达到高峰，高达 92.33～109.18 m³/亩。随着玉米籽粒的成熟，玉米植株后期逐渐枯萎，需水量渐渐变小。

The growth and development period of corn includes 6 stages, i. e. sowing – seedling, seedling – jointing, jointing – tasseling, tasseling – filling, filling – maturing, and maturing – harvest. Due to physiological and meteorological differences from growing stages, the crop water requirement can vary greatly. But in terms of water requirement at each stage, it changes in a similar way as for three different levels, i. e. requiring most water during tasseling-filling. Table 70 shows that, in the entire growth and development period, corn requires small amount of water before the growth, and gradually increases to the tasseling-filling stage, peaking at 92.33 – 109.18 m^3/mu. With the ripening of kernel, the plant begins to wither in the late period, so does the water requirement.

玉米全生育期不同生长阶段的需水量大小依次排列为:抽雄—灌浆期 > 出苗—拔节期 > 拔节—抽雄期 > 灌浆—成熟期 > 成熟—收割期 > 播种—出苗期。

The water requirement of corn at different stages can be arranged from large to small as: tasseling – filling > seedling – jointing > jointing – tasseling > filling – maturing > maturing – harvest > sowing – seedling.

从玉米生长特征方面考虑,抽雄—灌浆期,是玉米整个生育阶段中的一个关键时期,在这个生育阶段中,玉米籽粒日渐饱满,这时需要大量的水量,进行光合作用,合成有机物,形成干物质,这个生育阶段玉米的需水强度也最大。如果在这个时期玉米缺水,将直接影响玉米籽粒的饱满度和千粒重。

In terms of growth characteristics, tasseling – filling is a critical stage for corn in its entire growth and development period, during which kernel is to become plump, thus requiring abundant water to form organics and dry matters by means of photosynthesis. The water requirement intensity also peaks at this stage. If the amount of water cannot meet the requirement at this stage, the plumpness and the thousand kernel weight can be directly affected.

从气象因素方面来考虑,抽雄—灌浆期处于7月。此时,气温较高,同时,玉米植株也最茂盛,叶片较大,蒸腾量最高,这就需要大量的水维持自身的水量平衡,以及通过蒸腾散发水分来降低植株的温度,避免被灼伤。

In terms of meteorological factors, tasseling – filling stage falls in July. At this time, the temperature is high and the corn is most luxuriant with largest leaves and highest transpiration. Thus a large amount of water is required to balance its own water content and to lower the surface temperature of plant by evaporating water through transpiration in case of being scorched.

表70　　　　　　　　玉米各生长阶段需水量分析表(1989—1991年)　　　　　　单位:m³/亩

阶段		播种—出苗	出苗—拔节	拔节—抽雄	抽雄—灌浆	灌浆—成熟	成熟—收割
起止日期		9/5~15/5	16/5~24/6	25/6~8/7	9/7~29/7	30/7~14/8	15/8~7/9
水平	天数	7	40	14	21	16	23
1989年	1	23.87	108.13	36.31	168.32	31.85	14.40
	2	4.87	100.52	46.72	127.48	28.86	16.65
	3	21.74	86.30	46.86	138.91	14.75	4.69
1990年	1	21.76	32.90	74.32	63.93	48.46	—
	2	—	24.14	73.80	69.12	43.68	2.46
	3	1.16	32.12	59.40	66.30	49.71	7.04
1991年	1	22.37	64.69	35.51	95.28	64.42	67.48
	2	8.89	62.79	27.54	91.21	57.52	60.30
	3	4.55	57.97	16.25	71.79	37.92	39.72
3年平均	1	22.67	68.57	48.71	109.18	48.24	27.27
	2	4.59	62.48	49.35	95.95	43.36	26.47
	3	9.15	58.84	40.84	92.33	34.13	17.15

Table 70　Analysis of Corn Water Requirement at Different Stages (1989 to 1991) Unit: m³/mu

Stage		Seeding – Seedling	Seedling – Jointing	Jointing – Tasseling	Tasseling – Filling	Filling – Maturing	Maturing – Harvesting
Start and End Date		9/5 – 15/5	16/5 – 24/6	25/6 – 8/7	9/7 – 29/7	30/7 – 14/8	15/8 – 7/9
Level	Days	7	40	14	21	16	23
1989年	1	23.87	108.13	36.31	168.32	31.85	14.40
	2	4.87	100.52	46.72	127.48	28.86	16.65
	3	21.74	86.30	46.86	138.91	14.75	4.69
1990	1	21.76	32.90	74.32	63.93	48.46	—
	2	—	24.14	73.80	69.12	43.68	2.46
	3	1.16	32.12	59.40	66.30	49.71	7.04
1991	1	22.37	64.69	35.51	95.28	64.42	67.48
	2	8.89	62.79	27.54	91.21	57.52	60.30
	3	4.55	57.97	16.25	71.79	37.92	39.72
Average of the Three Years	1	22.67	68.57	48.71	109.18	48.24	27.27
	2	4.59	62.48	49.35	95.95	43.36	26.47
	3	9.15	58.84	40.84	92.33	34.13	17.15

我们将3个水平的平均需水量值点绘需水量累积曲线(见图14),从图中可看出,6

月以前即生育前期,曲线较平缓,需水量较小,7月到8月下旬这段时间,曲线的斜度较大,此时需水量值较大,8月以后,即生育后期,曲线又逐渐平缓,需水量的变化量不大,3个水平的需水量累积曲线均呈"S"形,体现出作物生育期需水规律,即生育前期需水量较小,中期需水量大,后期又逐渐变小。

The curve of accumulated water requirement is plotted in accordance with average water requirement of 3 levels (see Figure 14). It can be observed that the curve is gentle and water requirement is low before June, i.e. the early stages of growth and development period; then it becomes steep and water requirement becomes higher from July to late August; and the curve becomes flat and the variation of water requirement is small after August, i.e. the late stages of growth and development period. The curves of 3 levels are all in "S" shape, which present the law of water requirement in growth and development period, i.e. water requirement is low at early stages, large at middle stages and back to small at late stages.

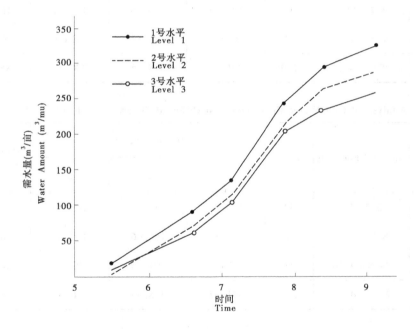

图14 玉米生产期累积需水量变化过程线

Figure 14 Changing Curves of Spring Wheat's Water Requirement Accumulated at Growth and Development Period

(四)甜菜不同生长阶段的需水规律

2.2.1.4 Law of Beet Water Requirement at Different Growing Stages

甜菜全生育期分前期、中期、后期3个阶段,前期为5月25日~7月7日,历时42 d;中期为7月8日~8月23日,历时46 d;后期历时41 d。3个水平的阶段需水量均表现为生长中期最高,占全生育期需水总量的47%左右(见表71),生长前期次之,后期略少。

The entire growth and development period of beet include 3 stages, i. e. early stage, middle stage and late stage. The early stage starts from May 25th to July 7th, 42 days altogether; the middle stage from July 8th to August 23rd, 46 days altogether; and the late stage lasts for 41 days. The middle stage requires most water among all 3 levels, contributing to about 47% of the whole required amount of water (see Table 71), followed by the early stage and the late stage.

表71　　　　　　　甜菜各生育阶段需水量分析表(1989—1991年)　　　　单位:m³/亩

生育期	起止日期	天数	不同水平			平均需水量
			1	2	3	
前期	25/5 ~ 7/7	42	106.98	90.95	103.40	100.44
中期	8/7 ~ 23/8	46	195.54	156.46	150.62	167.54
后期	24/8 ~ 4/10	41	88.65	101.54	76.69	88.96
合计	25/5 ~ 4/18	129	391.17	348.95	330.71	356.94

Table 71　　Analysis of Beet Water Requirement at Different Stages (1989 to 1991)　Unit: m³/mu

Growth and Development Period	Start and End Date	Days	Different Levels			Average Water Requirement
			1	2	3	
Initial Stage	25/5 - 7/7	42	106.98	90.95	103.40	100.44
Middle Stage	8/7 - 23/8	46	195.54	156.46	150.62	167.54
Later Stage	24/8 - 4/10	41	88.65	101.54	76.69	88.96
Total	25/5 - 4/18	129	391.17	348.95	330.71	356.94

甜菜生长中期需水量最大，因为这一时期是甜菜生长的重要时间，而甜菜生育阶段的后期，主要是甜菜糖分的形成期，甜菜需要一定的水分来进行光合作物，以形成糖，因而这一阶段的需水量也很重要，它决定甜菜的含糖量，也是甜菜生长的一个重要需水阶段。

Reason for this is that middle stage is the critical period for beet growth. The late stage is time to produce sugar. Beet requires water for photosynthesis to produce sugar. Thus water which decides the sugar content of beet, is important at this stage, and this is also an important water requirement stage of beet.

（五）作物最大需水生长阶段分析
2.2.1.5 Analysis of Max Water Requirement Stage of Crops

通过1989—1991年3年对春小麦、苜蓿、玉米及甜菜这4类作物不同阶段的需水规律试验研究,我们发现这4类作物的主要需水量期有所不同。春小麦的最大需水量阶段为成熟期;苜蓿的最大需水量生长阶段为生长中期蔓枝伸长期;玉米的最大需水量生长阶段为中后期抽雄—灌浆期;甜菜的最大需水量生长阶段为中期(见表72)。

Through the experimental research from 1989 to 1991 on the water requirement law of spring wheat, alfalfa, corn and beet at different stages, it is found that their major water requirement stages are different. Spring wheat requires most amount of water at the maturing stage, while alfalfa at the branch stretching stage in middle period, corn at the tasseling-filling stage in mid-late period, and beet at the middle stage (see Table 72).

表72　　　　　　　　作物最大需水生长阶段分析表

时段	时段			
	春小麦	苜蓿	玉米	甜菜
生长阶段	成熟期	蔓枝伸长期	抽雄—灌浆期	中期
起止日期	7月19日~8月13日	8月1日~8月31日	7月15日~8月20日	8月23日~10月4日

Table 72　　Analysis of Maximum Water Requirement Stage of Crops

Stages	Crops			
	Spring Wheat	Alfalfa	Corn	Beet
Growth Period	Maturing	Branch Extension	Tasseling-filling	Middle Stage
Start and End Date	July 19th to August 13rd	August 1st to August 31st	July 15th to August 20th	August 23rd to October 4th

（六）作物不同生长阶段需水规律研究小结
2.2.1.6 Summary of Crop Water Requirement at Different Growth Stages

通过1989—1991年3年对春小麦、苜蓿、玉米和甜菜不同生长阶段需水规律的试验研究,现将其归纳如下:

Through the experimental research from 1989 to 1991 on the water requirement law of spring wheat, alfalfa, corn and beet at different stages, the main results are summarized as follows:

春小麦各生长阶段的需水量变化由大到小依次排列为:灌浆—成熟期 > 抽穗—灌浆

期>拔节—孕穗期,该3个阶段为小麦生育中后期,阶段最小量较高,而苗期需水量较少。

The water requirement of spring wheat at different stages can be arranged from large to small as: filling-maturing > heading-filling > jointing-booting. These three stages, belonging to the mid-late period of wheat, require large amount of water, while the seedling stage requires a small amount of water.

苜蓿全生育期的三茬生长阶段内每一茬都出现一个峰值,即蔓枝伸长期。各阶段的需水量变化由大到小依次排列为:蔓枝伸长期>开花成熟期>苗期,并且苜蓿的第二茬生育期最短,仅43 d,次为第三茬,为56 d。

The growth peak of alfalfa occurs at the branch stretching stage in all three croppings. The water requirement of alfalfa at different stages can be arranged from large to small as: branch stretching > flowering and maturing > seedling. The second cropping experiences shortest growth and development period, 43 days, followed by the third cropping with 56 days.

玉米各生长阶段的需水变化由大到小排列为:抽雄—灌浆期>出苗—拔节期>拔节—抽雄期>灌浆—成熟期,而播种—出苗和成熟收割期需水量较少。

The water requirement of corn at different stages can be arranged from large to small as: tasseling-filling > seedling-jointing > jointing-tasseling > filling-maturing. The sowing-seedling and the maturing-harvest require the least amount of water.

甜菜在其生长阶段,以中期需水量最高,后期需不量较少。

Beet requires most amount of water at the middle stage but relatively small at the late stage.

二、作物各生育阶段需水量占全生育期需水量比重
2.2.2 Proportion of Water Requirement of Each Stage in the Entire Growth and Development Period

分析作物各生育期占全生育期需水量百分比(称阶段需水模数)主要用以研究各类作物不同阶段需水量在全生育期内的变化过程。

The analysis on the proportion (called the stage water requirement modulus) of each stage in the entire period is mainly used to study the changing process of water requirement in respect to different crops at different stages.

(一)春小麦各生育期需水量占全生育期需水量百分比
2.2.2.1 Percentage of Water Requirement of Every Growth and Development Period for Spring Wheat

春小麦的全生育期分7个不同阶段,我们对7个不同生育阶段的需水量状况分析表

明,3 个不同水平的需水量均表现为在小麦成熟期的需水量占全生育期需水量的百分比最大(见表73),变化在 23.51%~30.96% 之间。其中,1 号水平的需水量所占比重最大为 30.96%,3 号水平的需水量所占比重为 27.68%,介于 1 号水平和 2 号水平之间。

The entire growth and development period of spring wheat includes 7 stages. Through the water requirement analysis of these 7 stages, it is found that the maturing stage accounts for the largest proportion in all 3 levels (see Table 73), varying between 23.51% – 30.96%. Maturing stage of Level #1 requires the largest amount of water with 30.96%, followed by Level #3 with 27.68% and Level #2.

表73　春小麦不同生长期需水量占全生育期需水百分比分析表　　单位:%

水平	苗期	分蘖	拔节	孕穗	抽穗	灌浆	成熟
1	3.36	13.61	9.81	13.41	10.95	21.14	30.96
2	6.46	14.33	9.50	17.78	12.00	16.43	23.51
3	5.41	10.81	11.41	13.19	9.97	20.94	27.08
平均值	5.08	12.92	10.24	14.79	10.97	19.50	27.38

Table 73　Analysis of Proportion of Water Requirement at Each Stage in the Entire Growth and Development Period of Spring Wheat　　Unit:%

Level	Seedling	Tillering	Jointing	Booting	Heading	Filling	Maturing
1	3.36	13.61	9.81	13.41	10.95	21.14	30.96
2	6.46	14.33	9.50	17.78	12.00	16.43	23.51
3	5.41	10.81	11.41	13.19	9.97	20.94	27.08
Average	5.08	12.92	10.24	14.79	10.97	19.50	27.38

由表 73 可见,在春小麦全生育期内,除成熟期需水量所占比重最大外,小麦灌浆期和孕穗期也是十分重要的需水时段,这期间保证供水对作物的生长发育及产量是非常重要的。

Table 73 shows that, except for the maturing stage with largest water requirement, the filling and the booting stages also need large amount of water. It is important for the growth and development and yield of spring wheat to ensure water supply during these two stages.

(二)苜蓿各生育期需水量占全生育期需水量百分比
2.2.2.2　Percentage of Water Requirement of Every Growth and Development Period for Alfalfa

苜蓿三茬各生育期的需水量占全生育期需水量各不相同。由表 74 可见,三茬各生育阶段需水量均以苜蓿蔓枝伸长期所占百分比最大。由此可见,在苜蓿蔓枝伸长期水分的及时供给对苜蓿的正常生长有极为重要的作用。而在苜蓿苗期的需水量相对蔓枝伸长和开花成熟期来说,所占比重较小,表现为三茬各阶段需水量占全生育期需水量变化在1.24%~7.43%之间,均未超过10%。

  Proportions of each stage in the entire growth and development period in respect to water requirement of three alfalfa croppings are different. Table 74 shows that, no matter which cropping, the branch stretching stage requires most amount of water. Thus timely water supply in branch stretching stage is vital to the growth of alfalfa. The water requirement in seedling stage of alfalfa, compared to branch stretching and flowering and maturing, is relatively low, with proportion varying from 1.24% - 7.43% in all three croppings, not exceeding 10%.

表74 苜蓿各生育期需水量占全生育期需水量百分比分析表 单位:%

水平	第一茬			第二茬			第三茬		
	苗期	蔓枝伸长期	开花成熟期	苗期	蔓枝伸长期	开花成熟期	苗期	蔓枝伸长期	开花成熟期
1	1.84	17.2	12.79	4.95	18.89	7.02	5.09	26.21	6.02
2	1.88	23.11	12.73	5.25	16.2	5.52	4.93	24.72	5.70
3	1.24	16.09	14.45	7.43	15.34	6.0	5.3	25.73	8.42
平均值	1.65	18.8	13.32	5.88	16.81	6.81	5.11	25.55	6.71

Table 74 Analysis of Proportion of Water Requirement of Each Stage in the Entire Growth and Development Period of Alfalfa Unit:%

Level	1st Cropping			2nd Cropping			3rd Cropping		
	Seedling	Branch Extension	Flowering & Maturing	Seedling	Branch Extension	Flowering & Maturing	Seedling	Branch Extension	Flowering & Maturing
1	1.84	17.2	12.79	4.95	18.89	7.02	5.09	26.21	6.02
2	1.88	23.11	12.73	5.25	16.2	5.52	4.93	24.72	5.70
3	1.24	16.09	14.45	7.43	15.34	6.0	5.3	25.73	8.42
Average	1.65	18.8	13.32	5.88	16.81	6.81	5.11	25.55	6.71

  同时,由表74反映的3个不同水平生育阶段占全生育期需水量的百分比数值来看,也均表现出苜蓿在蔓枝伸长阶段的需水量所占比重最大这一共同特点,由此进一步说明,蔓枝伸长阶段是苜蓿生长发育的需水临界期,此阶段需求水分较大,如不能满足,会对苜蓿的生长发育产生很大影响。

  Meanwhile, according to the proportion shown in Table 74, the branch stretching stage of alfalfa requires most amount of water in all 3 levels. Thus it can be inferred that the stretching stage is the critical period of water requirement for the growth and development of alfalfa. If water supply is insufficient, the growth and development of alfalfa can be greatly affected.

(二)玉米各生育期需水量占全生育期需水量百分比
2.2.2.3 Percentage of Water Requirement of Every Growth and Development Period for Corn

玉米在生育期共分6个需水阶段,而各生育阶段的需水量所占全生育期需水量百分比是不同的,由表75可见,各生育阶段的需水量大体表现为玉米生长前期和后期所占百分比较小,如在播种—出苗期,3个不同水平的阶段需水量仅占全生育期需水量总数的1.42%~7.21%;成熟—收割期3个不同水平的阶段需水量也只占全生育期需水量的7.30%~9.04%。而玉米抽雄—灌浆期和出苗—拔节期为全生育期需水量最高阶段,其中抽雄—灌浆期的需水量占全生育期需水量的34.27%,这期间保证玉米生长用水对玉米的发育极为重要。

The growth and development period of corn includes 6 water requirement stages with different water requirement proportions. Table 75 shows that the early and the late stages account for relatively small proportion. For example, in the sowing - seedling stage, the required amount of water only accounts for 1.42% - 7.21% of the whole requirement in 3 levels; and the maturing - harvest stage accounts for 7.30% - 9.04%. The tasseling - filling and the seedling - jointing stages require most amount of water. The tasseling - filling stage accounts for 34.27% of the entire growth and development period. Thus it is vital for the corn to ensure its water supply during this stage.

表75　　玉米各生育期需水量占全生育期需水量百分比分析表　　单位:%

水平		播种—出苗	出苗—拔节	拔节—抽雄	抽雄—灌浆	灌浆—成熟	成熟—收割
1989年	1	6.23	28.24	9.48	43.96	8.32	3.76
	2	1.41	29.13	16.44	36.94	11.26	4.82
	3	6.94	27.55	14.96	44.34	4.71	1.50
1990年	1	9.00	13.63	30.79	26.49	20.17	—
	2	—	9.93	34.46	36.64	17.96	2.89
	3	0.5	13.78	25.40	31.78	25.54	3.01
1991年	1	6.4	18.5	10.65	27.42	18.41	19.29
	2	2.87	20.88	8.88	29.40	18.55	19.43
	3	1.99	25.38	7.11	31.52	16.60	17.39
3年平均	1	7.21	20.12	16.97	32.62	15.63	7.68
	2	1.42	21.08	19.92	34.32	15.92	9.04
	3	3.11	22.24	15.80	35.88	15.62	7.30

Table 75　　Analysis of Proportion of Water Requirement of Each Stage
in the Entire Growth and Development Period of Corn　　　Unit:%

Level		Sowing – Seedling	Seedling – Jointing	Jointing – Tasseling	Tasseling – Filling	Filling – Maturing	Maturing – Harvesting
1989	1	6.23	28.24	9.48	43.96	8.32	3.76
	2	1.41	29.13	16.44	36.94	11.26	4.82
	3	6.94	27.55	14.96	44.34	4.71	1.50
1990	1	9.00	13.63	30.79	26.49	20.17	—
	2	—	9.93	34.46	36.64	17.96	2.89
	3	0.5	13.78	25.40	31.78	25.54	3.01
1991	1	6.4	18.5	10.65	27.42	18.41	19.29
	2	2.87	20.88	8.88	29.40	18.55	19.43
	3	1.99	25.38	7.11	31.52	16.60	17.39
Average of the Three Years	1	7.21	20.12	16.97	32.62	15.63	7.68
	2	1.42	21.08	19.92	34.32	15.92	9.04
	3	3.11	22.24	15.80	35.88	15.62	7.30

（四）甜菜各生育期需水量占全生育期需水量百分比

2.2.2.4 Percentage of Water Requirement of Every Growth and Development Period for Beet

甜菜的生育期为前期、中期、后期3个阶段。从表76可以看出，通过1989—1991年3年的试验研究，3个不同水平的阶段需水量均以中期所占百分比最高，达44.84%~49.99%。前期需水量所占百分比次之，后期所占百分比最小。

The growth and development period of beet includes 3 stages, i.e. the early stage, the middle stage and the late stage. Table 76 shows that, according to the experimental research from 1989 to 1991, the middle stage requires most amount of water in all 3 levels, accounting for 44.84% –49.99%, followed by the early stage and the late stage.

表76　　　　甜菜各生育期需水量占全生育期需水量百分比分析表　　　　单位：%

生育期	时段	不同水平			平均
	起止日期	1	2	3	
前期	5月25日~7月7日	27.35	26.06	28.97	27.46
中期	7月8日~8月23日	49.99	44.84	46.94	47.26
后期	8月24日~10月4日	22.66	29.10	21.49	24.42

Table 76　Analysis of Proportion of Water Requirement of Each Stage in the Entire Growth and Development Period of Beet　　Unit: %

Period		Different Levels			Average
Growth and Development Period	Start and End Date	1	2	3	
Initial Stage	May 25th to July 7th	27.35	26.06	28.97	27.46
Middle Stage	July 8th to August 23rd	49.99	44.84	46.94	47.26
Later Stage	August 24th to October 4th	22.66	29.10	21.49	24.42

(五) 作物各生育阶段需水量所占需水总量百分比研究小结
2.2.2.5 Summary of the Proportion of Crops Water Requirement at Every Growth and Development Stage in Total Water Requirement

通过对春小麦、苜蓿、玉米和甜菜各生育阶段需水量所占全生育期需水量比重的试验研究,现将其基本特点归纳如下:

Through the experimental research on the proportion of each stage in the entire growth and development period in respect to water requirement of spring wheat, alfalfa, corn and beet, the main results are summarized as follows:

春小麦在灌浆—成熟期需水量占全生育期需水量百分比最高,达27.38%,次为小麦的抽穗—灌浆期,为19.50%,二者之和达46.88%,约为小麦全生育需水总量的1/2。

The filling – maturing stage of spring wheat requires most amount of water, accounting for 27.38%, followed by the heading – filling stage, accounting for 19.50%. The sum of the two is 46.88%, almost 1/2 of the total water requirement in the growth and development period.

从苜蓿打三茬各阶段需水量所占百分比分析,第三茬的苜蓿蔓枝伸长期需水量所占比例最高,约占苜蓿(打三茬)全生育期需水总量的25.55%,次为第一茬,占18.8%。

The branch stretching stage of the third alfalfa cropping requires most amount of water, accounting for 25.55% of the entire three croppings, followed by the first cropping, accounting for 18.8%.

玉米在全生育期内的需水高峰期出现在抽雄—灌浆期,阶段需水量占全生育期需水总量的 34.27%,次为出苗—拔节期,为 21.15%,二者之和达 55.42%,超过小麦全生育期需水总量的 1/2。

The tasseling - filling stage of corn requires most amount of water, accounting for 34.27%, followed by the seedling - jointing stage, accounting for 21.15%. The sum of the two is 55.42%, more than 1/2 of the total water requirement in the growth and development period.

甜菜在全生育期内以其生长中期需水量所占比重最大,平均为 47.26%,约为总需水量的一半,而苜蓿生育前期和后期的需水量所占比例接近,分别为 27.46% 和 24.42%。

The middle stage of beet requires most amount of water, accounting for 47.26% on average, which is almost 1/2 of the total water requirement. The early and the late stages are close in proportion, 27.46% and 24.42% respectively.

作物需水量最高值及生长阶段分析表见表 77。

The analysis of max value of water requirement and growth stage is shown in Table 77.

表 77　作物需水量最高值及生长阶段分析表

项目		春小麦	苜蓿	玉米	甜菜
最高值	百分比	27.38	25.55	34.27	47.26
	生长阶段	灌浆—成熟期	第三茬蔓枝伸长期	抽雄—灌浆期	生长中期
次高值	百分比	19.50	18.8	21.15	27.46
	生长阶段	抽穗—灌浆期	第一茬蔓枝伸长期	出苗—拔节期	生长前期

Table 77　Analysis of Maximum Water Requirement and Growth Stage of Crop

Items		Spring Wheat	Alfalfa	Corn	Beet
Peak Value	Percentage	27.38	25.55	34.27	47.26
	Growing Stage	Filling-Maturing Stage	Branch Stretching of the 3rd Cropping	Tasseling-filling Stage	Middle Growth Stage
Sub-peak Value	Percentage	19.50	18.8	21.15	27.46
	Growing Stage	Heading-Filling Stage	Branch Stretching of the 1st Cropping	Seedling-Jointing Stage	Early Growth Stage

三、作物需水强度的变化规律
2.2.3 Changing law of Crop Water Requirement Intensity

(一)春小麦需水强度的变化规律
2.2.3.1 Changing Law of Water Requirement Intensity of Spring Wheat

从本区春小麦各不同生长阶段分析需水强度的变化规律可以发现,小麦在其全生育期要经过从播种—苗期到成熟期 7 个阶段,在小麦生长的 7 个阶段中,以小麦的抽穗—灌浆期(6 月 20 日～7 月 19 日)需水强度最大,日需水量达 7.54～10.85 m³/亩,阶段需水量为 45.25～65.34 m³/亩,该期间是春小麦的需水临界期(见表 78)。次为小麦孕穗期—抽穗期,大致变化在 6 月 12 日～6 月 19 日,需水强度为 2.69～4.94 m³/(亩·d)。

According to the change law of spring wheat water requirement intensity at different stages of this plot, the growth and development period of wheat include 7 stages from sowing-seedling to maturing. The water requirement intensity at heading-filling stage (June 20th to July 19th) is the highest, daily water requirement being 7.54 – 10.85 m³/mu and the stage water requirement being 45.25 – 65.34 m³/mu. Thus this is the critical period of water requirement for spring wheat (see Table 78). The booting-heading stage, from June 12th to June 19th in general, follows with an intensity of 2.69 – 4.94 m³/(mu·day).

小麦全生育期需水变化规律与气象因素、土壤湿度和农业措施以及小麦本身的生理状态和品种有关。

The change law of spring wheat water requirement in the entire growth and development period links to meteorological factors, soil moisture, agricultural measures, physiological status and wheat varieties.

分析小麦 3 个不同水平的阶段需水强度可见,它们有着基本相似的变化规律。小麦出苗至分蘖期需水强度从 0.47 m³/(亩·d)增加至 2.34 m³/(亩·d),至拔节期又降至 1.70 m³/(亩·d),孕穗期升至 2.96 m³/(亩·日),抽穗期又升至 4.23 m³/(亩·d)。灌浆期是小麦的需水临界期,需水强度升至 10.89 m³/(亩·d),灌浆后,小麦停止生长到成熟期降至 3.83 m³/(亩·d)。从气候因素看,总的趋势是由于气温升降引起空气饱和差的变化,从而导致需水强度的增减,并且与小麦生长发育阶段关系极为密切。小麦抽穗—灌浆期处于营养生长阶段,作物生理活动旺盛,叶面积增加快,所以小麦蒸腾量骤增,并且这期间(6 月 20 日～6 月 24 日)气温高,日照时间长。乳熟后,植株逐渐死亡,蒸腾量减退,所以需水强度减少。

It is found that the wheat of 3 different levels shares a similar change law in respect to water requirement intensity. From seedling to tillering, the water intensity increases from 0.47

m³/(mu · day) to 2.34 m³/(mu · day), then decreases to 1.70 m³/(mu · day) at the jointing stage, and then increases to 2.96 m³/(mu · day) at the booting stage and 4.23 m³/(mu · day) at the heading stage. The intensity at the filling stage, a critical period of water requirement for wheat, increases to 10.89 m³/(mu · day), then the wheat stops growing and drops to 3.83 m³/(mu · day) at the maturing stage. From the meteorological aspect, the general trend is that the temperature variation causes changes in vapor pressure deficit, which leads to the variation of water requirement intensity and closely links to the growing stage of wheat. The heading-filling stage of wheat is for the vegetative growth, during which the physiological activities are vigorous and leaf area is fast expanding, plus the high temperature and long daytime during this stage (from June 20th to June 24th). Thus the wheat transpiration soars. The plant dies out after milk ripeness and transpiration decreases. Thus water requirement intensity becomes weak.

小麦各生育阶段需水量累积过程用曲线表示呈"S"形(见图15),3个不同水平累积过程用公式分别表示为：

The curves of accumulated water requirement of spring wheat in all 3 levels are in "S" shape (see Figure 15). The accumulating process of the 3 levels can be expressed as:

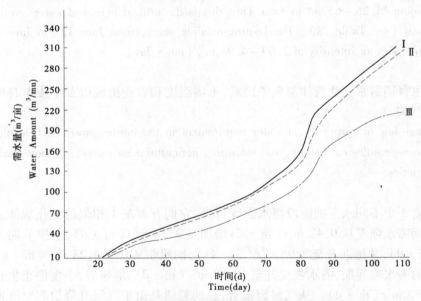

图15 小麦全生育期累积需水量变化过程线

Figure 15 Changing Process of Accumulated Water Requirement of Spring Wheat in the Growth and Development Period

1号水平 $E = -1.03387 + 0.005497T + 0.0259T^2$

Level #1 $E = -1.03387 + 0.005497T + 0.0259T^2$

相关系数 $R = 0.99064$

Correlation coefficient $R = 0.99064$

2 号水平 $E = 314.419/[1 + \mathrm{EXP}(3.95456 + 0.05426)T)]$

Level #2 $E = 314.419/[1 + \mathrm{EXP}(3.95456 + 0.05426)T)]$

相关系数 $R = 0.99104$

Correlation coefficient $R = 0.99104$

3 号水平 $E = 260.059/[1 + \mathrm{EXP}(4.11898 + 0.0505)T)]$

Level #3 $E = 260.059/[1 + \mathrm{EXP}(4.11898 + 0.0505)T)]$

相关系数 $R = 0.98998$

Correlation coefficient $R = 0.98998$

式中　E——各阶段累积需水量，$m^3/$亩；

　　　T——作物生长的时间，d。

Where　E = accumulated water requirement of each stage, m^3/mu

　　　　T = the growth period of crop, d.

表 78　春小麦生育期内不同阶段需水强度分析表

生育期	时段 起止日期	天数	1 日需水量	1 模系数	2 日需水量	2 模系数	3 日需水量	3 模系数	平均需水强度
苗期	3/4~24/4	22	0.47	3.36	0.85	6.46	0.53	5.41	0.62
分蘖	25/4~12/5	18	2.34	13.62	2.30	14.33	1.30	10.81	1.98
拔节	13/5~29/5	17	1.69	9.81	1.53	9.50	1.37	11.41	1.53
孕穗	30/5~12/6	14	2.96	13.41	3.67	17.78	2.13	13.79	2.92
抽穗	13/6~19/6	7	4.23	10.95	4.94	12.00	2.69	9.97	3.95
灌浆	20/8~24/6	5	10.85	21.14	2.92	16.43	7.54	28.94	8.77
成熟	25/6~19/7	25	3.83	30.96	2.72	23.51	2.39	27.68	2.98

Table 78　Analysis of Spring Wheat Water Requirement Intensity at Different Stages of Growth and Development Period

Growth and Development Period	Period Start and End Date	Days	1 Daily Water Requirement	1 Modulus	2 Daily Water Requirement	2 Modulus	3 Daily Water Requirement	3 Modulus	Average Water Requirement Intensity
Seedling Period	3/4 – 24/4	22	0.47	3.36	0.85	6.46	0.53	5.41	0.62
Tillering	25/4 – 12/5	18	2.34	13.62	2.30	14.33	1.30	10.81	1.98
Jointing Period	13/5 – 29/5	17	1.69	9.81	1.53	9.50	1.37	11.41	1.53
Booting	30/5 – 12/6	14	2.96	13.41	3.67	17.78	2.13	13.79	2.92
Heading	13/6 – 19/6	7	4.23	10.95	4.94	12.00	2.69	9.97	3.95
Filling Period	20/8 – 24/6	5	10.85	21.14	2.92	16.43	7.54	28.94	8.77
Maturing Period	25/6 – 19/7	25	3.83	30.96	2.72	23.51	2.39	27.68	2.98

（二）苜蓿需水强度的变化规律
2.2.3.2 Changing Law of Alfalfa Water Requirement Intensity

苜蓿生育期内不同生长阶段的需水强度的变化规律较为明显，从表 79 可看出，在苜蓿全生育期内二、三茬的蔓枝伸长期是苜蓿的需水临界期，日需水量达 3.53 ~ 5.74 m³/(亩·日)，阶段需水量为 70.51 ~ 159.43 m³/亩。第一茬蔓枝伸长期的日需水量较小是由于 4 月气温低、日照短，植株的叶面积小等原因，而进入 5、6 月，气温升高，植株生长旺盛，苜蓿需水强度增加。

The change law of alfalfa water requirement intensity at different growing stages has an evident pattern. Table 79 shows that the branch stretching stage of the 2nd and the 3rd croppings is the critical period of water requirement, daily water requirement being 3.53 to 5.74 m³/(mu·day) and stage water requirement being 70.51 - 159.43 m³/mu. The daily water requirement of the 1st cropping is relatively low in April, because the temperature is low, daytime is short, and leaf area is small. When it comes to May and June, the temperature rises and plants become luxuriant, the intensity increases accordingly.

表 79　　　　　　　苜蓿生育期内不同阶段需水强度分析表　　　　　　单位:m³/(亩·日)

生育期	生育阶段	起止日期	天数	时段 1 日需水数	模系数	2 日需水量	模系数	3 日需水量	模系数	平均需水强度
第一茬	苗期	27/3 - 20/4	25	0.45	1.84	0.41	1.88	0.23	1.24	0.36
	蔓枝伸长期	21/4 - 25/5	35	2.99	7.20	3.57	23.11	2.11	16.09	2.89
	开花成熟期	26/5 - 8/6	14	5.56	12.79	4.91	12.73	4.74	14.45	5.07
第二茬	苗期	9/6 - 21/6	13	2.27	4.85	2.18	5.25	2.62	7.43	2.36
	蔓枝伸长期	22/6 - 11/7	20	5.74	18.89	4.36	16.2	3.53	15.34	4.54
	开花成熟期	17/7 - 21/7	10	4.27	7.02	2.98	5.52	2.76	6.0	3.34
第三茬	苗期	23/7 - 31/7	10	3.09	5.09	2.66	4.93	2.44	5.3	2.73
	蔓枝伸长期	1/8 - 31/8	31	5.14	6.21	4.31	24.72	3.81	25.73	4.42
	开花成熟期	1/9 - 15/9	15	2.53	6.24	2.05	5.70	2.50	8.42	2.39
全生育期		27/3 - 15/9	173	3.50	100	3.24	100	2.66	100	3.13

Table 79   **Analysis of Alfalfa Water Requirement Intensity at Different Stages of Growth and Development Period**   Unit: m³/(mu · day)

Growth and Development Period	Growth and Development Period	Start and End Date	Days	1 Daily Water Requirement	1 Modulus	2 Daily Water Requirement	2 Modulus	3 Daily Water Requirement	3 Modulus	Average Water Requirement Intensity
1st Cropping	Seedling Period	27/3 – 20/4	25	0.45	1.84	0.41	1.88	0.23	1.24	0.36
1st Cropping	Branch Extension Period	21/4 – 25/5	35	2.99	7.20	3.57	23.11	2.11	16.09	2.89
1st Cropping	Flowering & Maturing Period	26/5 – 8/6	14	5.56	12.79	4.91	12.73	4.74	14.45	5.07
2nd Cropping	Seedling Period	9/6 – 21/6	13	2.27	4.85	2.18	5.25	2.62	7.43	2.36
2nd Cropping	Branch Extension Period	22/6 – 11/7	20	5.74	18.89	4.36	16.2	3.53	15.34	4.54
2nd Cropping	Flowering & Maturing Period	17/7 – 21/7	10	4.27	7.02	2.98	5.52	2.76	6.0	3.34
3rd cropping	Seedling Period	23/7 – 31/7	10	3.09	5.09	2.66	4.93	2.44	5.3	2.73
3rd cropping	Branch Extension Period	1/8 – 31/8	31	5.14	6.21	4.31	24.72	3.81	25.73	4.42
3rd cropping	Flowering & Maturing Period	1/9 – 15/9	15	2.53	6.24	2.05	5.70	2.50	8.42	2.39
Whole Growth and Development Period		27/3 – 15/9	173	3.50	100	3.24	100	2.66	100	3.13

同时从表79可以看出,苜蓿需水强度在整个生育期内共出现3个高峰值,从第一茬的蔓枝伸长期至开花成熟期达到第一个高峰值,即1号水平5.56 m³/(亩·d),2号水平4.91 m³/(亩·d),以及3号水平的4.74 m³/(亩·d)。在第二茬的蔓枝伸长期出现第二高峰值,1号水平为5.74 m³/(亩·d),2号水平4.36 m³/(亩·d),3号水平3.53 m³/(亩·d)。在第三茬蔓枝伸长期又出现第三个高峰值,1号水平为5.14 m³/(亩·d),2号水平为4.31 m³/(亩·d),3号水平为3.81 m³/(亩·d)。从全生育期的需水状况,蔓枝伸长期需水强度最大,是苜蓿对水分十分敏感期,此阶段枝叶繁茂,叶面蒸腾量大,是苜蓿三茬的需水临界期,苜蓿三茬平均各生育阶段的需水强度有着共同的规律,即由幼苗期至蔓枝伸长期逐渐增加,由2.25 m³/(亩·d),逐渐增至3.84 m³/(亩·d),开花成熟期为4.15 m³/(亩·d)。从气象因素分析,苜蓿全生育期的需水强度与气温升降引起空气饱和差的变化,也有一定的关系。

Table 79 shows that there are three peaks in the entire growth and development period. The first peak occurs between the branch stretching stage and the flowering & maturing stage of the 1st cropping, i.e. 5.56 m³/(mu · day) of Level #1, 4.91 m³/(mu · day) of Level #2, and 4.74 m³/(mu · day) of Level #3. The second peak occurs at the branch stretching stage of the 2nd cropping, i.e. 5.74 m³/(mu · day) of Level #1, 4.36 m³/(mu · day) of Level #2, and 3.53 m³/(mu · day) of Level #3. The third peak occurs at the branch stretching stage of the 3rd cropping, i.e. 5.14 m³/(mu · day) of Level #1, 4.31 m³/(mu · day) of Level #2, and 3.81 m³/(mu · day) of Level #3. The branch stretching stage requires most amount of water in the entire period. Very sensitive to water, with luxuriant leaves and huge transpiration, alfalfa at this stage is in the critical period of water requirement of three croppings. Three croppings share a similar pattern of water requirement intensity, i.e. gradually increases from seedling to branch stretching, from 2.25 m³/(mu · day) to 3.84 m³/(mu · day), and achieved 4.15 m³/(mu · day) at flowering & maturing stage. From the meteorological aspect, there is certain connection between alfalfa water requirement intensity of the entire growth and development period and changes in vapor pressure deficient caused by temperature variation.

(三)玉米需水强度的变化规律
2.2.3.3　Changing Law of Corn Water Requirement Intensity

玉米在生育期内不同阶段的需水强度变化较大,3个不同水玉的玉米需水强度均表现为在玉米的抽雄—灌浆期,即7月9日~7月29日之间的需水强度最大,其中1号水平为5.20 m³/(亩·d),2号水平为4.60 m³/(亩·d),3号水平为4.40 m³/(亩·d)。以1号水平的阶段需水强度最大;次为玉米的拔节—抽雄期(6月25日~7月8日)。3个不同水平的平均需水强度为3.31 m³/(亩·d),以2号水平的需水强度最高,达3.53 m³/(亩·d)。

The corn water requirement intensity varies largely at different stages of the growth and development period. The tasseling-filling stage, i.e. from July 9th to July 29th, requires the largest intensity in all 3 levels, 5.20 m³/(mu · day) for Level #1, 4.60 m³/(mu · day) for Level #2, and 4.40 m³/(mu · day) for Level #3. Level #1 corn requires the largest intensity at the stage, followed by the jointing-tasseling stage (from June 25th to July 8th). The average water requirement intensity of 3 levels is 3.31 m³/(mu · day). Level #2 has the largest intensity with 3.53 m³/(mu · day) on average.

玉米在生育期内的需水强度变化总趋势表现为单峰形式,即从玉米播种—出苗到玉米抽雄—灌浆,阶段需水强度逐渐增大,并达峰值,玉米灌浆以后,需水强度又逐渐减小。
The corn water requirement intensity has one peak in the entire period. The intensity increases from the sowing – seedling stage, achieves the peak at the tasseling – filling stage, and then decreases.

从玉米生长发育看,玉米灌浆期,正是玉米的籽粒物质转化时期,作物生理活动强盛,需水强度大,因此这时期的供水必须保证,它对玉米的籽粒和千粒重量影响极大。同时这时期气温升高,日照时期长,有助于作物的生长发育。玉米生育期内需水强度的次高值区为玉米拔节—抽雄期,这期间气温高,正值玉米生长发育阶段,玉米全生育期需水强度变化过程详见表80。

From the perspective of corn growth and development, the filling stage is time for kernel transformation. The corn is vigorous in physiological activities and large in water requirement intensity. Thus water supply must be ensured at this stage due to its great influence to kernel and thousand kernel weight. Meanwhile, rising temperature and prolonging daytime at this stage contribute to the growth and development of crops. The sub-peak value of corn water requirement intensity occurs at the jointing-tasseling stage, an important stage featured by high temperature. The intensity changing process of corn is shown in Table 80.

表80　　　　　玉米生育期内不同阶段需水强度分析表　　　　单位:m³/(亩·d)

生育期	时段		1	2	3	平均需水强度
	起止日期	天数	日需水量	日需水量	日需水量	
播种—出苗	9/5～15/5	73.24	0.66	1.31	1.74	
出苗—拔节	16/5～24/6	40	1.71	1.56	1.47	1.58
拔节—抽雄	25/6～8/7	14	3.48	3.53	2.92	3.31
抽雄—灌浆	9/7～29/7	21	5.20	4.60	4.40	4.73
灌浆—成熟	30/7～14/8	16	3.02	2.71	2.13	2.62
成熟—收割	15/8～7/9	23	1.19	1.15	0.75	1.03

Table 80　　Analysis of Corn Water Requirement Intensity at Different Stages of Growth and Development Period　　Unit: m³/(mu · day)

Period Growth and Development Period	Start and End Date	Days	1 Daily Water Requirement	2 Daily Water Requirement	3 Daily Water Requirement	Average Water Requirement Intensity
Seeding – Seedling	9/5-15/5	73.24	0.66	1.31	1.74	
Seedling – Jointing	16/5-24/6	40	1.71	1.56	1.47	1.58
Jointing – Tasseling	25/6-8/7	14	3.48	3.53	2.92	3.31
Tasseling – Filling	9/7-29/7	21	5.20	4.60	4.40	4.73
Filling – Maturing	30/7-14/8	16	3.02	2.71	2.13	2.62
Maturing – Harvesting	15/8-7/9	23	1.19	1.15	0.75	1.03

（四）甜菜需水强度的变化规律
2.2.3.4　Changing Law of Beet Water Requirement Intensity

分析甜菜生育期内需水强度的变化可以看出,甜菜在生长中期(7月8日~8月23日)的需水强度最大,3个不同水平的平均需水强度为3.64 m³/(亩·d),其中1号水平最大,为4.25 m³/(亩·d);2号水平次之,为3.40 m³/(亩·d);3号水平最小,为3.27 m³/(亩·d)。需水强度的变化过程线呈现为单峰形式。

According to the water requirement intensity change of beet, it is found that the intensity is the highest at the middle stage (from July 8th to August 23rd). The average intensity of the 3 levels is 3.64 m³/(mu · day), in which Level #1 is the highest with 4.25 m³/(mu · day), followed by Level #2 with 3.40 m³/(mu · day) and Level 3 with 3.27 m³/(mu · day). There is one peak in the changing process of water requirement intensity in the growth and development period.

从甜菜生长发育分析,在甜菜生长前期(5月15日~7月7日),为播种、出苗到幼叶形成时期,其中在幼苗形成期的气温较高,因而需保持一定土壤湿度而不至板结,保证顺利出苗,所以这期间需要一定水量,3个水平的平均需水强度为2.39 m³/(亩·d);甜菜生长中期(7月8日~8月23日),枝叶茂盛,叶面积增长迅速,叶面蒸腾量大,加之气温高,甜菜需水强度增大,为生育期内最高值;而甜菜生长后期(8月24日~10月4日),为甜菜块根成熟及糖分积累过程,这期间需水强度大幅度下降,阶段需水量为2.17 m³/(亩·d),为甜菜全生育期需水强度最低时期(见表81)。由上述分析可见,保证甜菜生长中期的需水强度是非常重要的。甜菜全生育期的累积需水量过程基本呈直线上升趋势(见图16)。

From the perspective of beet growth and development, the early stage (from May 15th to July 7th) is time for sowing, seedling and cotyledon emerging. Due to high temperature in the seedling stage, soil must contain enough water for sake of not hardening. Thus enough water must be supplied during this period. The average intensity of the three levels is 2.39 m³/(mu·day). The middle stage (from July 8th to August 23rd) is the peak of water requirement intensity, for its luxuriant leaves, large leaf area, high transpiration and temperature; and the late stage (from August 24th to October 4th) is time for root maturing and sugar accumulation, during which water requirement intensity drops to 2.17 m³/(mu·day), lowest of the entire growth and development period (see Table 81). In conclusion, it is vital for beet to meet its water requirement intensity at the middle stage. The entire accumulated water requirement process of beet is a straight rising line in general (see Figure 16).

表81　　　　甜菜生育期不同阶段需水强度分析表　　　　单位:m³/(亩·d)

生育期	时段		1	2	3	平均需水强度
	起止日期	天数	日需水量	日需水量	日需水量	
前期	15/5~7/7	42	2.55	2.17	2.46	2.39
前期	8/7~23/8	46	4.25	3.40	3.27	3.64
后期	24/8~4/10	41	2.16	2.48	1.87	2.17
合计	25/5~4/10	129	3.03	2.71	2.56	2.77

Table 81　　Analysis of Beet Water Requirement Intensity at Different Stages of Growth and Development Period　　Unit:m³/(mu·day)

Growth and Development Period	Period		1	2	3	Average Water Requirement Intensity
	Start and End Date	Days	Daily Water Requirement	Daily Water Requirement	Daily Water Requirement	
Initial Stage	15/5 - 7/7	42	2.55	2.17	2.46	2.39
Initial Stage	8/7 - 23/8	46	4.25	3.40	3.27	3.64
Later Stage	24/8 - 4/10	41	2.16	2.48	1.87	2.17
Total	25/5 - 4/10	129	3.03	2.71	2.56	2.77

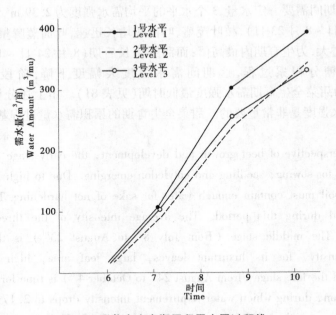

图 16 甜菜全生育期累积需水量过程线

Figure 16 Entire Accumulated Water Requirement Process of Beet

（五）作物不同阶段需水强度变化规律研究小结

2.2.3.5 Summary of Changing Law of Crop Water Requirement Intensity at Different Growing Stages

通过对春小麦、苜蓿、玉米及甜菜生育期不同阶段需水强度变化规律的研究，现将主要结果归纳如下：

Through the experimental research on the change law of water requirement intensity of spring wheat, alfalfa, corn and beet at different stages, the main results are summarized as follows：

春小麦生育期最大需水强度出现在小麦抽穗—灌浆期（6月19日～6月24日），需水强度达 8.77 m³/(亩·d)，需水强度变化过程线呈单峰形式。

The peak of spring wheat water requirement intensity occurs at the heading – filling stage (from June 19th to June 24th) with 8.77 m³/(mu · day). There is one peak in the entire changing process.

苜蓿生育期最大需水强度出现在蔓枝伸长期，并且需水强度与气温变化关系较大。

The peak of alfalfa water requirement intensity occurs at the branch stretching stage and relates to air temperature.

玉米生育期最大需水强度出现在抽雄—灌浆期(7月9日~7月29日),需水强度达4.73 m³/(亩·d),需水强度变化过程线呈单峰形式。

The peak of corn occurs at the tasseling – filling stage (from July 9th to July 29th) with 4.73 m³/(mu·day). There is one peak in the entire changing process.

甜菜生育期的最大需水强度出现在甜菜生长中期,即7月8日~8月23日,需水强度为3.64 m³/(亩·d),需水强度变化过程线亦呈单峰形式(见表82)。

The peak of beet occurs at the middle stage (from July 8th to August 23rd) with 3.64 m³/(mu·day). There is one peak in the entire changing process (see Table 82).

表82　　　　　　　　作物需水强度变化分析表　　　　　　　　单位:m³/(亩·d)

	项目	春小麦	苜蓿	玉米	甜菜
最大值	需水强度	8.77	5.07	4.73	3.64
	生长阶段	抽穗—灌浆	第一茬开花成熟期	抽雄—灌浆	生长中期
	日期	6月20日~6月24日	5月26日~6月8日	7月9日~7月29日	7月8日~8月23日
次大值	需水强度	3.95	4.54	3.31	2.39
	生长阶段	孕穗—抽穗	第二茬蔓枝伸长期	拔节—抽雄	生长后期
	日期	6月13日~6月19日	6月22日~7月11日	6月25日~8月7日	5月15日~7月7日

Table 82　　　　　　Analysis of Crops Water Requirement Intensity　　　　Unit:m³/(mu·day)

	Item	Spring Wheat	Alfalfa	Corn	Beet
Peak Value	Water Requirement Intensity	8.77	5.07	4.73	3.64
	Growing Stage	Heading – Filling	Flowering & Maturing of the 1st Cropping	Tasseling – Filling	Middle Growth Stage
	Date	June 20th to June 24th	May 26th to June 8th	July 9th to July 29th	July 8th to August 23rd
Sub-peak Value	Water Requirement Intensity	3.95	4.54	3.31	2.39
	Growing Stage	Booting – Heading	Branch Stretching of the 2nd Cropping	Jointing – Tasseling	Late Growth Stage
	Date	June 13th to June 19th	June 22nd to July 11th	June 25th to August 7th	May 15th to July 7th

· 150 ·

第三节 作物生理需水量分析
2.3 Analysis on Physiological Water Requirement of Crops

作物的生理需水量分析主要是通过对植株叶面的蒸腾、棵间土壤蒸发量以及叶面积系数与需水量的关系等方面进行测试研究的。而作物全生育的植株蒸腾量与棵间蒸发量的变化关系是随着下垫面和作物不同生长阶段的发育状况不同而变化着。

The physiological analysis of crop water requirement mainly focuses on the relationship between water requirement, leaf transpiration, soil water evaporation and leaf area index. The relationship between plant transpiration and soil water evaporation changes to the underlying surface and the growing stage of crop.

一、作物的植株蒸腾与棵间蒸发量
2.3.1 Plant Transpiration and Soil Water Evaporation

（一）春小麦的植株蒸腾量
2.3.1.1 Plant Transpiration of Spring Wheat

春小麦生理需水量在全生育期的不同生育时期是不同的。小麦一生中植株蒸腾量以6月为高，其中又以6月中旬最大。3个不同水平的小麦植株蒸腾量在6月中旬达到21.67~64.58 m³/亩，占该期间作物需水总量的54.38%~69.40%，比较3个不同水平的小麦植株蒸腾量，3号水平的蒸腾量最小，为21.67 m³/亩；1号水平的次之，为33.14 m³/亩，2号水平的最大，达41.23 m³/亩（见表83）。春小麦生育阶段的植株蒸腾量变化过程见图17。由图17可见，尽管3个不同水平的植株蒸腾量均以6月中旬出现峰值，但各不同水平的植株蒸腾量变化过程不尽相同，表现为1号水平的变化曲线呈单峰形式，只是在6月上旬有一个小小的波动；2号水平的植株蒸腾量在全生育阶段呈峰谷相间形式出现，有4次峰值和4次谷值；3号水平的植株蒸腾量变化有2次峰值，第一次出现在5月中旬，第二次6月中旬，且在最后还有一点翘尾。

The physiological water requirement of spring wheat differs from growing stages. The plant transpiration of spring wheat is the highest in June and peaks in mid-June. The plant transpiration of 3 different levels is between 21.67 to 64.58 m³/mu in mid-June, accounting for 54.38% – 69.40% of the total water requirement in this stage. Comparing these 3 levels, Level #3 produces least transpiration, 21.67 m³/mu, followed by Level #1, 33.14 m³/mu, and Level #2, 41.23 m³/mu (see Table 83). The changing process of spring wheat plant transpiration is shown in Figure 17. Although plant transpiration of the 3 different levels peaks in mid-July, their changing processes are different from one another. The changing curve of

第二章 作物需水量与需水规律
2 Water Requirement of Crops and Its Law

Level #1 has one peak and fluctuates a bit in early June. Level #2 has several fluctuations, 4 peaks and 4 valleys. Level #3 has 2 peaks, in mid-May and mid-June respectively, and a little rising trend at last.

表83　　　　　　　　　春小麦不同水平的植株蒸腾量分析表

月	旬	需水量(m^3/亩)			植株蒸腾量(m^3/亩)			蒸腾量占需水量(%)		
		1	2	3	1	2	3	1	2	3
4	上	1.16	0.48	3.62						
	中	5.49	3.55	11.74						
	下	31.39	36.59	14.05						
5	上	18.86	14.97	7.62	11.69	7.80	0.45	61.98	52.10	5.91
	中	37.78	20.72	33.36	22.77	5.71	18.35	60.27	27.56	55.01
	下	41.42	44.25	25.21	27.75	30.58	11.54	67.00	69.11	45.78
6	上	45.89	35.09	32.57	26.05	15.25	12.73	56.77	43.46	39.09
	中	51.32	59.41	39.85	33.14	41.23	21.67	64.58	69.40	54.38
	下	56.70	40.29	29.82	28.85	12.44	1.97	50.88	30.80	6.61
7	上	25.93	28.92	14.90	11.27	14.26	0.08	43.36	49.31	0.54
	中	3.22	4.87	3.36	1.19	2.34	1.13	39.96	58.32	39.58
合计		319.06	289.14	216.1	200.5	170.73	97.69	62.84	59.05	45.21

Table 83　　Analysis of Spring Wheat Plant Transpiration at Different Levels

Month	Every 10 Days	Requirement (m^3/mu)			Plant Transpiration (m^3/mu)			Transpiration/Water Requirement (%)		
		1	2	3	1	2	3	1	2	3
4	Early	1.16	0.48	3.62						
	Middle	5.49	3.55	11.74						
	Late	31.39	36.59	14.05						
5	Early	18.86	14.97	7.62	11.69	7.80	0.45	61.98	52.10	5.91
	Middle	37.78	20.72	33.36	22.77	5.71	18.35	60.27	27.56	55.01
	Late	41.42	44.25	25.21	27.75	30.58	11.54	67.00	69.11	45.78
6	Early	45.89	35.09	32.57	26.05	15.25	12.73	56.77	43.46	39.09
	Middle	51.32	59.41	39.85	33.14	41.23	21.67	64.58	69.40	54.38
	Late	56.70	40.29	29.82	28.85	12.44	1.97	50.88	30.80	6.61
7	Early	25.93	28.92	14.90	11.27	14.26	0.08	43.36	49.31	0.54
	Middle	3.22	4.87	3.36	1.19	2.34	1.13	39.96	58.32	39.58
Total		319.06	289.14	216.1	200.5	170.73	97.69	62.84	59.05	45.21

从作物生长发育各不同时期分析春小麦的植株蒸腾量变化可以看出(见表84)，小麦的植株蒸腾量在分蘖时期较大，达 13.18～31.90 m^3/亩，占阶段需水总量的 56.45%～

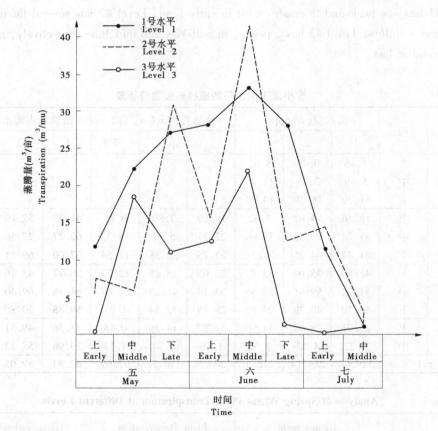

图 17　春小麦各生育期植株蒸腾量变化过程线
Figure 17　Changing Process of Spring Wheat Plant Transpiration at Different Stages

75.81%；然后骤然降低，后又随植株叶面积的增大而逐渐增加，至小麦抽穗—灌浆期，植株蒸腾量大幅度增加，由 7.0～20.16 m³/亩增加到 30.47～50.56 m³/亩，这时的植株蒸腾量达作物需水量的 67.34%～77.32%。到了小麦灌浆—成熟期，植株蒸腾量又进一步增大到 33.18～69.03 m³/亩，达到最高峰值区，然而这期间的蒸腾量所占作物需水总量百分比却较小麦抽穗—灌浆期低。

　　From the perspective of changes of spring wheat plant transpiration at different stages (see Table 84), it is found that transpiration at the tillering stage is the highest, 13.18 – 31.90 m³/mu, accounting for 56.45% – 75.81% of the stage water requirement; then drops sharply, followed by gradual increase with the expanding of leaf area; by the heading-filling stage, the plant transpiration soars from 7.0 – 20.16 m³/mu to 30.47 – 50.56 m³/mu, accounting for 67.34% – 77.32% of the water requirement. By the filling – maturing stage, the transpiration peaks to 33.18 – 69.03 m³/mu, but the proportion to total water requirement at this stage is smaller than that at the heading – filling stage.

第二章 作物需水量与需水规律
2 Water Requirement of Crops and Its Law

表84　　　　　　　春小麦各生育期植株叶面蒸腾量分析表

生育阶段	天数	需水量(m³/亩)			蒸腾量(m³/亩)			蒸腾量占需水量(%)		
		1	2	3	1	2	3	1	2	3
苗期	22	10.37	18.67	11.70						
分蘖	18	42.07	41.42	23.35	31.90	31.25	13.18	75.81	75.45	56.45
拔节	10	30.33	27.46	24.65	6.03	3.16	0.35	19.88	11.51	1.42
孕穗	14	41.43	51.40	29.79	13.45	23.43	1.81	32.46	45.58	6.08
抽穗	8	33.85	34.70	21.54	19.31	20.16	7.00	57.05	50.10	32.50
灌浆	6	65.34	47.52	45.25	50.56	32.74	30.47	77.32	68.90	67.34
成熟	25	95.67	67.9	59.82	69.03	41.33	33.10	72.15	60.81	55.47
合计	111	319.06	209.14	216.10	200.50	170.73	97.69	62.04	59.05	45.21

Table 84　　Analysis of Spring Wheat Leaf Transpiration at Different Stages

Growth and Development Period	Days	Water Requirement (m³/mu)			Transpiration (m³/mu)			Transpiration/Water Requirement (%)		
		1	2	3	1	2	3	1	2	3
Seedling	22	10.37	18.67	11.70						
Tillering	18	42.07	41.42	23.35	31.90	31.25	13.18	75.81	75.45	56.45
Jointing	10	30.33	27.46	24.65	6.03	3.16	0.35	19.88	11.51	1.42
Booting	14	41.43	51.40	29.79	13.45	23.43	1.81	32.46	45.58	6.08
Heading	8	33.85	34.70	21.54	19.31	20.16	7.00	57.05	50.10	32.50
Filling	6	65.34	47.52	45.25	50.56	32.74	30.47	77.32	68.90	67.34
Maturing	25	95.67	67.9	59.82	69.03	41.33	33.10	72.15	60.81	55.47
Total	111	319.06	209.14	216.10	200.50	170.73	97.69	62.04	59.05	45.21

(二)春小麦棵间蒸发量
2.3.1.2 Soil Water Evaporation in Spring Wheat Plot

春小麦的棵间蒸发在本地区主要发生在5～7月3个月，从小麦全生育期分析，棵间土壤蒸发量最大值出现在6月下旬(见表85)，从5月至7月中旬，小麦的棵间地面蒸发量变化过程线呈峰谷相间形式，而在6月下旬达到最高峰值后，呈直线衰减，曲线直线下降(见图18)。

Soil water evaporation in spring wheat plot mainly occurs between May to July. From the perspective of the entire growth and development period, the peak evaporation occurs in late June (see Table 85). With peaks and lows from May to mid-July, the soil water evaporation goes straightly downward sharply after achieving the highest peak in late June (see Figure 18).

表85　　　　　　　春小麦不同水平的棵间蒸发量分析表

月	旬	棵间蒸发量 (m³/亩)	棵间蒸发量占需水量%			平均值(%)
			1	2	3	
4	上	—	—	—	—	
	中	—	—	—	—	
	下	—	—	—	—	
5	上	7.17	38.02	47.90	94.09	60.00
	中	15.01	39.73	72.44	44.99	53.39
	下	13.67	33.00	30.89	54.22	39.37
6	上	19.84	43.23	56.54	60.91	53.56
	中	18.18	35.42	30.60	45.62	37.21
	下	27.85	49.12	69.12	93.39	70.54
7	上	14.66	56.55	50.69	98.39	68.54
	中	2.03	63.04	41.68	60.42	55.05
合计		118.41	358.11	399.86	552.03	436.67

Table 85　　Analysis of Soil Water Evaporation in Spring Wheat Field at Different Levels

Month	Every 10 Days	Soil Water Evaporation (m³/mu)	Soil Water Evaporation/Water Requirement (%)			Average (%)
			1	2	3	
4	Early	—	—	—	—	
	Middle	—	—	—	—	
	Late	—	—	—	—	
5	Early	7.17	38.02	47.90	94.09	60.00
	Middle	15.01	39.73	72.44	44.99	53.39
	Late	13.67	33.00	30.89	54.22	39.37
6	Early	19.84	43.23	56.54	60.91	53.56
	Middle	18.18	35.42	30.60	45.62	37.21
	Late	27.85	49.12	69.12	93.39	70.54
7	Early	14.66	56.55	50.69	98.39	68.54
	Middle	2.03	63.04	41.68	60.42	55.05
Total		118.41	358.11	399.86	552.03	436.67

再则，从春小麦全生育期各不同阶段分析其棵间蒸发量的变化可以发现(见表86)，小麦的棵间土壤蒸发出现过2次高峰期，一次在小麦拔节—孕穗期，变化在24.30~27.98 m³/亩之间，另一次则出现在小麦灌浆—成熟期，变化在14.78~26.64 m³/亩之间。

From the perspective of different stages (see Table 86), it is found that there are two peaks during the entire evaporation, which are at the jointing-booting (varying between 24.30-27.95 m³/mu) and the filling-maturing stage (14.78-26.64 m³/mu) respectively.

表86　　　　　　　　　春小麦在生育期棵间蒸发量分析表

生育阶段	天数	棵间蒸发 (m^3/亩)	棵间蒸发量占需水量(%)			平均值(%)
			1	2	3	
苗期	22					
分蘖	18	10.17	24.17	24.55	43.55	30.77
拔节	18	24.30	80.12	88.49	98.58	89.07
孕穗	14	27.98	67.54	54.44	93.32	71.77
抽穗	8	14.54	42.95	41.81	67.50	50.75
灌浆	6	14.78	22.62	31.10	32.66	28.79
成熟	25	26.64	27.85	39.19	44.53	37.19
合计	111	118.41	265.25	279.58	380.14	308.32

Table 86　　Analysis of Soil Water Evaporation in Spring Wheat Field at Different Stages

Growth and Development Period	Days	Soil Water Evaporation (m^3/mu)	Soil Water Evaporation/Water Requirement (%)			Average (%)
			1	2	3	
Seedling Period	22					
Tillering Period	18	10.17	24.17	24.55	43.55	30.77
Jointing Period	18	24.30	80.12	88.49	98.58	89.07
Booting Period	14	27.98	67.54	54.44	93.32	71.77
Heading Period	8	14.54	42.95	41.81	67.50	50.75
Filling Period	6	14.78	22.62	31.10	32.66	28.79
Maturing Period	25	26.64	27.85	39.19	44.53	37.19
Total	111	118.41	265.25	279.58	380.14	308.32

由上述分析可见，春小麦全生育期内的植株蒸腾量为256 mm，占总需水量的60%左右，其中以6月中旬最高，占69.40%；小麦全生育期的棵间土壤蒸发量为177.5 mm，占全部需水总量

的 40%，其中以 6 月下旬最高，约占阶段需水总量的 70.54%。

In conclusion, plant transpiration of spring wheat in the entire growth and development period is 256 mm, accounting for 60% of the total water requirement, and peaks in mid-June, accounting for 69.40% of the stage water requirement; soil water evaporation of spring wheat in the entire period is 177.5 mm, accounting for 40% of the entire water requirement, and peaks in late June, accounting for 70.54% of the stage water requirement.

（三）玉米的植株蒸腾量
2.3.1.3　Plant Transpiration of Corn

分析玉米全生育期不同生育阶段的植株蒸腾量变化可以发现(见表 87)，3 个不同水平的玉米植株蒸腾量变化是不同的。1 号水平的玉米全生育期的植株蒸腾量以 8 月中旬最高，达 44.15 m³/亩，占阶段需水量的 87.6%，7 月下旬次之；2 号水平的植株蒸腾量则以 7 月上旬最高，为 39.72 m³/亩，占阶段需水量的 73.7%，7 月下旬次之；3 号水平的植株蒸腾量以 7 月下旬最高，为 34.3 m³/亩，占阶段需水量的 74.8%，而 8 月中旬次之。比较 3 个不同水平的植株蒸腾量，由大到小依次为：1 号水平 > 2 号水平 > 3 号水平，分别为 251.54、188.52 m³/亩和 150.07 m³/亩。

From the perspective of different stages (see Table 87), it is found that the plant transpiration of corn changes to levels. The transpiration of Level #1 corn peaks in the mid-August, 44.15 m³/mu, accounting for 87.6% of the stage water requirement, followed by the late July; the transpiration of Level #2 peaks in the early-July, 39.72 m³/mu, accounting for 73.7%, followed by the late July; Level #3 peaks in the late July, 34.3 m³/mu, accounting for 74.8%, followed by the mid-August. The plant evaporation of 3 different levels can be arranged from large to small as: Level #1 > Level #2 > Level #3, 251.54, 188.52 and 150.07 m³/mu respectively.

分析 3 个不同水平的玉米全生育期植株蒸腾变化过程还可以发现，其变化过程线均呈多峰形式，其中 2 号水平和 3 号水平的植株蒸腾量变化过程线呈峰谷相间形式(见图 19)。

From the perspective at different levels, it is found that each of the three changing process of plant transpiration has several peaks. Level #2 and Level #3 have several peaks and lows (see Figure 19).

在分析玉米植株蒸腾量及其所占作物需水量百分比时则发现，3 个不同水平玉米植株蒸腾量所占作物阶段需水量百分比均以 8 月中旬最高，1 号水平、2 号水平和 3 号水平所占百分比分别达 87.6%、83.4% 和 80.3%，次为 7 月下旬，变化在 74.20% ~ 77.34% 之间。

From the perspective of plant transpiration and its proportion in the water requirement, it is found that the proportion of 3 levels peaks in mid-August, accounting for 87.6%, 83.4% and 80.3% for Level #1, #2 and #3 respectively. The sub-peak is in late July, varying from 74.20% to 77.34%.

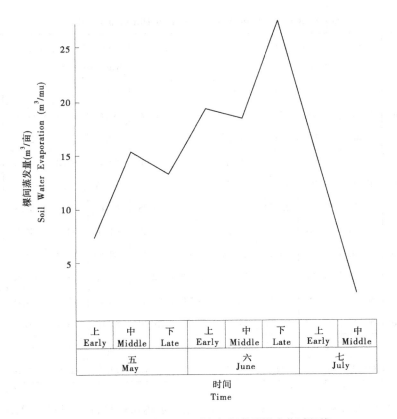

图 18 春小麦不同时段棵间蒸发量变化过程线

Figure 18　Changing Process of Soil Water Evaporation of Spring Wheat at Different Stages

表 87　　　　　　　　　　玉米不同水平的植株蒸腾量分析表

月	旬	需水量(m^3/亩)			植株蒸腾量(m^3/亩)			植株蒸腾量占需水量百分比(%)			平均值(%)
		1	2	3	1	2	3	1	2	3	
5	中	13.83	6.85	14.67	8.1	1.12	8.94	58.6	16.4	60.9	47.97
	下	8.46	5.16	5.06	4.45	1.15	1.05	52.6	22.3	20.8	31.9
6	上	7.79	12.07	10.46	4.07	8.35	6.74	52.7	69.2	64.4	62.1
	中	12.97	9.45	15.41	6.93	5.05	9.37	53.43	36.7	60.8	50.3
	下	47.78	37.05	26.78	37.95	27.22	16.95	79.42	37.5	63.3	72.1
7	上	36.35	35.90	32.18	22.17	39.72	18.0	61.0	73.7	55.9	63.5
	中	32.40	21.83	26.43	22.18	11.61	16.21	68.5	53.2	61.3	361.01
	下	51.02	44.73	45.86	39.46	33.17	34.3	77.34	74.2	74.8	75.5
8	上	50.02	21.2	21.43	38.06	9.24	9.47	76.1	43.6	44.0	54.6
	中	50.37	37.55	61.61	44.15	31.33	25.39	87.6	83.4	80.3	83.8
9	上	23.39	28.43	12.53	15.61	20.7	4.75	66.7	72.6	37.9	46.6
合计		345.21	282.19	247.7	251.54	188.52	150.07	72.9	66.8	60.6	66.0

Table 87　　　　　　Analysis of Corn Plant Transpiration at Different Levels

Month	Every 10 Days	Water Requirement (m³/mu)			Plant Transpiration (m³/mu)			Plant Transpiration/ Water Requirement (%)			Average (%)
		1	2	3	1	2	3	1	2	3	
5	Middle	13.83	6.85	14.67	8.1	1.12	8.94	58.6	16.4	60.9	47.97
	Late	8.46	5.16	5.06	4.45	1.15	1.05	52.6	22.3	20.8	31.9
6	Early	7.79	12.07	10.46	4.07	8.35	6.74	52.7	69.2	64.4	62.1
	Middle	12.97	9.45	15.41	6.93	5.05	9.37	53.43	36.7	60.8	50.3
	Late	47.78	37.05	26.78	37.95	27.22	16.95	79.42	37.5	63.3	72.1
7	Early	36.35	35.90	32.18	22.17	39.72	18.0	61.0	73.7	55.9	63.5
	Middle	32.40	21.83	26.43	22.18	11.61	16.21	68.5	53.2	61.3	361.01
	Late	51.02	44.73	45.86	39.46	33.17	34.3	77.34	74.2	74.8	75.5
8	Early	50.02	21.2	21.43	38.06	9.24	9.47	76.1	43.6	44.0	54.6
	Middle	50.37	37.55	61.61	44.15	31.33	25.39	87.6	83.4	80.3	83.8
9	Early	23.39	28.43	12.53	15.61	20.7	4.75	66.7	72.6	37.9	46.6
Total		345.21	282.19	247.7	251.54	188.52	150.07	72.9	66.8	60.6	66.0

同时,进一步分析玉米全生育期内不同阶段的植株蒸腾量可见,在玉米生育前期,植株小,植株的叶面蒸腾量所占比重较小;随着玉米植株成长,叶面积长大,植株的叶面蒸腾量加大;但到后期,随着玉米籽粒的逐渐成熟,玉米枝叶的逐渐枯萎,叶面蒸腾量减小。

From the perspective of different stages, the plant itself is small at early stages. Thus the proportion of transpiration is small. As the plant grows up, the leaf area expands and the transpiration increases. The kernel gradually matures and leaves are withered at late stages. Thus leaf transpiration decreases.

(四)玉米的棵间蒸发量

2.3.1.4　Soil Water Evaporation in Corn Plot

玉米的棵间蒸发量在玉米全生长期,经历过一个由高到低、再由低到高的变化过程。在玉米生育前期,植株小,玉米的棵间蒸发相对较大,到5月下旬至6月上旬,这期间玉米的棵间土壤蒸发量减小,6月中旬以后,随气温升高,玉米棵间蒸发量又上升,至7月上旬达最高值,为14.18 m³/亩(见表88),7月上旬至8月上旬的棵间蒸发量变化在10.22~14.18 m³/亩。玉米全生育期棵间蒸发量的变化过程线呈不规则双峰形式(见图20)。

The soil water evaporation in the corn plot changes from high to low and then from low to high. The plant is small at early stages and the soil water evaporation is huge, then the evaporation decreases from late May to Early June. As the air temperature rises after mid-June, the evaporation

2 Water Requirement of Crops and Its Law

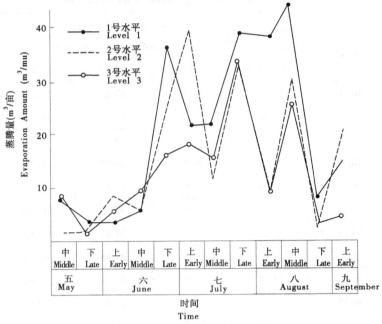

图 19 玉米不同水平的植株蒸腾量变化过程线

Figure 19 Changing Process of Corn Plant Transpiration at Different Levels

rises again and peaks in early July with 14.18 m³/mu (see Table 88). The evaporation from early June to early August varies between 10.22 to 14.18 m³/mu. The change process of soil water evaporation in the corn plot is irregular and has two peaks (see Figure 20).

表 88 玉米棵间蒸发量分析表

月	旬	棵间蒸发 (m³/亩)	棵间蒸发量占需水量百分比(%)			平均值 (%)
			1	2	3	
5	中	5.73	41.43	83.65	39.06	54.71
	下	4.01	47.40	77.71	79.25	68.12
6	上	3.72	47.75	30.82	35.56	38.04
	中	6.04	46.57	63.92	39.20	49.90
	下	9.83	20.57	26.53	36.71	27.94
7	上	14.18	39.01	39.50	44.06	40.86
	中	10.22	31.54	46.82	38.67	39.01
	下	11.56	22.66	25.84	25.21	24.57
8	上	11.96	23.91	56.42	55.81	45.38
	中	6.22	12.35	16.56	19.68	16.20
	下	2.42	23.31	62.37	45.49	43.72
9	上	7.78	33.26	27.37	62.09	40.91
合计		93.67	389.76	557.51	520.79	489.36

Table 88   Analysis of Soil Water Evaporation in the Corn Plot

Month	Every 10 Days	Soil Water Evaporation (m^3/mu)	Soil Water Evaporation/Water Requirement (%)			Average (%)
			1	2	3	
5	Middle	5.73	41.43	83.65	39.06	54.71
	Late	4.01	47.40	77.71	79.25	68.12
6	Early	3.72	47.75	30.82	35.56	38.04
	Middle	6.04	46.57	63.92	39.20	49.90
	Late	9.83	20.57	26.53	36.71	27.94
7	Early	14.18	39.01	39.50	44.06	40.86
	Middle	10.22	31.54	46.82	38.67	39.01
	Late	11.56	22.66	25.84	25.21	24.57
8	Early	11.96	23.91	56.42	55.81	45.38
	Middle	6.22	12.35	16.56	19.68	16.20
	Late	2.42	23.31	62.37	45.49	43.72
9	Early	7.78	33.26	27.37	62.09	40.91
Total		93.67	389.76	557.51	520.79	489.36

由上述分析试验可知,玉米全生育期的生理需水情况表现为:玉米全生育期的植株蒸腾量为 150.07~251.54 m^3/亩,约占需水量的 67%;棵间土壤蒸发量为 93.67 m^3/亩,约占需水量的 33%。

In conclusion, the physiological water requirement of corn in the entire growth and development period is: the total plant transpiration varies between 150.07 to 251.54 m^3/mu, accounting for 67% of the water requirement; and the total soil water evaporation is 93.67 m^3/mu, accounting for 33%.

(五)甜菜的植株蒸腾量

2.3.1.5 Plant Transpiration of Beet

甜菜的叶面蒸腾量在本区全生育期内主要发生在 5—9 月 5 个月。由表 90 可知,3 个不同水平的甜菜叶面蒸腾量均以 7 月下旬最高,其中 1 号水平最高,为 57.23 m^3/亩;2 号水平次之,为 50.95 m^3/亩;3 号水平最低,为 47.45 m^3/亩,分别占需水量的 93.47%,92.72% 和 92.23(见表 89)。

The leaf transpiration of beet in this plot mainly occurs from May to September. Table 90 shows that late July is the peak of leaf transpiration for all 3 levels. Level #1 is the highest with

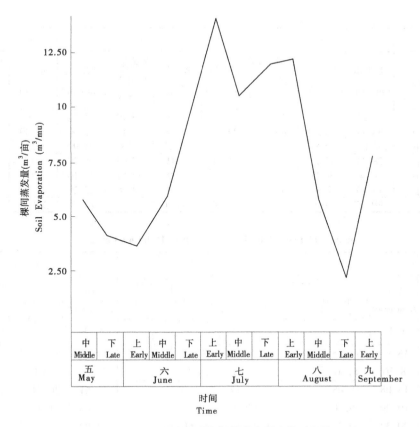

图 20 玉米全生育期棵间蒸发量变化过程线

Figure 20 Changing Process of Soil Water Evaporation of Corn in the Growth and Development Period

57.23 m³/mu, followed by Level #2 with 50.95 m³/mu, and Level #3 with 47.45 m³/mu. They account for 93.47%, 92.72% and 92.23% of the water requirement respectively (see Table 89).

表 89 甜菜不同水平的叶面蒸腾量分析表

时间		1			2			3		
月	旬	需水量 (m³/亩)	蒸腾量 (m³/亩)	蒸腾量占需水量 (%)	需水量 (m³/亩)	蒸腾量 (m³/亩)	蒸腾量占需水量 (%)	需水量 (m³/亩)	蒸腾量 (m³/亩)	蒸腾量占需水量 (%)
5	中	13.95	13.95		5.61	5.61		11.28	11.28	
	下	8.30	8.30		11.35	11.35		9.04	9.04	
6	上	18.08	18.08		13.12	13.12		20.51	20.51	
	中	19.20	19.20		17.10	17.10		11.40	11.40	
	下	30.32	30.32		24.69	24.69		29.02	29.02	
7	上	24.47	18.20	74.58	27.25	21.03	77.17	31.65	25.43	80.35
	中	37.95	32.84	86.53	20.23	15.12	74.74	17.09	11.98	70.10
	下	60.23	57.23	93.47	54.95	50.95	92.72	51.45	47.45	92.23

续表 89

时间		1			2			3		
月	旬	需水量 (m^3/亩)	蒸腾量 (m^3/亩)	蒸腾量占需水量 (%)	需水量 (m^3/亩)	蒸腾量 (m^3/亩)	蒸腾量占需水量 (%)	需水量 (m^3/亩)	蒸腾量 (m^3/亩)	蒸腾量占需水量 (%)
8	上	38.29	18.23	47.61	29.95	9.94	33.19	25.49	5.48	21.50
	中	38.44	33.10	86.11	29.02	22.68	78.15	35.64	30.30	85.02
	下	45.07	34.84	77.30	55.50	45.27	81.57	41.99	31.76	75.64
9	上	28.33	15.21	53.69	21.14	8.02	37.94	16.79	3.67	21.86
	中	8.14	9.25		18.30	2.29	12.51	17.10	1.09	6.37
	下	19.40	7.39	38.09	21.24	9.23	43.46	12.26	0.25	2.04
合计		391.17	306.99		348.95	256.40		330.71	238.66	

Table 89   **Analysis of Beet Plant Transpiration at Different Levels**

Time		1			2			3		
Month	Every 10 Days	Water Requirement (m^3/mu)	Transpiration (m^3/mu)	Transpiration/ Water Requirement (%)	Water Requirement (m^3/mu)	Transpiration (m^3/mu)	Transpiration/ Water Requirement (%)	Water Requirement (m^3/mu)	Transpiration (m^3/mu)	Transpiration/ Water Requirement (%)
5	Middle	13.95	13.95		5.61	5.61		11.28	11.28	
	Late	8.30	8.30		11.35	11.35		9.04	9.04	
6	Early	18.08	18.08		13.12	13.12		20.51	20.51	
	Middle	19.20	19.20		17.10	17.10		11.40	11.40	
	Late	30.32	30.32		24.69	24.69		29.02	29.02	
7	Early	24.47	18.20	74.58	27.25	21.03	77.17	31.65	25.43	80.35
	Middle	37.95	32.84	86.53	20.23	15.12	74.74	17.09	11.98	70.10
	Late	60.23	57.23	93.47	54.95	50.95	92.72	51.45	47.45	92.23
8	Early	38.29	18.23	47.61	29.95	9.94	33.19	25.49	5.48	21.50
	Middle	38.44	33.10	86.11	29.02	22.68	78.15	35.64	30.30	85.02
	Late	45.07	34.84	77.30	55.50	45.27	81.57	41.99	31.76	75.64
9	Early	28.33	15.21	53.69	21.14	8.02	37.94	16.79	3.67	21.86
	Middle	8.14	9.25		18.30	2.29	12.51	17.10	1.09	6.37
	Late	19.40	7.39	38.09	21.24	9.23	43.46	12.26	0.25	2.04
Total		391.17	306.99		348.95	256.40		330.71	238.66	

把甜菜全生育期叶面蒸腾量的变化过程绘制成图,由图 21 可见,1 号水平的全生育期叶面蒸腾量变化过程呈三峰现象。2 号水平的全生育期叶面蒸腾量变化过程表现为,在甜菜生育前期,叶面蒸腾量是个逐渐增大过程,至 6 月下旬,达到第一次峰值;随后渐减,至 7 月下旬骤然升至最高峰。3 号水平的叶面蒸腾量在全生育期的变化过程表现为,8 月中旬以前,呈峰、谷相间形式,而在此以后,叶面蒸腾量呈单一衰减。

Figure 21, plotted in accordance with the changing process of leaf transpiration, shows that Level #1 beet experiences three peaks. The transpiration of Level #2 beet rises at the early stage and peaks in late June, then gradually decreases until a sharp rise in late July and achieves the highest peak. The transpiration of Level #3 experiences several ups and downs until a progressive decrease since mid-August.

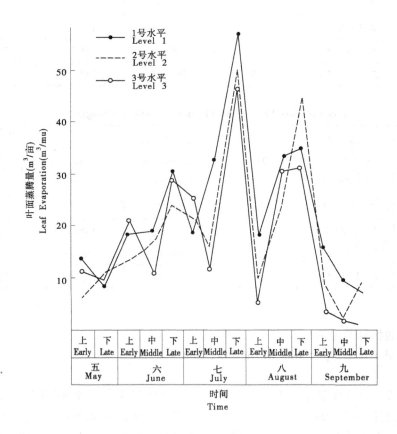

图 21 甜菜生育期叶面蒸腾量变化过程线

Figure 21 Changing Curves of Beet's Leaf Evaporation at Growth and Development Period

从全生育期不同阶段叶面蒸腾量分析,甜菜的叶面蒸腾量主要集中在生长中期的 7 月 8 日~8 月 23 日,达 148.15 m³/亩,占需水量的 75% 以上,而后期 8 月 24 日—10 月 4 日的植株蒸腾量仅占 35% 左右(见表 90)。

From the perspective of different stages, the leaf transpiration mainly occurs in July 8th to August 23rd of the middle stage with 148.15 m³/mu, accounting for over 75% of the stage water requirement, while the transpiration in the late stage, August 24th to October 4th, only accounts for 35% (see Table 90).

表90　　　　　　　　甜菜不同生育阶段的植株蒸腾量分析表

生育阶段	起止日期（日/月）	天数	阶段需水量（m³/亩）	植株蒸腾量（m³/亩）	蒸腾量占需水量(%)
前期	25/5~7/7	42	106.98		
中期	8/7~23/8	46	195.54	148.15	75.76
后期	24/8~4/10	41	88.65	30.49	34.39
合计	25/5~4/10	129	391.17	285.62	73.02

Table 90　　　Analysis of Beet Plant Transpiration at Different Stages

Growth and development period	Start and End Date (day/month)	Days	Stage Water Requirement (m³/mu)	Plant Transpiration (m³/mu)	Transpiration/Water Requirement (%)
Initial Stage	25/5 – 7/7	42	106.98		
Middle Stage	8/7 – 23/8	46	195.54	148.15	75.76
Later Stage	24/8 – 4/10	41	88.65	30.49	34.39
Total	25/5 – 4/10	129	391.17	285.62	73.02

（六）甜菜的棵间蒸发量
2.3.1.6　Soil Water Evaporation in the Beet Plot

甜菜的棵间土壤蒸发量主要发生在7—9月3个月（见表91），棵间蒸发量最高值出现在8月上旬，变化在20.01 m³/亩左右，占需水量的52.26%~78.50%。比较3个不同水平的甜菜棵间蒸发量占需水量百分比则发现，3号水平的棵间蒸发量在后期所占比重很大，9月上、中、下3旬的棵间蒸发量占需水量百分比分别为78.14%、93.62%和97.96%。表明甜菜后期的棵间蒸发量较大。这一特征从表92上也可以证实。

The soil water evaporation of beet mainly occurs from July to September (see Table 91) and peaks in early August, varying around 20.01 m³/mu and accounting for 52.26% to 78.50%. Contrast on the proportion of soil water evaporation among 3 levels shows that the evaporation at the late stage of Level #3 accounts for large proportion, 78.14%, 93.62% and 97.96% of the water requirement in early, middle and late September respectively. It means the soil water evaporation of beet is quite high at the late stage. Table 92 displays the same result.

表91　　甜菜不同水平全生育期棵间蒸发量分析表

时间			1		2		3	
月	旬	棵间蒸发 (m³/亩)	需水量 (m³/亩)	蒸发量占需水量(%)	需水量 (m³/亩)	蒸发量占需水量(%)	需水量 (m³/亩)	蒸发量占需水量(%)
5	中		13.95		5.61		11.28	
	下		8.30		11.35		9.04	
6	上		18.08		13.12		20.51	
	中		19.20		17.1		11.40	
	下		30.32		24.69		29.02	
7	上	6.22	24.47	25.42	27.25	22.82	31.65	19.25
	中	5.11	37.95	13.46	20.23	25.25	17.09	29.90
	下	4.00	61.23	6.53	54.95	7.27	51.45	7.77
8	上	20.01	38.29	52.26	29.95	66.81	25.49	78.50
	中	5.34	38.44	13.89	29.02	19.06	35.64	14.98
	下	10.23	45.07	22.69	55.50	18.43	41.99	24.36
9	上	13.12	28.33	46.31	21.14	62.06	16.79	78.14
	中	16.01	8.14		18.3	87.47	17.1	93.62
	下	12.01	19.40	61.90	21.24	50.54	12.26	97.96
合计		92.05	391.17	24.29	348.95	26.38	330.71	27.83

Table 91　　Analysis of Soil Water Evaporation in the Beet Plot at Different Levels

Time			1		2		3	
Month	Every 10 Days	Soil Water Evaporation (m³/mu)	Water Requirement (m³/mu)	Evaporation/ Water Requirement (%)	Water Requirement (m³/mu)	Evaporation/ Water Requirement (%)	Water Requirement (m³/mu)	Evaporation/ Water Requirement (%)
5	Middle		13.95		5.61		11.28	
	Late		8.30		11.35		9.04	
6	Early		18.08		13.12		20.51	
	Middle		19.20		17.1		11.40	
	Late		30.32		24.69		29.02	
7	Early	6.22	24.47	25.42	27.25	22.82	31.65	19.25
	Middle	5.11	37.95	13.46	20.23	25.25	17.09	29.90
	Late	4.00	61.23	6.53	54.95	7.27	51.45	7.77
8	Early	20.01	38.29	52.26	29.95	66.81	25.49	78.50
	Middle	5.34	38.44	13.89	29.02	19.06	35.64	14.98
	Late	10.23	45.07	22.69	55.50	18.43	41.99	24.36
9	Early	13.12	28.33	46.31	21.14	62.06	16.79	78.14
	Middle	16.01	8.14		18.3	87.47	17.1	93.62
	Late	12.01	19.40	61.90	21.24	50.54	12.26	97.96
Total		92.05	391.17	24.29	348.95	26.38	330.71	27.83

· 166 ·

表92 甜菜全生育期内不同阶段棵间蒸腾量分析表

生育阶段	起止日期	天数	阶段需水量（m³/亩）	棵间蒸腾量（m³/亩）	占需水量（%）
前期	25/5 ~ 7/7	42	106.98	—	—
中期	8/7 ~ 23/8	46	195.54	47.39	24.23
后期	24/8 ~ 4/10	41	88.65	58.16	65.61
合计	25/5 ~ 4/10	129	391.17	105.55	26.98

Table 92 Analysis of Soil Water Evaporation in the Beet Plot at Different Stages

Growth and Development Period	Start and End Date	Days	Stage Water Requirement (m³/mu)	Soil water evaporation (m³/mu)	Evaporation/water requirement (%)
Initial Stage	25/5 – 7/7	42	106.98	—	—
Middle Stage	8/7 – 23/8	46	195.54	47.39	24.23
Later Stage	24/8 – 4/10	41	88.65	58.16	65.61
Total	25/5 – 4/10	129	391.17	105.55	26.98

由图22可见，甜菜在生育期的棵间地面蒸发量呈不规则双峰现象，最高峰出现在8月上旬，次高峰出现在9月中旬，随甜菜生长发育的不同阶段而变化，而在7月下旬甜菜的叶面蒸腾量呈最大值时，棵间蒸发量则呈底谷，作物的生理需水量要用于满足甜菜的叶面蒸腾。

Figure 22 shows that the soil water evaporation of beet is irregular and has two peaks, the highest in early August and the sub-peak in mid-September. The evaporation changes to different stages of beet. When the leaf transpiration peaks in late July, the evaporation drops to the low point. Thus the physiological water requirement is mainly used to satisfy the leaf transpiration at this time.

由上述分析可见，甜菜在本区全生育期的植株（叶面）蒸腾量占需水量的69%，以7月下旬的植株蒸腾量最大，达47.45~57.23 m³/亩；甜菜的棵间土壤蒸发量占需水量的31%左右。

In conclusion, the plant (leaf) transpiration of beet in this plot accounts for 69% of the water requirement, peaking in late July with 47.45 – 57.23 m³/mu, and the soil water evaporation accounts for 31% of total water requirement.

通过对春小麦、玉米、甜菜等作物生理需水量的观测试验，现就不同作物的植株叶面蒸腾量和棵间土壤蒸发量测试结果归纳见表93。

Through the observation of physiological water requirement of spring wheat, corn and beet, the test results of the leaf transpiration and the evaporation are summarized in Table 93.

第二章 作物需水量与需水规律
2　Water Requirement of Crops and Its Law

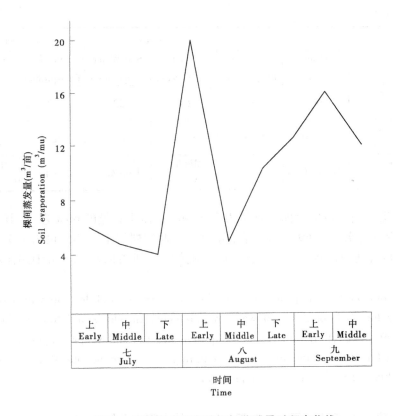

图 22　甜菜各生育期的棵间蒸腾量过程变化线

Figure 22　Changing Curves of Soil Water Evaporation in the Beet Plot at Growth and Development Period

表 93　　　　　　　　　不同作物生理需水量分析总结一览表　　　　　　　单位：m^3/亩

项目		春小麦		玉米		甜菜	
		植株蒸腾量	棵间蒸发量	植株蒸腾量	棵间蒸发量	植株蒸腾量	棵间蒸发量
全生育期	mm	256	177.5	274.21	137.99	428.46	158.22
	m^3/亩	170.65	118.41	182.79	92.05	285.62	105.55
占需水量(%)		60	40	67	33	73	27
最大值(%)		69.40	70	83.8	68	92.8	90
出现时段		6月中旬	5月下旬	8月中旬	5月下旬	7月下旬	9月中、下旬

Table 93　　Summary of Physiological Water Requirement of Different Crops　　Unit: m^3/mu

Item		Spring Wheat		Corn		Beet	
		Plant Transpiration	Soil Water Evaporation	Plant Transpiration	Soil Water Evaporation	Plant Transpiration	Soil Water Evaporation
Whole Growth and Development Period	mm	256	177.5	274.21	137.99	428.46	158.22
	m^3/mu	170.65	118.41	182.79	92.05	285.62	105.55

Continue Table 93

Item	Spring Wheat		Corn		Beet	
	Plant Transpiration	Soil Water Evaporation	Plant Transpiration	Soil Water Evaporation	Plant Transpiration	Soil Water Evaporation
Evaporation/water requirement (%)	60	40	67	33	73	27
Maximum (%)	69.40	70	83.8	68	92.8	90
Emergence Period	Middle-June	Late-May	Middle-August	Late-May	Late-July	Middle and late September

春小麦全生育期的植株叶面蒸腾量为 256 mm,占需水总量的 60% 左右,其中以 6 月中旬最高,占阶段需水量的 69.40%;春小麦全生育期的棵间土壤蒸发量为 177.5 mm,约占需水量的 40%,其中以 5 月下旬的棵间土壤蒸发量所占阶段需水量比例最高,约占 70%。

The plant transpiration of spring wheat in the entire growth and development period is 256 mm, accounting for 60% of the total water requirement, in which mid-June is the highest, accounting for 69.40% of the stage water requirement; the soil water evaporation is 177.5 mm, accounting for 40% of the total water requirement, in which late May is the highest, accounting for 70% of the stage water requirement.

玉米全生育期的植株叶面蒸腾量为 274.2 mm,约占需水总量的 67%,所占阶段需水量百分比以 8 月中旬最高,达 83.8%。蒸腾强度为 5.04 mm/d;玉米全生育期的棵间土壤蒸发量为 137.99 mm,约占需水量的 33%,所占阶段需水量百分比以 5 月下旬最高,达 68%。

The plant transpiration of corn is 274.2mm, accounting for 67% of the entire period, in which mid-August is the highest, accounting for 83.8% of the stage water requirement, and the transpiration intensity is 5.04 mm/d; the soil water evaporation of the entire period is 137.99 mm, accounting for 33%, in which late May is the highest, accounting for 68% of the stage water requirement.

甜菜全生育期的植株叶面蒸腾量为 285.62 mm,占需水总量的 73%。其中以 7 月下旬最高,占阶段需水量的 92.8%;甜菜全生育期的棵间土壤蒸发量为 158.22 mm,约占总需水量的 27%,其中以甜菜生长后期的 9 月中、下旬最高,约占阶段需水量的 90%。

The plant transpiration of beet in the entire period is 285.62 mm, accounting for 73% of the total water requirement, in which late July records the highest, accounting for 92.8% of the stage water requirement; the evaporation is 158.22 mm, accounting for 27% of the total water requirement, in which middle and late September are the highest, accounting for 90% of the stage water requirement.

二、作物叶面积系数与需水量变化分析
2.3.2 Analysis of Correlation Between Leaf Area Index and Water Requirement

植株叶面积系数的大小与需水量有直接的关系。试验表明,在通常情况下,叶面积愈大,植株蒸腾量愈大,需水量也愈大,而不同作物的植株生长发育及叶面积系数不同,因此,其相互之间的关系也各有所不同。

The leaf area index is directly linked to water requirement. The experiment indicates that, in general, the larger the leaf area, the higher the water requirement. Different crops have different growth and development characteristics and different leaf area index. Thus the correlation between them is different.

(一)春小麦的叶面积系数与需水量变化分析
2.3.2.1 Analysis of Correlation Between Leaf Area Index of Spring Wheat and Water Requirement

1. 春小麦叶面积变化与时间的关系
I Correlation between leaf area and time

春小麦生育期内的叶面积大小的变化随生长阶段(时间)的变化而异,由图23可见,在小麦生育初期(5月中旬),叶面积生长速度很快,斜率较大;而5月中旬至6月上旬,虽然叶面积仍然是增大过程,但生长速度放慢,曲线斜率变缓,至6月10日以前,小麦叶面积变化达到峰值,其中1号水平的叶面积最大值接近65 cm²;次为3号水平,为54 cm²。随后叶面积开始变小。3个不同水平的小麦叶面积变化均表现为基本相似的特点。

The leaf area of spring wheat changes to the growing stage (time). Figure 23 shows that leaf area expands fast and the slope is big at the early stages (mid-May); the leaf area expands in a slow speed and the slope becomes flat from mid-May to early June. The leaf area achieves peak on June 10th, in which that of Level #1 achieves 65 cm² and that of Level #3 achieves 54 cm². Then the area of leaves begins to diminish. Three levels present the same area changing characteristics.

2. 春小麦叶面积系数变化规律
II Change law of spring wheat leaf area index

春小麦的叶面积系数在生育初期(5月上旬)较小,仅为0.02~0.03,随着小麦的生长发育,叶面积系数不断增加,至6月上旬和中旬达到最大值。其中1号水平的叶面积系数最大值为0.25,出现在6月上旬;2号水平的为0.27,出现在5月下旬;3号水平为0.25,出现在6月中旬。3个不同水平的叶面积系数最大值均在0.25~0.27之间,差异

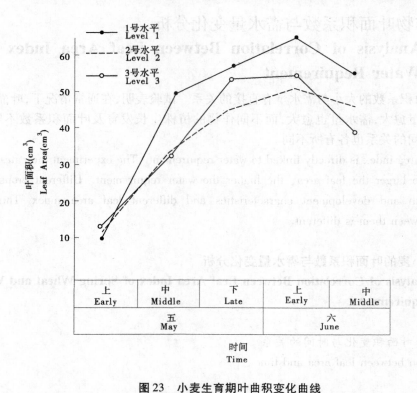

图23 小麦生育期叶曲积变化曲线
Figure 23 Changing Curves of Wheat Leaf Area at Growth and Development Period

The leaf area index of spring wheat is small in the early stages (early May), only between 0.02 – 0.03. With the growth and development of wheat, the index increases continuously, reaching the maximum in early and middle June. The maximum leaf area index of Level #1 is 0.25, which appears in early June; that of Level #2 is 0.27 in late May; that of Level #3 is 0.25 in mid-June. The maximum values of the three different levels are varying from 0.25 to 0.27, thus presenting a small difference (see Table 94).

表94　　　　　　　春小麦不同水平的叶面积系数与日需水强度分析表

时间		1		2		3	
月	旬	需水强度 (m³/(亩·d))	叶面积系数	需水强度 (m³/(亩·d))	叶面积系数	需水强度 (m³/(亩·d))	叶面积系数
5	上	1.89	0.02	1.50	0.02	0.70	0.03
	中	3.78	0.17	2.07	0.18	3.34	0.15
	下	4.14	0.22	4.43	0.27	2.52	0.22
6	上	4.59	0.25	3.51	0.25	3.26	0.24
	中	5.13	0.24	5.94	0.26	3.99	0.25
	下	5.67	0.22	4.02	0.21	2.98	0.17

Table 94 **Analysis of Spring Wheat Leaf Area Index and Daily Water Requirement Intensity at Different Levels**

Time		1		2		3	
Month	Every 10 Days	Water Requirement Intensity (m³/(mu·day))	Leaf Area Index	Water Requirement Intensity (m³/(mu·day))	Leaf Area Index	Water Requirement Intensity (m³/(mu·day))	Leaf Area Index
5	Early	1.89	0.02	1.50	0.02	0.70	0.03
	Middle	3.78	0.17	2.07	0.18	3.34	0.15
	Late	4.14	0.22	4.43	0.27	2.52	0.22
6	Early	4.59	0.25	3.51	0.25	3.26	0.24
	Middle	5.13	0.24	5.94	0.26	3.99	0.25
	Late	5.67	0.22	4.02	0.21	2.98	0.17

3. 春小麦叶面积系数与需水强度

Ⅲ Spring wheat leaf area index and water requirement intensity

小麦的叶面积系数与其需水强度有着较好的相关过程,即叶面积系数较大时段的需水强度亦较大(见表94)。由表94可见,小麦叶面积系数较大的6月中旬,需水强度也为最大值。因为,叶面积愈大,植株蒸腾量就愈大,需水量也就愈大。比较3个不同水平的叶面积系数与需水强度可以发现,它们虽然略有差异,但总的变化趋势是基本一致的。

The leaf area index of wheat has a good correlation with water requirement intensity, that is, the water requirement intensity is large when the leaf area index is large (see Table 94). Table 94 shows that, in mid-June, the water requirement index peaks when the leaf area index is large. This is because the larger the leaf area, the greater the plant transpiration, so is the water requirement. By comparing these 3 levels, we can find that they have a similar trend in general with slight differences.

我们对小麦的叶面积系数和需水强度建立回归方程,进行相关分析,3 个不同水平的小麦叶面积系数与需水强度的函数关系如下:

A regression equation for the correlation between leaf area index and water requirement intensity of wheat is established for correlation analysis. The functional relationship between the leaf area index and the water requirement intensity of three different levels is as follows:

$$E = 1.5388 + 16.54417S - 10.40721S^2$$

相关系数 $R = 0.90411$

Correlation coefficient $R = 0.90411$

2 号水平: $E = 1.62337 - 9.1843S + 80.95425S^2$
Level #2: $E = 1.62337 - 9.1843S + 80.95425S^2$

相关系数 $R = 0.85556$

Correlation coefficient $R = 0.85556$

3 号水平: $E = 3.65744 - 0.0869S$
Level #3: $E = 3.65744 - 0.0869S$

相关系数 $R = 0.9054$

Correlation coefficient $R = 0.9054$

式中　E——日需水强度,m³/(亩·d);
　　　S——叶面积系数。

Where　E = daily water requirement intensity, m³/(mu·day)
　　　　S = leaf area index

以上 3 个函数关系式的相关系数较高,均在 0.85 以上,3 条曲线形态表现为:1 号水平和 2 号水平的叶面积系数与日需水强度的变化曲线呈抛物线相关,3 号水平的呈直线相关(见图 24)。

The correlation coefficient of the above three functional equations is high, all above 0.85. The forms of curve show: the leaf area index of Level #1 and #2 is parabolically related to the daily water requirement intensity, and Level #3 is linearly related (see Figure 24).

(二)苜蓿的株高变化与需水量变化分析
2.3.2.2 Analysis of Correlation Between Plant Height and Water Requirement

苜蓿打三茬株高的变化状况(见图 25)。在第一茬苜蓿的生长初期,即 4 月苜蓿的返

第二章 作物需水量与需水规律
2 Water Requirement of Crops and Its Law

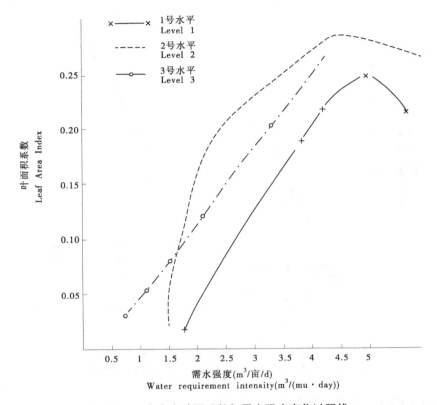

图 24 春小麦叶面系数与需水强度变化过程线

Figure 24 Spring Wheat's Leaf Area Index and its Changing Curves of Water Requirement Intensity

青—现蕾阶段,株高生长速度较慢,3 个不同水平的苜蓿株高均不超过 5 cm,这时,苜蓿的需水量较少。当进入 5 月蔓枝伸长阶段,株高生长速度迅速加快,3 个不同水墙都均以直线形式上升,斜率较大,这时苜蓿需水量出现第一次高值区;对比第二茬和第三茬苜蓿株高生长过程可以发现,第二茬苜蓿的株高生长速度较快,日需水强度增大,最高达 5.3 m³/(亩·d)(1 号水平)。这期间正值 6 月底至 7 月中旬,气温高,日照时间长,苜蓿的植株蒸腾量较大。第三茬苜蓿株高变化与需水量状况较第一茬大,但较第二茬略小。

Here outlines the changing process of alfalfa plant height during three croppings (see Figure 25). Alfalfa grows at a quite low speed at the early stage of the first cropping, i.e. the turning green-budding stage in April. At the time, three levels are no taller than 5 cm and require small amount of water. Alfalfa grows rapidly in the branch stretching stage in May. Three levels present a straight rise with a steep slope and come into the first high-value zone; the second cropping grows faster than the third cropping, and its daily water requirement intensity increases up to 5.3 m³/(mu·day) (Level #1). It owes to the high temperature and long daytime during the stage, late June to mid-July. Thus the plant transpiration is huge. In terms of plant height and water requirement, the third cropping differs greater from the first cropping than the second cropping.

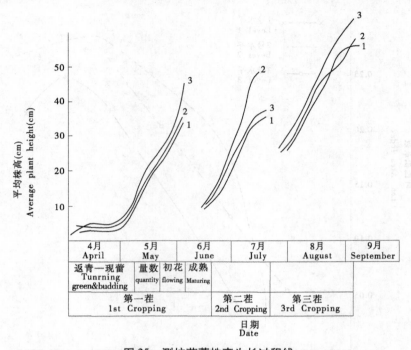

图 25　测坑苜蓿株高生长过程线
Figure 25　Height Growing Process of Test Pit Alfalfa

（三）玉米的叶面积系数与需水量变化分析
2.3.2.3　Analysis of Correlation between Leaf Area Index of Corn and Water Requirement

1. 玉米的叶面积变化与时间的关系
Ⅰ　Correlation between leaf area and time

玉米的叶面积随时间变化大致经过了 4 个阶段:6 月中、下旬的均匀增大过程;7 月上旬至中旬的快速增大过程;7 月下旬的缓慢增大过程,至 7 月底,叶面积增为最大;8 月以后的衰减过程。3 个不同水平的玉米叶面积变化的时间与关系均表现出基本相似的特点,其中,在同期生长过程中,1 号水平的叶面积最大,超过 4 000 cm²;2 号水平次之,为 3 800 cm²;而 3 号水平的叶面积最小(见图 26),最大叶面积仅为 3 250 cm²,均出现在 7 月下旬。

The leaf area of corn goes through 4 stages over time: expanding in constant speed in middle and late June; expanding in fast speed in early and middle July; then slows in late July and peaks in the end of the month. The leaf area shrinks since August. The time and relationship of leaf area change of 3 different levels show similar characteristics. In terms of the same period of growth, the leaf area of Level #1 is the largest, exceeding 4 000 cm², followed by Level #2 with 3 800 cm² and Level #3 with only 3 250 cm² at best (see Figure 26) which appears in late July.

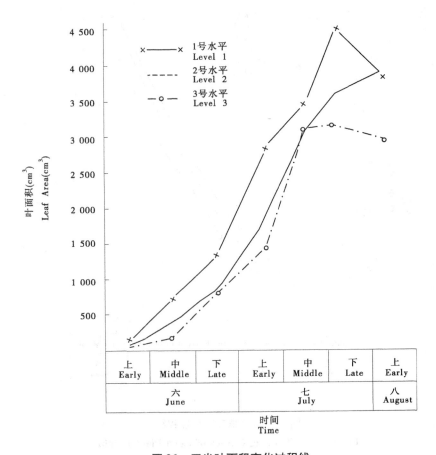

图 26 玉米叶面积变化过程线
Figure 26 Changing Curves of Corn Leaf Area

同时分析观测玉米株高变化过程也发现,玉米株高生长在抽雄以前变化很快,而到了灌浆期以后,株高停止生长,3 个不同水平均表现出一致的特点,见图 27。

In addition, through the observation of corn's plant height change, it is found that the plant height grows rapidly before the tasseling, and stops after the filling stage. The three different levels show the same characteristics, as shown in Figure 27.

2. 玉米叶面积系数变化规律

Ⅱ Change law of corn leaf area index

玉米叶面积系数在生育初期最小,仅为 0.01,随着玉米的生长发育,叶面积系数逐渐增大,到 7 月底至 8 月初,叶面积系数达最高值,变化在 1.62~2.09 之间。

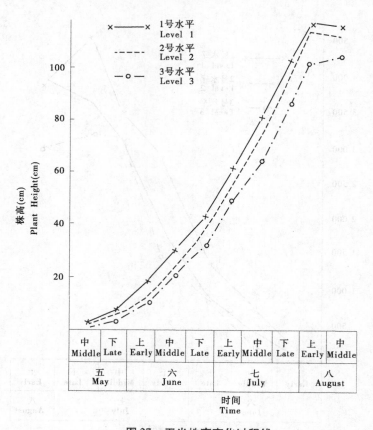

图 27 玉米株高变化过程线

Figure 27 Changing Curves of Corn Plant Height

The leaf area index is the smallest at the early stages with only 0.01. With the growth and development of corn, the leaf area index gradually increases. By late July to early August, the leaf area index reaches the peak, varying from 1.62 to 2.09.

其中 1 号水平的叶面积系数最高值出现在 7 月 7 日,2 号水平和 3 号水平均出现在 8 月初。自此以后,叶面积系数又降低至 0.5 左右,最低值为 0.35(3 号水平),出现在 9 月上旬(见表 95)。

The highest index of Level #1 appears in July 7, and Level #2 and #3 appear in early August. Since then, the leaf area index shrinks to around 0.5, and the lowest is 0.35 (Level #3) which appears in early September (see Table 95).

表95　　玉米不同水平的叶面积系数的需水强度分析表

时间		1		2		3	
月	旬	需水强度 ($m^3/(亩·d)$)	叶面积系数	需水强度 ($m^3/(亩·d)$)	叶面积系数	需水强度 ($m^3/(亩·d)$)	叶面积系数
5	下	0.84	0.01	0.52	0.01	0.97	0.01
6	上	0.77	0.09	1.21	0.05	1.05	0.06
	中	1.30	0.24	1.47	0.18	1.54	0.16
	下	4.76	0.59	3.70	0.38	2.68	0.43
7	上	3.84	1.14	5.40	0.85	3.21	0.66
	中	3.27	1.51	2.18	1.37	2.64	1.16
	下	5.10	2.09	4.47	1.91	6.02	1.57
8	上	5.04	2.04	2.10	2.00	2.15	1.62
	中	3.20	0.53	3.76	0.44	3.16	0.42
	下	1.08	0.43	0.39	0.38	0.53	0.41
9	上	1.71	0.43	2.85	0.39	1.24	0.35

Table 95　Analysis of Corn Leaf Area Index and Daily Water Requirement Intensity at Different Levels

Time		1		2		3	
Month	Every 10 Days	Water Requirement Intensity ($m^3/(mu·day)$)	Leaf Area Index	Water Requirement Intensity ($m^3/(mu·day)$)	Leaf Area Index	Water Requirement Intensity ($m^3/(mu·day)$)	Leaf Area Index
5	Late	0.84	0.01	0.52	0.01	0.97	0.01
6	Early	0.77	0.09	1.21	0.05	1.05	0.06
	Middle	1.30	0.24	1.47	0.18	1.54	0.16
	Late	4.76	0.59	3.70	0.38	2.68	0.43
7	Early	3.84	1.14	5.40	0.85	3.21	0.66
	Middle	3.27	1.51	2.18	1.37	2.64	1.16
	Late	5.10	2.09	4.47	1.91	6.02	1.57
8	Early	5.04	2.04	2.10	2.00	2.15	1.62
	Middle	3.20	0.53	3.76	0.44	3.16	0.42
	Late	1.08	0.43	0.39	0.38	0.53	0.41
9	Early	1.71	0.43	2.85	0.39	1.24	0.35

（四）甜菜的叶面积系数与需水量变化分析
2.3.2.4 **Analysis of Correlation between Leaf Area Index of Beet and Water Requirement**

1. 甜菜叶面积变化与时间关系
I Correlation between leaf area and time

由图 28 可见，甜菜自 6 月中旬至 8 月是一个叶面积不断增大的过程，其中 1 号水平的叶面积在 8 月中旬达到最大值，达 5 000 cm²，随后开始减小，而 2 号水平和 3 号水平在 8 月下旬和 9 月初才达到最大值。

Figure 28 shows that the leaf area of beet is expanding from mid–June to August. Level # 1 reaches to its peak in mid-August with 5 000 cm², and then begins to decrease, while not until late August and early September do Level #2 and Level #3 reach their maximum.

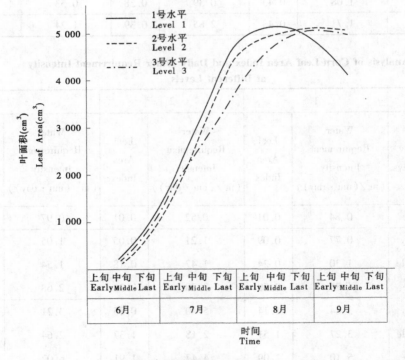

图 28 甜菜生育期叶面积变化过程线
Figure 28 Changing Curves of Beet Leaf Area at Growth and Development Period

甜菜株高生长过程表现为：6 月上旬至 8 月是甜菜株高增大过程，其中 1 号水平在 8 月中旬达最大值，为 46 cm 高；2 号水平在 8 月下旬达最高值，为 44.5 cm 高；3 号水平更晚，至 9 月初株高才达到最大值，为 43 cm 高。

The growth process of beet height is as follows: the height of beet grows from early June to August. Level #1 reaches the maximum in mid – August with 46 cm. Level #2 is in late August with 44.5 cm; and Level #3, even slower, reaches the maximum in early September with 43 cm.

3 个不同水平的株高变化表现为 1 号水平 > 2 号水平 > 3 号水平(见图 29)。

The height change can be ranked as Level #1 > Level #2 > Level #3 (see Figure 29).

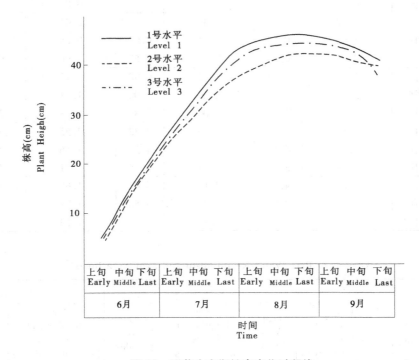

图 29　甜菜生育期株高变化过程线

Figure 29　Changing Curves of Beet Height at growth and Development Period

2. 甜菜叶面积系数变化规律

Ⅱ　Change law of beet leaf area index

甜菜的叶面积系数变化规律较强,从 6 月中旬至 8 月下旬一直呈增大过程,3 个不同水平的甜菜叶面积系数均表现出相似的特点,其中 1 号水平的叶面积系数最大,为 5.25;2 号水平和 3 号水平的甜菜叶面积系数接近,分别为 4.88 和 4.91(见表 96)。

The leaf area index of beet has a strong change rule. It has been increasing from mid – June to late August. The leaf area indexes of three levels show similar characteristics during this period. Index of Level #1 is the largest with 5.25. Level #2 and #3 are close, 4.88 and 4.91 respectively (see Table 96).

表96　甜菜需水强度与叶面积系数分析表

时间		1		2		3	
月	旬	需水强度 ($m^3/(亩·d)$)	叶面积系数	需水强度 ($m^3/(亩·d)$)	叶面积系数	需水强度 ($m^3/(亩·d)$)	叶面积系数
5	上	—		—		—	
	中	1.39		0.56		1.13	
	下	0.75		1.03		0.82	
6	上	1.02		1.31		2.05	
	中	1.92	0.15	1.71	0.07	0.65	0.16
	下	3.03	0.04	2.47	0.57	2.90	0.36
7	上	2.45	1.13	2.72	0.98	3.17	1.08
	中	3.79	2.20	2.2	2.00	1.71	2.21
	下	5.57	3.19	5.00	2.83	4.68	2.79
8	上	3.83	2.80	2.99	4.32	2.55	3.54
	中	3.84	4.15	2.00	4.34	3.56	4.22
	下	4.09	5.25	5.04	4.88	3.81	4.91
9	上	2.83	4.72	2.11	4.82	1.68	4.77
	中	0.81	3.87	1.83	3.58	1.71	3.31
	下	1.94		2.17		1.23	

Table 96　Analysis of Beet Water Requirement and Leaf Area Index

Time		1		2		3	
Month	Every 10 Days	Water Requirement Intensity ($m^3/(mu·day)$)	Leaf Area Index	Water Requirement Intensity ($m^3/(mu·day)$)	Leaf Area Index	Water Requirement Intensity ($m^3/(mu·day)$)	Leaf Area Index
5	Early	—		—		—	
	Middle	1.39		0.56		1.13	
	Late	0.75		1.03		0.82	
6	Early	1.02		1.31		2.05	
	Middle	1.92	0.15	1.71	0.07	0.65	0.16
	Late	3.03	0.04	2.47	0.57	2.90	0.36
7	Early	2.45	1.13	2.72	0.98	3.17	1.08
	Middle	3.79	2.20	2.2	2.00	1.71	2.21
	Late	5.57	3.19	5.00	2.83	4.68	2.79
8	Early	3.83	2.80	2.99	4.32	2.55	3.54
	Middle	3.84	4.15	2.00	4.34	3.56	4.22
	Late	4.09	5.25	5.04	4.88	3.81	4.91
9	Early	2.83	4.72	2.11	4.82	1.68	4.77
	Middle	0.81	3.87	1.83	3.58	1.71	3.31
	Late	1.94		2.17		1.23	

3. 甜菜叶面积系数与需水量变化
Ⅲ Change in leaf area index and water requirement

由表 96 可见,甜菜生育期内各旬日需水强度的变化与叶面积系数变化有着较为密切的关系。需水强度随甜菜叶面积系数增加,植株的叶面蒸腾量增大而增强,日需水强度最大值的出现时间与叶面积系数最大值出现时间相一致。

Table 96 shows that there is a close relationship between the change of daily water requirement intensity and the change of leaf area index in every ten days of a month. The water requirement intensity increases with the increase of leaf area index and leaf transpiration of the plant. The maximum daily water requirement intensity appears at the same time to the maximum leaf area index.

由图 30 可清楚地看到,甜菜的阶段需水强度与叶面积系数的变化过程密切相关。3 个不同水平的阶段需水强度在达到最高值时,其叶面积系数也均为最高值,曲线变化过程也表现出基本相似的形态(见图 30)。

Figure 30 clearly shows that the stage water requirement intensity of beet is closely related to the changing process of leaf area index. When the water requirement intensity of 3 levels reaches the highest, the leaf area coefficient peaks, and the changing processes of curves also show a similar shape (see Figure 30).

(五)作物叶面积系数与需水量变化研究小结
2.3.2.5 Summary of Correlation between Leaf Area Index and Water Requirement

通过对本区春小麦、苜蓿、玉米和甜菜叶面积系数与需水量变化的试验研究,现将结果归纳如下:

Through the experimental research on the change of leaf area coefficient and water requirement of spring wheat, alfalfa, corn and beet in this Plot, the results are summarized as follows:

不同作物的叶面积最大值出现时间不同:小麦叶面积最大值出现在 6 月上旬;玉米叶面积最大值出现在 7 月下旬;甜菜叶面积最大值出现在 8 月下旬。3 个不同水平的作物叶面积变化比较,均以 1 号水平的叶面积大,生长速度快(见表 97)。

The maximum leaf area of crops differs from occurrence time: the maximum leaf area of wheat comes out in early June, while corn leaf in late July, and beet in late August. In terms of the changes of leaf area of three different levels of crops, Level #1 is the largest in leaf area and the fastest in growth rate (see Table 97).

不同作物叶面积系数与需水量的变化过程均相关较好,表现出相似的变化规律。即

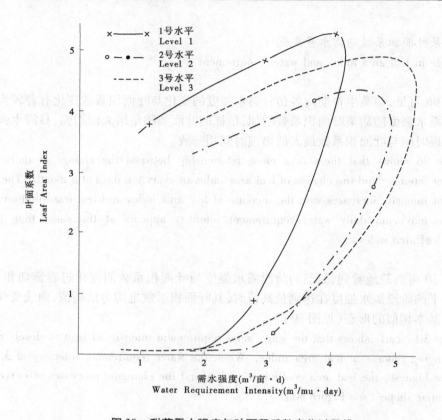

图30 甜菜需水强度与叶面积系数变化过程线
Figure 30 Changing Process of Beet Water Requirement Intensity and Leaf Area Index

作物需水强度随叶面积系数的增加而增大,叶面积系数较大时的作物需水强度也较大。

The close correlation between the leaf area index and the water requirement is found in different crops, showing a similar changing pattern. That's to say, crop water requirement intensity increases with the increase of leaf area index, and the crop water requirement intensity is large when the leaf area index is large.

表97　　　　　　　　作物叶面积(株高)最大值出现时间分析统计表

作物	春小麦	苜蓿(株高)			玉米		甜菜
		第一茬	第二茬	第三茬	叶面积	株高	
1	6月上旬	6月上旬	7月中旬	9月上旬	7月下旬	8月上旬	8月中旬
2	6月上旬	6月上旬	7月中旬	9月上旬	8月上旬	8月上旬	8月下旬
3	6月上旬	6月上旬	7月中旬	9月上旬	7月下旬	8月中旬	9月上旬

Table 97　Analysis of Occurrence Time of Maximum Crop Leaf Area (Plant Height)

Crop	Spring Wheat	Alfalfa (Plant Height)			Corn		Beet
		1st Cropping	2nd Cropping	3rd Cropping	Leaf Area	Plant Height	
1	Early-June	Early-June	Mid-July	Early September	Late-July	Early-August	Mid-August
2	Early-June	Early-June	Mid-July	Early September	Early-August	Early-August	Late-August
3	Early-June	Early-June	Mid-July	Early September	Late-July	Mid-August	Early-September

第四节　作物潜在腾发量及作物系数
2.4　Potential Evapo-Transpirative Capacity of Crops and Its Coefficient

作物的潜在蒸发蒸腾量是指在作物水肥供给充足，作物完全覆盖地面，茎秆高度均一的条件下，单位时间内所消耗的最大可能的总水量。但实际上，不是所有作物的各个生育阶段都能达到这个水平。因此具体应用时，必须按照作物的情况和土壤条件加以折算修正，才能得出实际的蒸发蒸腾量。

The potential evapotranspiration of crops refers to the maximum possible total water consumption per unit time under conditions that water and fertilizer are supplied sufficiently, crops completely cover the ground, and the stem height is uniform. But in fact, not all crops can reach this level at any stage of growth and development period. Therefore, it must be corrected according to the crop and soil conditions in specific application, so as to figure out the actual evapotranspiration.

作物在蒸发过程中，无论是体内液态水的输送，还是田间腾发面上水分的汽化和扩散均需克服一定阻力，这种阻力越大，需要消耗的能量也越大，由此可见，作物需水量的大小与腾发消耗能量有较密切的关系。腾发过程中的能量消耗，主要是以耗能形式进行的。

During the evaporation process of the crop, whether it is the transport of inner liquid water or the vaporization and diffusion of water on the evapotranspiration surface of the field, a certain resistance must be overcome. The greater the resistance, the greater the energy will be consumed. Thus the amount of water requirement is closely related to the energy consumed for evapotranspiration. The energy consumed during the evapotranspiration is mainly carried out in the form of energy consumption.

一、作物潜在腾发量的求解计算
2.4.1 Working Out Crop Potential Evapotranspiration

这里主要采用彭曼公式对作物潜在腾发量进行求解。彭曼公式是以太阳辐射热能平衡为基础的一种方法,着重考虑气象条件作为对蒸发的影响,运用彭曼公式确定潜在腾发量是作物需水量研究的一种基本方法。

Penman-Monteith equation is adopted to work out the potential evapotranspiration of crops. The equation, based on the solar radiation energy and heat energy balance, focuses on the effect of meteorological conditions on evaporation. Using Penman-Monteith equation to determine the potential evapotranspiration is a basic method in crop water requirement research.

该方法是利用当地实测的有关气象资料,从热量平衡原理推导出基本公式,再用水汽扩散理论求解公式中难于用热量平衡方程本身求解的参数,从而探求作物需水量的气象诸因素之间的关系,从而求出作物潜在蒸腾量(ET_0)。

Local measured meteorological data are used in this method to derive the basic formula through the heat balance principle, and then figure out parameters difficult to solve with the heat balance formula itself by using the water vapor diffusion theory, so as to explore the relationship between the crop water requirement and meteorological factors and determine the potential evapotranspiration (ET_0) of the crop.

这里我们用1989—1992年4年实际观测的气象资料,采用彭曼公式对作物潜在蒸腾量(ET_0)进行求解,其结果见表98。

Here is the meteorological data observed during 1989 to 1992. The Penman-Monteith equation is used to figure out the potential evapotranspiration (ET_0) of the crop. Results are shown in Table 98.

第二章 作物需水量与需水规律
2 Water Requirement of Crops and Its Law

表98　　　　　　　　作物生长期气象资料及 ET_0 计算统计表

月份	4	5	6	7	8	9	10
日照时数(h)	309.6	353.1	355.7	302.5	314.5	301.9	231.5
气压值(mb)	960.0	956.5	950.2	948.3	952.5	958.4	964.1
蒸发值(mm)	162.5	259.8	294.8	256.9	266.8	167.2	93.1
平均气温(℃)	9.4	18.3	24.4	24.8	21.5	17.0	9.0
湿度(%)	55.2	52.8	50.2	61.1	64.9	55.5	73.2
地温(℃)	10.3	20.1	27.3	27.4	25.4	19.1	9.7
最低温度(℃)	0.7	9.5	19.9	17.7	15.9	10.9	0.6
最高温度(℃)	16.4	24.6	29.0	19.0	28.2	22.1	14.9
风速(m/s)	3.9	3.95	4.7	3.7	3.75	2.85	3.3
ET_0(mm)	105.19	110.77	173.82	162.35	133.19	79.72	32.61
ET_0(m³/亩)	70.13	73.85	115.89	108.23	88.79	53.15	21.74
ET/d		1.02	2.13	4.04	2.34	1.93	

Table 98　　Statistics of Meteorological Data and ET_0 Results during the Growth Period of the Crop

Month	4	5	6	7	8	9	10
Sunshine Duration (hour)	309.6	353.1	355.7	302.5	314.5	301.9	231.5
Atmospheric Pressure (mb)	960.0	956.5	950.2	948.3	952.5	958.4	964.1
Evaporation (mm)	162.5	259.8	294.8	256.9	266.8	167.2	93.1
Average Air Temperature (℃)	9.4	18.3	24.4	24.8	21.5	17.0	9.0
Humidity (%)	55.2	52.8	50.2	61.1	64.9	55.5	73.2
Ground Temperature (℃)	10.3	20.1	27.3	27.4	25.4	19.1	9.7
Minimum Temperature (℃)	0.7	9.5	19.9	17.7	15.9	10.9	0.6
Maximum Temperature (℃)	16.4	24.6	29.0	19.0	28.2	22.1	14.9
Wind Speed (m/s)	3.9	3.95	4.7	3.7	3.75	2.85	3.3
ET_0(mm)	105.19	110.77	173.82	162.35	133.19	79.72	32.61
ET_0(m³/mu)	70.13	73.85	115.89	108.23	88.79	53.15	21.74
ET/day		1.02	2.13	4.04	2.34	1.93	

通过计算求出,本区4~10月的潜在蒸腾量为797.65 mm,即531.77 m³/亩。

The potential transpiration of the Prefecture from April to October is calculated as being 797.65 mm, that is, 531.77 m³/mu.

由潜在蒸腾蒸发量(ET_0)和作物实际需水量(ET)可以根据如下公式求出作物系数K_c值:

With the potential evapotranspiration (ET_0) and the actual crop water requirement (ET), the crop coefficient K_c can be calculated according to the following formula:

$$K_c = ET/ET_0$$

二、作物的潜在腾发量及作物系数
2.4.2 Crop Potential Evapotranspiration and Crop Coefficient

(一)春小麦的潜在腾发量及作物系数
2.4.2.1 Potential Evapotranspiration and Coefficient of Spring Wheat

本区春小麦的生育期为4月3日至7月19日,从而求出生育期内春小麦的潜在腾发量为:

The growth and development period of spring wheat in the Prefecture is from April 3rd to July 19th, so the potential evapotranspiration is:

$$ET_0(小麦) = 482.27 \text{ mm} = 321.51 \text{ m}^3/亩$$
$$ET_0(\text{wheat}) = 482.27 \text{ mm} = 321.51 \text{ m}^3/\text{mu}$$

小麦在生育期内的实际需水量为289.14 m³/亩,由此并根据小麦的潜在腾发量,求算本区小麦作物系数(K_c)为:

The actual water requirement during the growth and development period is 289.14 m³/mu. Therefore, according to the potential evapotranspiration of wheat, the crop coefficient (K_c) of wheat in this Prefecture is calculated to be:

$$K_c = ET(小麦)/ET_0(小麦) = 289.14/321.51 = 0.899$$
$$K_c = ET(\text{wheat})/ET_0(\text{wheat}) = 289.14/321.51 = 0.899$$

(二)苜蓿的潜在腾发量及作物系数
2.4.2.2 Potential Evapotranspiration and Coefficient of Alfalfa

本区苜蓿打三茬的全生育期为3月底至9月中旬,从而求出苜蓿生育期内的潜在腾发量为 725.18 mm。苜蓿3个不同水平的作物系数 K_c 分别为：1号水平 1.257,2号水平 1.117,3号水平为 0.950(见表99)。1号水平最大,2号水平次之,3号水平最小。

In this Prefecture, the entire growth and development period of alfalfa for three croppings is from late March to mid-September, so the potential evapotranspiration during the entire period is 725.18 mm. The crop coefficients, K_c, of alfalfa of three levels are: 1.257 for Level #1, 1.117 for Level #2, and 0.950 for Level #3 (see Table 99). Level #1 is the biggest, followed by Level #2 and #3.

表99 苜蓿生育期内潜在腾发量和作物系数

作物名称	水平	需水量		潜在腾发量 (mm)	K_c 值
		m³/亩	mm		
苜蓿	1	608.23	911.89	725.18	1.257
	2	540.38	810.16	725.18	1.117
	3	459.53	688.95	725.18	0.950

Table 99 Alfalfa's Potential Evapotranspiration and Crop Coefficient in the Growth and Development Period

Crop Name	Level	Water Requirement		Potential Evapotranspiration (mm)	K_c Value
		m³/mu	mm		
Alfalfa	1	608.23	911.89	725.18	1.257
	2	540.38	810.16	725.18	1.117
	3	459.53	688.95	725.18	0.950

(三)玉米的潜在腾发量及作物系数
2.4.2.3 Potential Evapotranspiration and Coefficient of Corn

本区玉米的全生育期为5月中旬到9月上旬,从而求出1989、1990和1991年苜蓿生育期内的潜在腾发量分别为:1989年为397.66 m³/亩,1990年394.0 m³/亩,1991年397.56 m³/亩,3年平均潜在腾发量为396.41 m³/亩。即:

The whole growth and development period of corn in this Prefecture is from mid-May to early September. Thus the potential evapotranspiration during the growth period of 1989, 1990 and 1991 is 397.66 m³/mu, 394.0 m³/mu, and 397.56 m³/mu respectively; and the average is 396.41 m³/mu. That is:

$$ET_0(玉米) = 396.41 \text{ m}^3/亩 = 594.61 \text{ mm}$$

$ET_0(\text{corn}) = 396.41 \text{ m}^3/\text{mu} = 594.61 \text{ mm}$

根据玉米生育期内的需水量(273.0 m³/亩),从而求出玉米的作物系数为:

According to the water requirement during the entire growth and development period of corn (273.0 m³/mu), the crop coefficient is:

$$K_c(\text{玉米}) = 273.0/396.41 = 0.689$$
$$K_c(\text{corn}) = 273.0/396.41 = 0.689$$

(四)甜菜的潜在腾发量及作物系数
2.4.2.4 Potential Evapotranspiration and Coefficient of Beet

本区甜菜的生育期为5月至10月初,从而求出甜菜生育期的潜在腾发量为570.82 mm。

In the Prefecture, the entire growth and development period of beet lasts from May to early October, so the potential evapotranspiration during the entire period is 570.82 mm.

根据3个不同水平甜菜的实际需水量求出,1号水平的作物系数为1.027,2号水平的为0.916,3号水平的为0.869(见表100)。

Based on the actual water requirement, the crop coefficients of beet of three levels are: 1.027 for Level #1, 0.916 for Level #2, and 0.869 for Level #3 (see Table 100).

表100　　　　　　　　　　苜蓿生长期内的潜在腾发量和作物系数

作物名称	水平	需水量		潜在腾发量 (mm)	K_c 值
		m³/亩	(mm)		
甜菜	1	391.17	586.46	570.82	1.027
	2	348.95	523.16	576.82	0.916
	3	330.71	495.82	570.82	0.869

Table 100　Beet's Potential Evapotranspiration and Crop Coefficient in the Growth and Development Period

Crop Name	Level	Water Requirement		Potential Evapotranspiration (mm)	K_c Value
		m³/mu	mm		
Beet	1	391.17	586.46	570.82	1.027
	2	348.95	523.16	576.82	0.916
	3	330.71	495.82	570.82	0.869

（五）作物潜在腾发量研究小结
2.4.2.5 Summary of Crop Potential Evapotranspiration

通过对春小麦、苜蓿、玉米及甜菜生育期潜在腾发量（ET_0）和作物系数（K_c）的分析测试，现归纳总结如下：

Through the analysis and test on the potential evapotranspiration (ET_0) and crop coefficient (K_c) in the growth and development period of spring wheat, alfalfa, corn and beet, results are summarized as follows:

用彭曼公式计算求出本区 4～10 月的潜在腾发量为 797.65 mm，即 531.77 m³/亩。

The potential evapotranspiration of the Prefecture from April to October is calculated, with Penman-Monteith equation, as being 797.65 mm, that is, 531.77 m³/mu.

春小麦生育期的潜在腾发量（ET_0）为 482.27 mm，即 321.51 m³/亩，作物系数（K_c）为 0.899。

The potential evapotranspiration (ET_0) in the growth and development period of spring wheat is 482.27 mm, that is 321.51 m³/mu, and the crop coefficient (K_c) is 0.899.

苜蓿生育期的潜在腾发量（ET_0）为 725.18 mm，即 483.45 m³/亩，作物系数（K_c）为 1.108。

The potential evapotranspiration (ET_0) in the growth and development period of alfalfa is 725.18 mm, that is 483.45 m³/mu, and the crop coefficient (K_c) is 1.108.

玉米生育期的潜在腾发量（ET_0）为 594.61 mm，即 396.41 m³/亩，作物系数（K_c）为 0.689。

The potential evapotranspiration (ET_0) in the growth and development period of corn is 594.61 mm, that is 396.41 m³/mu, and the crop coefficient (K_c) is 0.689.

甜菜生育期的潜在腾发量（ET_0）为 570.82 mm，即 380.55 m³/亩，作物系数（K_c）为 0.937（见表 101）。

The potential evapotranspiration (ET_0) in the growth and development period of beet is 570.82 mm, that is 380.55 m³/mu, and the crop coefficient (K_c) is 0.937 (see Table 101).

表 101　　　　　　不同作物潜在腾发量及作物系数分析表

项目		春小麦	苜蓿	玉米	甜菜
潜在腾发量	mm	482.27	725.18	594.61	570.82
	m³/亩	325.51	483.45	396.41	380.55
作物系数 K_c		0.899	1.108	0.689	0.937

Table 101   Analysis of Crop Potential Evapotranspiration and Crop Coefficients

Level		Spring Wheat	Alfalfa	Corn	Beet
Potential Evapotranspiration	mm	482.27	725.18	594.61	570.82
	m³/mu	325.51	483.45	396.41	380.55
Crop Coefficient K_c		0.899	1.108	0.689	0.937

第五节　作物需水量的数学模型
2.5 Mathematical Model of Water Requirement of Crops

一、作物需水量数学模型的建立
2.5.1 Mathematical Modeling of Water Requirement of Crops

作物生长与土壤、大气的关系受物理、化学、生物学机理的制约,是一个复杂的传递系统,构成土壤—植物—大气的连续系统,作物的需水量取决于作物生长发育和对水分需求的内部因子与外部因子,内部因子是指对需水规律有影响的那些生物学特性 B,这些特性与作物的种类 K 和品种 V 有关,同时也与作物的发育期 Φ 和生长状况 G 有关,天气条件 M(包括太阳辐射、气温、日照、风速和温度等)和土壤条件 S(包括土壤含水量、土壤质地、结构和地下水位等)属外部因子,各种不同的农业措施和排水措施只对作物需水量产生间接影响,作物需水和主要影响因子 B、M、S 之间多因子关系比较复杂,用函数形式表示为:

Constrained by physical, chemical or biological mechanisms, the relationship between crop growth and soil and atmosphere is a complex transmission system, which constitutes a continuous soil-plant-atmosphere system. How much water a crop requires depends on its growth and development, as well as internal and external factors that have a demand for water. Internal and external factors refer to those biological properties (B) that influence the water requirement, which are related to the kind (K) and the variety (V) of a crop, and also related to the development phase (φ) and the growth conditions (G) of a crop. The external factors refer to climate conditions (M, including solar radiation, air temperature, sunlight, wind speed and temperature, etc.) and soil conditions (S, including soil water content, soil texture, structure, water table, etc.). A variety of agricultural practices and drainage measures only have indirect impacts on the water requirement. The multifactorial relationship between crop water requirement and main impact factors ($B/M/S$) is complicated, and can be expressed in the following function equation:

$$E = f_1(M、S、B)$$

式中　E——作物日需水量；
　　　f_1——函数符号。

Where　E = daily water requirement of the crop
　　　　f_1 = function symbol.

为了分别考虑气象因素、土壤条件对需水量的影响，将上式分解成单独的一元函数来分析作物腾发量与影响因素的关系，气象因素对腾发量的影响，可表示为：

In order to separately consider the influence of meteorological factors and soil conditions on water requirement, the above equation is decomposed into a unary function to analyze the correlation between crop evapotranspiration and influencing factors. The influence of meteorological factors on evapotranspiration can be expressed as:

$$E = f_2(M) = f_2(T、V、N、D、H)$$

式中　T——气温，℃；
　　　V——风速，m/s；
　　　N——当地实际日照时数，h/d；
　　　D——日平均饱和差，mb；
　　　H——日平均相对湿度，%。

Where　T = air temperature, ℃
　　　　V = wind speed, m/s
　　　　N = local actual sunshine duration, h/d
　　　　D = daily average saturation deficit, mb
　　　　H = daily average relative humidity.

二、相关气象因子分析统计
2.5.2　Analysis and Statistics of Relevant Meteorological Factors

试验地区气候干燥，作物生长季节日照时间长，昼间气温较高，风频繁、风速大，太阳的短波辐射，使地温和植物叶面温度升高，有一部分又被反射到大气中，使气温增高，因此把气温饱和差、日照时数、风速等要素作为影响本地区作物需水量的主要气象因子。

The climate in the Prefecture is dry, the sunshine duration in the growing season is long, the air temperature is high in daytime, the wind is frequent and with high speed, and the short-wave radiation of the sun lifts the temperature of the ground and the leaf. Some of the radiation is reflected to the atmosphere, uprising the air temperature. Thus the air temperature saturation deficit, sunshine duration, and wind speed are regarded as the main meteorological factors that affect the crop water requirement in the Prefecture.

由表 102 可见,试验区作物生长期 4~10 月的气温变化以 7 月最高,月平均气温为 24.8 ℃,次为 6 月,为 24.4 ℃。而 4 月和 10 月的月平均气温较低,在 9 ℃左右;地温变化与气温的变化过程基本接近,最高地温亦出现在 7 月,达 27.4 ℃,6 月次之。

Table 102 shows that, in the growing season from April to October, the air temperature is the highest in July with a monthly average temperature of 24.8 ℃, followed by June with 24.4 ℃. The average monthly air temperature was relatively low in April and October with around 9 ℃. The change in ground temperature is similar to the air temperature. The highest ground temperature occurs in July with 27.4 ℃, followed by June.

表 102 试验区作物生育期内主要气象系数表

月份	日照时数(h)	气压值(mb)	蒸发(mm)	平均气温(℃)	湿度(1 K)	平均地温(℃)	最高气温(℃)	最低气温(℃)	风速(m/s)
4	309.6	960.0	162.5	9.4	55.2	10.3	16.4	0.7	3.9
5	353.1	965.5	259.1	18.3	52.1	21.1	24.6	9.5	3.95
6	355.7	950.2	294.1	24.4	50.2	27.3	29.1	19.9	4.7
7	302.5	948.3	256.9	24.1	61.15	27.4	29.1	17.3	3.7
8	314.6	952.5	236.1	21.5	64.9	25.4	28.2	15.9	3.75
9	301.9	958.4	167.2	17.11	55.5	19.1	22.1	10.9	2.85
10	231.5	964.1	93.1	9.1	73.2	9.7	14.9	0.6	3.3

Table 102 Major Meteorological Coefficients of the Prefecture in the Growth and Development Period of Crops

Month	Sunshine Duration (h)	Air Pressure (mb)	Evaporation (mm)	Average Temperature (℃)	Humidity (1 K)	Average Ground Temperature (℃)	Maximum Air Temperature (℃)	Minimum Air Temperature (℃)	Wind Speed (m/s)
4	309.6	960.0	162.5	9.4	55.2	10.3	16.4	0.7	3.9
5	353.1	965.5	259.1	18.3	52.1	21.1	24.6	9.5	3.95
6	355.7	950.2	294.1	24.4	50.2	27.3	29.1	19.9	4.7
7	302.5	948.3	256.9	24.1	61.15	27.4	29.1	17.3	3.7
8	314.6	952.5	236.1	21.5	64.9	25.4	28.2	15.9	3.75
9	301.9	958.4	167.2	17.11	55.5	19.1	22.1	10.9	2.85
10	231.5	964.1	93.1	9.1	73.2	9.7	14.9	0.6	3.3

本试验区作物生育期内的日照时数以 6 月最长,达 355.7 h,次为 5 月,为 353.1 h。其中 4~9 月的各月平均日照时数均在 300 h 以上,对作物的生长发育很有利。

The longest sunshine duration in the growth and development period of crops in the Prefecture is in June with 355.7 h, followed by May with 353.1 h. The monthly average sunshine duration is above 300 h from April to September, which is good for the growth and development of crops.

本试验区风速变化,以 6 月风速最大,平均为 4.7 m/s,在作物生育期内的各月平均风速除 6 月较大和 9 月较小(2.85 m/s)外,大都在 3~4 m/s 之间。

In the Prefecture, the highest wind speed is in June with 4.7 m/s. Except for the highest, June, and the lowest, September (2.85 m/s), the monthly average wind speed is between 3 – 4 m/s in other months.

三、作物需水量模型及模拟计算
2.5.3 Crop Water Requirement Model and Analog Calculation

根据上述作物需水量的数字计算模型和相关的气象因子、土壤含水率等方面的要素,分别就本区春小麦、苜蓿、玉米、甜菜需水量与气象因子(气温、日照时数、风速)的函数模式以及上述作物需水量与气温、饱和差、土壤含水率的函数模式建立如下:

According to the above numerical calculation model of the crop water requirement and related meteorological factors, soil water content, etc., a functional model of spring wheat, alfalfa, corn, beet for the water requirement and the meteorological factors (air temperature, sunshine duration, wind speed) and a model for the crop water requirement and the air temperature, saturation deficit and soil water content are built as follows:

作物需水量与气象因子的函数模式:
Functional model for the crop water requirement and meteorological factors:

$$E = F(T, N, V) \tag{1}$$

式中　E——需水强度,$m^3/(亩 \cdot day)$;
　　　T——气温,℃;
　　　N——日照时数,h/day;
　　　V——风速,m/s。

Where　E = the water requirement intensity, $m^3/(mu \cdot day)$
　　　　T = the air temperature, ℃
　　　　N = the sunshine duration, hour/day
　　　　V = the wind speed, m/s.

作物需水量与气温、日平均饱和差、土壤含水率的函数模式:

The function model for crop water requirement and air temperature, daily average saturation deficit and soil water content is expressed as:

$$E = f(T, D, W) \tag{2}$$

式中　　D——日平均饱和差，mb；
　　　　W——土壤含水率，%。

Where　D = daily average saturation deficit, mb
　　　　W = soil water content, %.

（一）小麦需水量的函数模式
2.5.3.1 Functional Model for Spring Wheat Water Requirement

1. 春小麦需水量与气象因子的函数关系
Functional correlation between spring wheat water requirement and meteorological factors

为进一步比较春小麦需水量与诸气象因子相互间的关系，我们首先就小麦需水量与日平均气温、日平均饱和差、平均日照时数、日平均水面蒸发量以及日平均地温等单因子的关系作了计算分析，其函数关系如下：

In order to further compare the correlation between the spring wheat water requirement and meteorological factors, the correlation between the wheat water requirement and single factors, such as daily average air temperature, daily average saturation deficit, average sunshine duration, daily average evaporation from water surface, and daily average ground temperature, is calculated and analysed. The functional correlation is as follows:

需水强度与日平均气温（T）的关系：
The correlation between water requirement intensity and the daily average air temperature (T):

$$E = -12.813\ 67 + 2.057\ 08T - 0.045\ 12T^2$$

相关系数 $R = 0.751$
Correlation coefficient $R = 0.751$

需水强度与日平均饱和差（D）的关系：
The correlation between water requirement intensity and the daily average saturation deficit (D):

$$E = -8.269\ 83 + 3.133\ 59D - 0.103\ 39D^2$$

相关系数 $R = 0.836$
Correlation coefficient $R = 0.836$

需水强度与平均日照时数(H)的关系：

The correlation between water requirement intensity and average sunshine duration (H):

$$E = 771.462\ 44 - 147.232\ 59H + 7.043\ 69H^2$$

相关系数 $R = 0.991$

Correlation coefficient $R = 0.991$

需水强度与日平均水面蒸发量(X)的关系：

The correlation between water requirement intensity and daily average evaporation from water surface (X):

$$E = 23.476\ 9 - 8.119\ 76X - 0.744\ 56X^2$$

相关系数 $R = 0.979$

Correlation coefficient $R = 0.979$

需水强度与日平均地温(C)的关系：

The correlation between water requirement intensity and the daily average ground temperature (C):

$$E = -14.785\ 59 + 1.949\ 09C - 0.037\ 27C^2$$

相关系数 $R = 0.836$

Correlation coefficient $R = 0.836$

式中　E——需水强度，$m^3/(亩 \cdot d)$。

Where　E = water requirement intensity, $m^3/(mu \cdot day)$.

通过建立春小麦需水量与各气象因子的函数关系式发现，它们均呈抛物线性相关，其中以平均日照时数、日平均水面蒸发量的相关关系最好，相关系数达 0.98～0.99 之间；其余的也均在 0.75 以上。

By establishing functional equations between the spring wheat water requirement and various meteorological factors, it is found that they are all parabolically related. The correlations with average sunshine duration and daily average evaporation from water surface are the best with coefficient being 0.98 to 0.99. The rest are all above 0.75.

我们将春小麦需水量与气温、日照时数和风速进行相关分析，得出下列函数式：

With the correlation analysis of the spring wheat water requirement with air temperature, sunshine duration and wind speed, the functional equation is writtened as:

$$E = 7.054\ 854\ 19 \times 10^{-9} T^{1.833} \cdot H^{7.532} \cdot V^{-1.92}$$

2. 春小麦需水量与气温、饱和差、土壤含水率的函数关系

II The functional relationship between spring wheat water requirement and air temperature, saturation deficit, soil water content

由函数模式(2)得春小麦需水量与月平均气温(℃)、月平均饱和差(mb)、月平均土壤含水率(%)的函数关系如下：

With the functional model (2), the functional relationship between spring wheat water requirement and monthly average air temperature (℃), monthly average saturation deficit (mb), and monthly average soil water content (%) is as follows:

$$E = 1\,194.496\,5 - 10.115\,3T - 0.006\,8D - 141.863W$$

相关系数 $R = 0.999$

Correlation coefficient $R = 0.999$

(二) 苜蓿需水量的函数模式
2.5.3.2 Functional Model for Alfalfa Water Requirement

1. 苜蓿需水量与气象因子的函数关系

I Functional correlation between the alfalfa water requirement and meteorological factors

苜蓿需水量与气象因子的函数关系式是由公式(2)进行回归分析得出：

With the regression analysis of equation (2), the functional equation for correlation between the alfalfa water requirement and meteorological factors is listed as follows:

$$E = 0.036\,8 T^{0.966\,9} \cdot H^{0.866\,1} \cdot V^{-0.311\,7}$$

相关系数 $R = 0.999$

Correlation coefficient $R = 0.999$

2. 苜蓿需水量与气温、饱和差、土壤含水率的函数关系

II The functional relationship between alfalfa water requirement and air temperature, saturation deficit, soil water content

由公式(1)进行回归分析，得出苜蓿需水量与气温、饱和差及土壤含水率的函数关系为：

With the regression analysis to equation (1), the functional correlation between alfalfa water requirement and air temperature, saturation deficit and soil water content is formed as:

$$E = 5.0399 + 0.3386T - 0.4249D - 0.7074W$$

相关系数 $R = 0.998$

Correlation coefficient $R = 0.998$

（三）玉米需水量的函数模式
2.5.3.3 Functional Model for Corn Water Requirement

1. 玉米需水量与气象因子的函数关系

I Functional correlation between corn water requirement and meteorological factors

玉米需水量与日照时数、平均气温、风速的函数关系式为：

The functional equation for correlation between corn water requirement and sunshine duration, average air temperature and wind speed is:

$$E = 0.0004 T^{3.9434} \cdot H^{-0.1361} \cdot V^{-2.4433}$$

相关系数 $R = 0.999$

Correlation coefficient $R = 0.999$

2. 玉米需水量与气温、饱和差、土壤含水率的函数关系

II The functional relationship between corn water requirement and air temperature, saturation deficit, soil water content

函数关系为：

The functional equation is:

$$E = 0.0331 T^{0.7081} \cdot D^{0.1175} \cdot W^{-2.4474}$$

相关系数 $R = 0.933$

Correlation coefficient $R = 0.933$

（四）甜菜需水量的函数模式
2.5.3.4 Functional Model for Beet Water Requirement

1. 甜菜需水量与气象因子的函数关系

I Functional correlation between the beet water requirement and meteorological factors

通过 3 年气象资料和试验结果得出的甜菜需水量值进行回归分析，得出甜菜需水量与平均气温、平均日照时数及风速的函数关系式如下：

A regression analysis is conducted with three-year meteorological data and test results, a functional equation for correlation between beet water requirement and average air temperature, average sunshine duration and wind speed is formed as:

$$E = 16.5637 + 0.0125T - 2.1424H + 2.2578V$$

相关系数 $R = 0.958$

Correlation coefficient $R = 0.958$

2. 甜菜需水量与气温、饱和差、土壤含水率的函数关系

II The functional relationship between beet water requirement, air temperature, saturation deficit and soil water content

甜菜需水量与平均气温、月平均日饱和差及土壤含水率之间的函数关系由公式(2)回归计算得出:

The functional relationship, calculated with regression analysis to equation (2), between beet water requirement and average air temperature, monthly average of daily saturation deficit, and soil water content is:

$$E = -7.8142 + 0.8425T - 0.1359D - 1.1331W$$

相关系数 $R = 0.888$

Correlation coefficient $R = 0.888$

式中　E——月平均需水强度,$m^3/(亩 \cdot d)$;
　　　T——月平均日气温,℃;
　　　H——月平均日照时数,h/d;
　　　V——平均风速,m/s;
　　　D——月平均日饱和差,mb;
　　　W——土壤含水率,%。

Where　E = the monthly water requirement, $m^3/(mu \cdot day)$
　　　T = the monthly average daily air temperature, ℃
　　　H = monthly average sunshine duration, h/day
　　　V = the average wind speed, m/s
　　　D = the monthly average daily saturation deficit, mb
　　　W = soil water content, %.

(五)作物需水量数学模型研究小结
2.5.3.5 Summary of Crop Water Requirement Mathematical Model

通过对春小麦、苜蓿、玉米及甜菜需水量数学模型的研究可以看出,这4类作物的需水强度与气象因子和土壤含水率之间存在着密切的关系。它既可以根据气温、日照时数

及风速等推算出作物需水量的函数关系式,也可以根据气温、饱和差及土壤含水率建立作物需水量的函数关系,由表 103 可以看出,这两个数学模型精度都较高,相关系数均在 0.888 以上,因而可以采用该模型来预测作物不同时期的需水状况。

Through research on the mathematical model for water requirement of spring wheat, alfalfa, corn and beet, it is found that there is a close correlation between water requirement intensity and meteorological factors and soil water content. It can deduce the functional equation for crop water requirement based on air temperature, sunshine duration and wind speed, and can also establish a functional relationship of crop water requirement based on air temperature, saturation deficit and soil water content. Table 103 shows that both mathematical models have high accuracy and correlation coefficients are above 0.888, so the models can be used to predict the crop water requirement in different periods.

表 103 作物需水量数学模型分析表

作物	函数关系式	相关系数 R
春小麦	$E = 7.0548519 \times 10^{-9} T^{1.833} \cdot H^{7.532} \cdot V^{-1.92}$	0.999
	$E = 1194.4965 - 10.1153T - 0.1168D - 141.863W$	0.999
苜蓿	$E = 0.0368 T^{0.967} \cdot H^{0.866} \cdot V^{-0.312}$	0.999
	$E = 5.0399 + 0.3386 - 0.4249D - 0.7074W$	0.998
玉米	$E = 0.0004 T^{3.9434} \cdot H^{0.1361} \cdot V^{-2.4433}$	0.999
	$E = 0.0331 R^{0.7081} \cdot D^{0.1175} \cdot W^{-2.4474}$	0.933
甜菜	$E = 16.5637 + 0.0125T - 2.1424H + 2.2578V$	0.958
	$E = -7.8142 + 0.8425T - 0.1359D - 1.1331W$	0.888

Table 103 Analysis of Mathematical Model for Crop Water Requirement

Crop	Functional Equation	Correlation Coefficient R
Spring Wheat	$E = 7.0548519 \times 10^{-9} T^{1.833} \cdot H^{7.532} \cdot V^{-1.92}$	0.999
	$E = 1194.4965 - 10.1153T - 0.1168D - 141.863W$	0.999
Alfalfa	$E = 0.0368 T^{0.967} \cdot H^{0.866} \cdot V^{-0.312}$	0.999
	$E = 5.0399 + 0.3386T - 0.4249D - 0.7074W$	0.998
Corn	$E = 0.0004 T^{3.9434} \cdot H^{0.1361} \cdot V^{-2.4433}$	0.999
	$E = 0.0331 R^{0.7081} \cdot D^{0.1175} \cdot W^{-2.4474}$	0.933
Beet	$E = 16.5637 + 0.0125T - 2.1424H + 2.2578V$	0.958
	$E = -7.8142 + 0.8425T - 0.1359D - 1.1331W$	0.888

第六节 作物产量与需水量分析
2.6 Crop Yields and Its Water Requirement

作物产量与需水量有着十分密切的关系。在新疆北部的荒漠戈壁地，农作物生产主要靠灌溉作业，因而正确把握作物产量与需水量的最佳关系，不仅提高了作物产量，而且又节约了水资源。研究作物产量与需水量之间关系，探索最优灌溉定额是本章中主要研究内容。

The crop yield is closely related to the crop water requirement. In the Gobi Desert in northern Xinjiang Autonomous Region, crop production mainly relies on irrigation, so correctly understanding the best correlation between crop yield and water requirement will not only improve crop yield, but also save water resources. This chapter mainly studies on the correlation between crop yield and water requirement and aims at finding out the optimal irrigation quota.

通过3年(1989—1991年)对春小麦、玉米、苜蓿、甜菜等作物进行不同水平、不同重复的产量、需水量测试分析，现将试验结果分述如下：

Through a three-year (1989 - 1991) test and analysis on crop yield and water requirement of spring wheat, corn, alfalfa and beet at different levels and different repetitions, results are summarized and stated as follows:

一、作物产量分析
2.6.1 Analysis of Crop Yield

通过3年(1989—1991年)对春小麦、苜蓿、玉米及甜菜的作物产量与需水量关系的试验分析，3年不同水平的作物产量各有差异，这其中同时也包括因自然环境变化造成的影响。现将春小麦、苜蓿、玉米及甜菜在试验期(1989—1991年)内产量状况分析如下：

Through the 3-year (1989 - 1991) experiment on the correlation between crop yield and water requirement of spring wheat, alfalfa, corn and beet, it is found that the crop yield at different levels in 3 years varies, which may be partially caused by natural environment changes. The analysis of yield of spring wheat, alfalfa, corn and beet during the experiment period (1989 - 1991) is as follows:

第二章 作物需水量与需水规律
2 Water Requirement of Crops and Its Law

（一）春小麦的产量变化分析
2.6.1.1 Analysis of Spring Wheat Yield Change

在试验期内(1989—1991年)春小麦的生产因自然环境的变化,不同年份的小麦产量也受到一定影响,如1989年,正值小麦生长期,受到冰雹灾害,1990年则又受到干热风影响,仅1991年属小麦生产正常年份。因冰雹和干热风的影响,1989、1990年小麦产量受到影响,最高产量分别只达到106.03 kg/亩和117.7 kg/亩,而1991年正常年份,3个不同水平的小麦产量为118.27~329.73 kg/亩,其中以2号水平的产量较高,达329.73 kg/亩。同时,再比较1989—1991年3年3个不同水平的春小麦产量的平均值,1号水平为154.32 kg/亩,2号水平为174.38 kg/亩,3号水平为123.39 kg/亩,所以2号水平居高(见表104)。

The yield of spring wheat changes to the natural environment in the experiment period (1989 – 1991). For example, the wheat suffered hail in the growing period in 1989 and dry – hot wind in 1990, only in 1991 did the wheat production smoothly go on. Due to the hail and dry-hot wind, the maximum yields in 1989 and 1990 were only 106.03 kg/mu and 117.7 kg/mu respectively. While in 1991, a normal year, the yields of 3 levels varied between 118.27 kg/mu to 329.73 kg/mu, among which Level #2 was the highest with 329.73 kg/mu. In terms of the average yield of 3 levels in three years, Level #1 was 154.32 kg/mu, Level #2 was 174.38 kg/mu, and Level #3 was 123.39 kg/mu. Thus Level #2 is the highest (see Table 104).

表104　　　春小麦不同水平产量变化分析表　　　单位:kg/亩

年份	1989	1990	1991	平均
1	106.03	17.60	259.34	154.32
2	77.50	117.7	329.7	174.9
3	74.63	107.9	188.27	123.39
平均	86.05	107.73	259.10	150.87

Table 104　　Analysis of Spring Wheat Yield Change at Different Levels　　Unit: kg/mu

Year	1989	1990	1991	Average
1	106.03	17.60	259.34	154.32
2	77.50	117.7	329.7	174.9
3	74.63	107.9	188.27	123.39
Average	86.05	107.73	259.10	150.87

(二) 苜蓿的产量变化分析
2.6.1.2 Analysis of Alfalfa Wheat Yield Change

通过1990—1991年2年对3个不同水平苜蓿产量的变化分析,发现1990年和1991年2年的苜蓿产量各不相同,其中以1990年产量略高。比较3个不同水平的小麦产量可见,1号水平在本次试验中产量最高,达716.74 kg/亩;2号水平次之,为669.9 kg/亩;3号水平较低,为616.1 kg/亩(见表105)。

When analysing the yield change of alfalfa at different levels in 1989 to 1991, it is found that the yield varies, and the yield in 1990 is higher. When comparing the yield of 3 different levels, it is found that Level #1 is the highest with 716.74 kg/mu, followed by Level #2 with 669.9 kg/mu and Level #3 with 616.1 kg/mu (see Table 105).

表105　苜蓿不同水平的产量变化分析表　　　　　　　　　　单位:kg/亩

年份	1990	1991	平均
1	719.18	714.30	716.74
2	708.37	631.47	669.42
3	661.79	570.4	616.10
平均	696.45	638.72	667.59

Table 105　Analysis of Alfalfa Yield Change at Different Levels　　Unit: kg/mu

Yera	1990	1991	Average
1	719.18	714.30	716.74
2	708.37	631.47	669.42
3	661.79	570.4	616.10
Average	696.45	638.72	667.59

(三) 玉米的产量变化分析
2.6.1.3 Analysis of Corn Wheat Yield Change

通过1989—1991年3年的试验对比,从时间上看,本次试验中,1990年玉米的产量较高,最高达350.3 kg/亩;1991年次之,为260.3 kg/亩;1989年最低,为219 kg/亩。比较3个不同水平的玉米产量可见,3年的玉米产量均为1号水平最高,变化在219.0~350.3 kg/亩之间,3号水平的产量最低,在133.3~171.2 kg/亩之间(见表106)。

By comparing experiments during 1989 to 1991, it is found that the yield of corn is the highest in 1990 with 350.3 kg/mu, followed by 1991 with 260.3 kg/mu and 1989 with 219 kg/mu. When comparing the yield of 3 different levels, it is found that Level #1 is the highest, varying between 219.0 to 350.3 kg/mu, and Level #3 is the lowest with 133.3 to 171.2 kg/mu (see Table 106).

表 106　　　　　玉米不同水平的产量变化分析表　　　　　单位:kg/亩

年份	1989	1990	1991	平均
1	219.0	350.3	260.3	276.53
2	152.0	242.6	224.1	206.23
3	133.3	171.2	156.7	153.73

Table 106　　　Analysis of Corn Yield Change at Different Levels　　　Unit: kg/mu

Year	1989	1990	1991	Average
1	219.0	350.3	260.3	276.53
2	152.0	242.6	224.1	206.23
3	133.3	171.2	156.7	153.73

（四）甜菜的产量变化分析
2.6.1.4　Analysis of Beet Yield Change

衡量甜菜的产量与质量有两个方面的因素：一是甜菜重量；二是甜菜的含糖量。
The yield and quality of beet can be assessed from two aspects: weight and sugar content.

由 1989—1991 年实测结果可知,1990 年 1 号水平的甜菜产量最高,达 1 964 kg/亩,1991 年 1 号水平的产量最低,为 1 514 kg/亩。比较 1989、1990 和 1991 年 3 年的甜菜平均产量,以 1990 年最高,为 1 745.78 kg/亩,1991 年与 1989 年 2 年的产量接近,分析为 1 669.5 和 1 694.5 kg/亩。同时由表 107 可见,3 个不同水平的甜菜产量比较,以 1 号水平的最高,为 1 757 kg/亩;2 号水平次之,为 1 701.94 kg/亩;3 号水平较低,为 1 650.89 kg/亩,见表 108。

The results measured during 1989 to 1991 show that the yield of Level #1 beet in 1 990 is the highest with 1 964 kg/mu and Level #1 in 1991 is the lowest with 1 514 kg/mu. Comparing the average yield in three years, it is found that the yield in 1990 is the highest with 1 745.78 kg/mu, and 1991 and 1990, close in yield, are 1 669.5 and 1 694.5 kg/mu respectively. Table 107 shows that, when comparing the yield of 3 different levels, Level #1 is the highest with 1 757 kg/mu, followed by Level #2 with 1 701.94 kg/mu and Level #3 with 1 650.89

kg/mu (see Table 107).

表107 甜菜不同水平的产量分析统计表 （单位：kg/亩）

年份	1989	1990	1991	平均
1	1 793.00	1 964.00	1 514.00	1 757.00
2	1 658.00	1 551.33	1 896.50	1 701.94
3	1 632.67	1 722.00	1 598.00	1 650.89
平均	1 694.56	1 745.78	1 669.50	1 703.08

Table 107 Analysis of Beet Yield Change at Different Levels Unit: kg/mu

Year	1989	1990	1991	Average
1	1 793.00	1 964.00	1 514.00	1 757.00
2	1 658.00	1 551.33	1 896.50	1 701.94
3	1 632.67	1 722.00	1 598.00	1 650.89
Average	1 694.56	1 745.78	1 669.50	1 703.08

比较不同水平甜菜的含糖量（见表108）可见，以1990年甜菜的含糖量较高，1990年2号水平的甜菜含糖量达27.70%，1991年的甜菜含糖量较低，仅为13.24%。比较不同水平的甜菜含糖量可见，以2号水平的甜菜含糖量最高，其3年的平均值为21.10%；3号水平最低，为18.58%；2号水平介于二者之间，为21.10%。

When comparing the sugar content of different years (see Table 108), it is found that the sugar content in 1990 is the highest, Level #2 being 27.70%, and that in 1991 is low, being 13.24%. When comparing the sugar content at different levels, it is found that Level #2 is the highest with a three-year average of 21.10%, followed by Level #2 with 21.10%, and Level #3 with 18.58%.

表108 甜菜不同水平含糖量分析表 单位：kg/亩

年份	1989	1990	1991	平均
1	23.91	24.81	13.20	20.64
2	22.25	27.70	13.35	21.10
3	21.86	20.71	13.18	18.58
平均	22.67	24.41	13.24	20.11

Table 108　　　　Analysis of Beet Sugar Content at Different Levels　　　　Unit: kg/mu

Year	1989	1990	1991	Average
1	23.91	24.81	13.20	20.64
2	22.25	27.70	13.35	21.10
3	21.86	20.71	13.18	18.58
Average	22.67	24.41	13.24	20.11

（五）作物不同水平产量变化研究小结
2.6.1.5　Summary of Crops Yield Change at Different Levels

通过1989—1991年3年间对春小麦、苜蓿、玉米及甜菜3个不同水平的作物产量分析,现归纳如下:
Through the analysis of yield of spring wheat, alfalfa, corn and beet at different levels in 3 years, the results are summarized as follows:

春小麦以1991年正常生育年的产量为高,3个不同水平的小麦平均产量为259.10 kg/亩,其中以1991年的2号水平产量最高,达329.7 kg/亩。
The yield of spring wheat is the highest in 1991 with an average yield of three different levels being 259.10 kg/mu, among which Level #2 was the highest, reaching 329.7 kg/mu.

苜蓿产量以1990年略高,其中以1990年的1号水平最高,达719.18 kg/亩。
The yield of alfalfa is higher in 1990, among which Level #1 is the highest with 719.18 kg/mu.

玉米产量以1990年1号水平最高,达350.3 kg/亩,3号水平的玉米产量均较低。
The yield of Level #1 corn in 1990 is the highest, reaching 350.3 kg/mu, and Level #3 is low.

甜菜产量以1990年1号水平最高,达1 964 kg/亩,其次为1991年的2号水平,为1 896.50 kg/亩,相差67.5 kg/亩;含糖量以1990年的2号水平最高,达27.70%（见表109）。
The yield of Level #1 beet in 1990 is the highest, reaching 1 964 kg/mu, followed by Level #2 in 1991 with 1 896.50 kg/mu, the difference being 67.5 kg/mu. The sugar content of Level #2 in 1990 is the highest, reaching 27.70% (see Table 109).

表 109　　　　　　　　　　　作物最高产量分析表

作物	春小麦	苜蓿	玉米	甜菜
最高产量（kg/亩）	329.7	719.18	350.3	1964
年份	1991	1990	1990	1990
水平号	2	1	1	1

Table 109　　　　　　　Analysis of Crops' Maximum Yield

Crop	Spring Wheat	Alfalfa	Corn	Beet
Maximum Yield (kg/mu)	329.7	719.18	350.3	1964
Year	1991	1990	1990	1990
Level	2	1	1	1

二、作物产量与需水量
2.6.2 Crop Yield and Water Requirement

作物产量与需水量之间有着极为密切的关系，它们二者呈抛物线关系，也就是说，当作物产量和需水量呈增长关系达到顶峰后，产量则会随需水量增加而下降。同时，农地也会因灌水量太多而造成土壤盐碱化或成为下潮地，致使成土壤肥力下降甚至弃耕。这种现象在新疆北部阿勒泰地区的荒漠瘠薄戈壁地尤为突出。该区土层较薄，下覆地层为第三系泥岩成为隔水层，因此该区土地极易因灌水太多而成为下潮地。所以正确掌握作物产量与需水量之间的关系，不但对提高作物效益，减少水资源浪费有直接实际意义，而且对预防土地退化也有重要的意义。

　　The crop yield and the water requirement have a close parabolic relationship, that is, after the relationship curve of yield and water requirement reach the peak, the yield will then decrease as the water requirement increases. Meanwhile, the soil can be salinized or become fluvo-aquic soil, which leads to the decrease of fertility or even abandonment of the field. Such phenomenon is especially prominent in the Altay Prefecture in northern Xinjiang. With thin soil layer and the tertiary mud rock as its underlayer and aquiclude, land in the Prefecture is very likely to become fluvo - aquic soil land due to excessive irrigation. Therefore, to correctly figure the correlation between crop yield and water requirement out will not only do good to the improvement of economic benefits and avoid waste of water, but also to prevent land deterioration.

(一)春小麦的产量与需水量
2.6.2.1 Yield and Water Requirement of Spring Wheat

春小麦产量与需水量之间有着极为密切的关系,我们把1989—1991年3年3个不同水平的试验结果进行回归分析,建立小麦需水量与产量的函数关系式如下:

The yield of spring wheat is closely correlated with the water requirement. When making a regression analysis to the experiment results of 3 levels during 1989 – 1991, a functional equation for correlation between wheat water requirement and yield can be established as followed:

$$ET = 1\,232.552\,74 + 10.024\,82W - 0.017\,85W^2$$

相关系数 $R = 0.999\,9$

Correlation coefficient $R = 0.999\,9$

式中 ET——需水量,m^3/亩;
W——作物产量,kg/亩。

Where ET = water requirement, m^3/mu
W = crop yield, kg/mu.

由计算可知,小麦产量与需水量之间呈幂相关,二者的相关曲线呈抛物线形式,即小麦需水量在289.14 m^3/亩以前较低时,其产量是一个增加过程,基本呈线性关系,当投入水量较大达289.14 m^3/亩时,春小麦产量达到最高,以后随水量增加,小麦产量反而开始下降(见图31)。

It can be seen from the calculation that the yield of wheat is power – related to the water demand, and the correlation curve of the two is parabolic. When the water requirement of wheat is before the point of 289.14 m^3/mu, the yield increases, which is generally a linear relationship. When the input water reaches 289.14 m^3/mu, the yield reaches the highest. Then, with the increase of water volume, the wheat yield declines (see Figure 31).

(二)苜蓿产量与需水量关系分析
2.6.2.2 Analysis of Alfalfa Yield and Water Requirement

根据本次试验结果,我们通过回归分析,建立苜蓿产量与需水量之间的函数关系式为:

With the experiment result and regression analysis, a functional equation is established for the correlation between alfalfa yield and water requirement:

$$ET = -2\,304.731\,59 + 106.034W - 0.023\,15W^2$$

相关系数 $R = 0.926\,3$

Correlation coefficient $R = 0.926\,3$

式中 ET——需水量,m^3/亩;

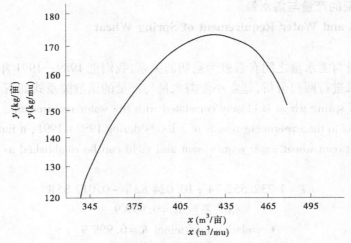

图 31 春小麦产量与需水量相关曲线
Figure 31 Correlation Curve of Spring Wheat Yield and Water Requirement

式中 W——作物产量,kg/亩。

Where ET = the water requirement, m³/mu
　　　 W = the crop yield, kg/mu.

苜蓿产量与需水量之间的变化关系与春小麦有着十分相似的特点,即二者的相关曲线也存在抛物线形。当投入水量为 540 m³/亩时,产量达到最高,为 670 kg/亩。而当水量超出此值后,苜蓿产量则因投入水量的增加反而下降(见图 32)。

The correlation of alfalfa is similar to the spring wheat, that is, the correlation curve is also parabolic. When the input water reaches 540 m³/mu, the yield is highest with 670 kg/mu. When the water exceeds the volume, the yield the alfalfa decreases (see Figure 32).

(三) 玉米产量与需水量关系分析
2.6.2.3 Analysis of Corn Yield and Water Requirement

根据 1989—1991 年试验结果,我们通过回归分析建立玉米产量与需水量的函数关系式为:

With the experimental result and regression analysis, a functional equation is established for the correlation between corn yield and water requirement:

$$W = -3\,057.7622 + 24.1431\,ET - 0.04374\,ET^2$$

相关系数 $R = 0.93607$

Correlation coefficient $R = 0.93607$

式中 ET——需水量,m³/亩;

第二章 作物需水量与需水规律
2 Water Requirement of Crops and Its Law

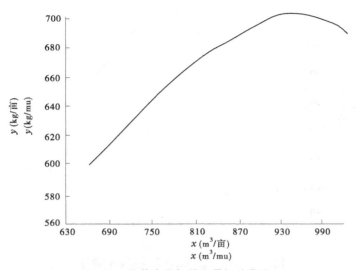

图 32 苜蓿产量与需水量相关曲线

Figure 32   Correlation between Alfalfa Yield and Irrigation Amount

式中 W——玉米产量, kg/亩。

Where ET = the water requirement, m³/mu

W = the corn yield, kg/mu.

由上式可见, 玉米产量和需水量之间的相关曲线呈抛物线形式, 玉米的产量开始阶段随投入水量的增大而增加, 当投入水量达 273.0 m³/亩时, 玉米产量达到顶峰, 而当投入水量超过 273.0 m³/亩时, 玉米产量反而下降(见图 33)。

Above equation shows that the correlation curve of corn yield and water requirement is parabolic. The yield of corn increases with the increase of input volume of water at the beginning. When the volume reaches 273.0 m³/mu, the yield is the highest. As the volume exceeds the value, the yield of corn decreases (see Figure 33).

(四) 甜菜产量与需水量关系分析
2.6.2.4   Analysis of Beet Yield and Water Requirement

根据 1989—1991 年 3 年对甜菜产量和需水量试验结果的分析, 我们进行回归分析, 得出本次试验甜菜产量与需水量之间的函数关系式为:

With the experiment result and regression analysis, a functional equation is established for the correlation between beet yield and water requirement:

$$W = -525\,84.480\,29 + 327.052\,05ET - 0.490\,79ET^2$$

相关系数 $R = 0.946\,51$

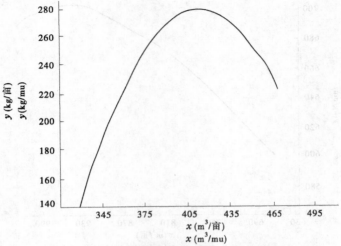

图33 玉米产量与灌水量关系变化曲线

Figure 33 Changing Relation Curve of Corn Yield and Irrigation Amount

Correlation coefficient $R = 0.94651$

式中 　ET——需水量，$m^3/$亩；
　　　W——甜菜产量，kg/亩。

Where 　ET = water requirement, $m^3/$mu
　　　W = beet yield, kg/mu.

甜菜产量与需水量之间关系仍然同上述小麦、苜蓿及玉米的相似，呈幂相关，曲线呈抛物线性形式。当投入水量在到达 391.17 $m^3/$亩以前时，甜菜产量的增加幅度最大，而当投入水量再增大时，产量增加速度减慢，甚至产量开始下降（见图34）。

Similar to wheat, alfalfa and corn, the correlation of beet yield and water requirement is also power‑related, and the curve is parabolic. When the input volume of water is before the point of 391.17 $m^3/$mu, the increasing trend of yield is most obvious. When the volume exceeds the value, the increasing speed slows down and the yield even to decrease (see Figure 34).

（五）作物产量与需水量关系分析研究小结
2.6.2.5 Summary of the Relation Between Crop Yield and Water Requirement

通过上述对春小麦、苜蓿、玉米及甜菜产量与需水量关系的分析，现将结果归纳如下：

Through the above analysis of yield and water requirement of spring wheat, alfalfa, corn and beet, results are summarized as follows:

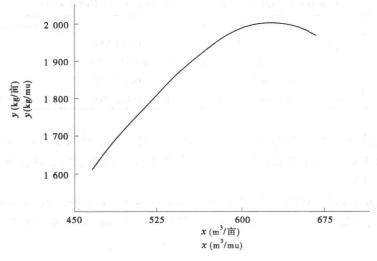

图 34 甜菜产量与灌水量关系变化曲线

Figure 34 Changing Relation Curve of Beet Yield and Irrigation Amount

春小麦、苜蓿、玉米及甜菜的产量与需水量的关系均呈幂相关,二者的相关曲线呈抛物线形式,即不同作物均存在一个临界需水量值,在到达该值前,作物产量与需水量呈线性关系,随水量增大而产量增加;当到达该值时,产量为最高值;当水量再增大时,在耕作条件不改变的情况下,产量反而下降,其函数关系式见表110。

The relationship between yield and water requirement of spring wheat, alfalfa, corn and beet is power – related. The correlation curves are all parabolic, that is, there exists a critical water requirement value for crops. Before reaching this value, the crop yield and the water requirement are linear, and the yield increases with the increase of water volume. When reaching this value, the yield is the highest; when the volume continues to increase, farming conditions keep unchanged, and the yield decreases. The functional equation is shown in Table 110.

表 110   作物产生与需水量之间函数关系式

作物	函数关系式	相关系数 R
春小麦	$ET = -1\,232.552\,74 + 10.024\,82W - 0.017\,85W^2$	0.999 9
苜蓿	$ET = -2\,304.731\,59 + 106.034W - 0.023\,15W^2$	0.926 3
玉米	$W = -3\,057.762\,2 + 24.143\,1ET - 0.043\,74ET^2$	0.936 1
甜菜	$W = -52\,584.480\,29 + 327.052\,05ET - 0.490\,79ET^2$	0.946 5

Table 110　　　　　Functional Equation of Crop Yield and Water Requirement

Crop	Functional Equation	Correlation Coefficient R
Spring Wheat	$ET = -1\,232.552\,74 + 10.024\,82W - 0.017\,85W^2$	0.999 9
Alfalfa	$ET = -2\,304.731\,59 + 106.034W - 0.023\,15W^2$	0.926 3
Corn	$W = -3\,057.762\,2 + 24.143\,1ET - 0.043\,74ET^2$	0.936 1
Beet	$W = -52\,584.480\,29 + 327.052\,05ET - 0.490\,79ET^2$	0.946 5

在本次试验中(1989—1991 年),获得春小麦的最优需水量为 289.14 m³/亩;苜蓿的最优需水量为 540 m³/亩;玉米的最优需水量为 273 m³/亩;甜菜的最优需水量为 391.17 m³/亩(见表 111)。

In the experiment (1989 - 1991), the optimal water requirement of spring wheat is 289.14 m³/mu, alfalfa is 540 m³/mu, corn is 273 m³/mu, and beet is 391.17 m³/mu (see Table 111).

表 111　　　　　作物最优需水量统计表

作物		春小麦	苜蓿	玉米	甜菜
需水量	m³/亩	289.14	540	273	391.17
	mm	433.71	810.0	409.3	586.8

Table 111　　　　　Statistics of Optimal Water Requirement of Crops

Crop		Spring Wheat	Alfalfa	Corn	Beet
Water Requirement	m³/mu	289.14	540	273	391.17
	mm	433.71	810.0	409.3	586.8

三、作物需水系数
2.6.3　Crop Water Requirement Coefficient

作物需水系数是反映作物产量与需水量之间关系的一个重要参数,在进行作物需水量分析和最优方案确定时,常以作物需水系数作为衡量指标和参选条件。需水系数 K 值的求解为:

The crop water requirement coefficient is an important parameter to reflect the relationship between crop yield and water requirement. When doing the crop water requirement analysis and determining the optimal scheme, the crop water requirement coefficient is usually used as a measurement index and selection criteria. The water requirement coefficient (K) is derived from:

$$K = \frac{ET}{W}$$

式中　K——需水系数，m³/kg；
　　　ET——需水量，m³/亩；
　　　W——产量，kg/亩。

Where　K = the water requirement coefficient, m³/kg
　　　　ET = the water requirement, m³/mu
　　　　W = the crop yield, kg/mu.

（一）春小麦需水系数分析
2.6.3.1　Analysis of Spring Wheat Water Requirement Coefficient

分析3个不同水平的小麦需水系数可见（见表112），1989—1991年的3年平均值以1号水平的需水系数最高，达2.48 m³/kg；3号水平的次之，为1.75 m³/kg；2号水平的最低，为1.70 m³/kg，即2号水平产1 kg小麦的耗水量最低。2号水平3个试验年度的平均单产较1号水平和3号水平都高，而需水系数又较1号水平和3号水平的都低，从既考虑作物高产，又省水的效果分析，2号水平为最优试验设计方案。

Through the analysis of the water requirement coefficients of three different levels of wheat (see Table 112), it is found that, based on the 3 - year average from 1989 to 1991, Level #1 has the highest water requirement coefficient of 2.48 m³/kg, followed by Level #3 with 1.75 m³/kg and Level #2 with 1.70 m³/kg, that is to say, Level #2 consumes least water to produce 1 kg wheat. The average per unit yield of Level #2 in 3 experimental years is higher than those of Level #1 and Level #3, and the water requirement coefficient is lower than those of Level #1 and Level #3. From the perspective of high yield with small requirement of water, Level 2 is the optimal experiment design.

表112　　　　　　　　　　　春小麦需水量分析表

	年份	1989年	1990年	1991年	平均
1	产量（kg/亩）	106.03	97.60	259.34	154.32
	需水量（m³/亩）	355.21	278.14	324.13	319.16
	需水系数（m³/kg）	3.35	2.85	1.25	2.48
2	产量（kg/亩）	77.50	117.70	329.73	174.98
	需水量（m³/亩）	312.39	266.71	315.30	289.14
	需水系数（m³/kg）	4.03	2.26	0.96	1.70
3	产量（kg/亩）	74.00	107.90	188.27	123.39
	需水量（m³/亩）	267.31	214.74	166.38	216.10
	需水系数（m³/kg）	3.61	1.99	0.88	1.75

Table 112   **Analysis of Spring Wheat Water Requirement**

	Year	1989	1990	1991	Average
1	Yield (kg/mu)	106.03	97.60	259.34	154.32
1	Water Requirement (m³/mu)	355.21	278.14	324.13	319.16
1	Water Requirement Coefficient (m³/kg)	3.35	2.85	1.25	2.48
2	Yield (kg/mu)	77.50	117.70	329.73	174.98
2	Water Requirement (m³/mu)	312.39	266.71	315.30	289.14
2	Water Requirement Coefficient (m³/kg)	4.03	2.26	0.96	1.70
3	Yield (kg/mu)	74.00	107.90	188.27	123.39
3	Water Requirement (m³/mu)	267.31	214.74	166.38	216.10
3	Water Requirement Coefficient (m³/kg)	3.61	1.99	0.88	1.75

（二）苜蓿需水系数分析

2.6.3.2 Analysis of Water Requirement Coefficient of Alfalfa

由表113可见，苜蓿的需水系数以1号水平最高，为0.849 m³/kg；2号水平次之，为0.805 m³/kg；3号水平的需水系数较低，为0.755 m³/kg。但从3个不同水平的苜蓿产量分析，1号、2号和3号水平的平均产量分别为716.74、669.9 kg/亩和616.1 kg/亩，由苜蓿产量和需水量关系曲线可见，苜蓿的产量在需水量540 m³/亩(810 mm)以前，增加幅度较大，即1 m³水可增加1.44 kg苜蓿；而当投入水量大于540 m³/亩时，产量这时也虽有增加，但幅度很小，1 m³水只产0.56 kg苜蓿，当投入水量达608 m³/亩(911.5 mm)时，苜蓿产量不再增加。因而，从考虑高产又省水的效果分析，2号水平的试验设计方案应为最优方案。

Table 113 shows that the water requirement coefficient of Level #1 is the highest with 0.849 m³/kg, followed by Level #2 with 0.805 m³/kg and Level #3 with 0.755 m³/kg. However, in terms of yield of three different levels, the average yields of Level #1, #2 and #3 are 716.74, 669.9 and 616.1 kg/mu respectively. It can be seen from the correlation curve between yield and water requirement that, before 540 m³/mu (810 mm), the increase of yield is large, that is, increase of 1 m³ water can increase 1.44 kg alfalfa; and when the input water is more than 540 m³/mu, the yield increases in a very small range, 1 m³ water for only 0.56 kg alfalfa. When the input water reaches 608 m³/mu (911.5 mm), the yield stops increasing. Therefore, from the perspective of high yield with water saving, experimental design scheme of Level #2 is the optimum.

表 113　　　　　　　　　　苜蓿需水系数分析表

	年份	1990 年	1991 年	平均
1	产量(kg/亩)	719.18	714.30	716.74
1	需水量(m³/亩)	566.62	649.83	608.23
1	需水系数(m³/kg)	0.788	0.910	0.849
2	产量(kg/亩)	708.37	631.47	669.92
2	需水量(m³/亩)	494.86	585.90	540.38
2	需水系数(m³/kg)	0.699	0.908	0.804
3	产量(kg/亩)	661.79	570.40	616.10
3	需水量(m³/亩)	422.66	496.59	459.63
3	需水系数(m³/kg)	0.64	0.87	0.755

Table 113　　　　　Analysis of Water Requirement Coefficient of Alfalfa

	Year	1990	1991	Average
1	Yield (kg/mu)	719.18	714.30	716.74
1	Water Requirement (m³/mu)	566.62	649.83	608.23
1	Water Requirement Coefficient (m³/kg)	0.788	0.910	0.849
2	Yield (kg/mu)	708.37	631.47	669.92
2	Water Requirement (m³/mu)	494.86	585.90	540.38
2	Water Requirement Coefficient (m³/kg)	0.699	0.908	0.804
3	Yield (kg/mu)	661.79	570.40	616.10
3	Water Requirement (m³/mu)	422.66	496.59	459.63
3	Water Requirement Coefficient (m³/kg)	0.64	0.87	0.755

(三)玉米需水系数分析
2.6.3.3 Analysis of Water Requirement Coefficient of Corn

由表 114 可见,不同年份玉米的需水系数变化很大。比较 1989—1991 年 3 年的作物需水系数,1989 年 2 号、3 号水平的作物需水系数达 2.27~2.36 m³/kg,1 号水平也高达 1.84 m³/kg。1989 年玉米需水系数偏大与自然因素有关(冰雹)。根据玉米产量与需水量的函数关系式可见,玉米的最优需水量为 273 m³/亩(409 mm)时,生产效益最高,玉米的需水系数为 0.98 m³/kg,该值接近试验方案中的 2 号水平。

Table 114 shows that the water requirement coefficient of corn varies greatly over years. Comparing the crop water requirement coefficient during 1989 – 1991, the water requirement coefficient of Level #2 and #3 ranged from 2.27 – 2.36 m³/kg while that of Level #1 was also as high as 1.84 m³/kg in 1989. The water requirement coefficient of corn in 1989 was larger due to a natural factor (the hailstorm). According to the functional relationship between corn yield and water requirement, when the optimal water requirement of corn was 273 m³/mu (409 mm), the production benefit recorded the highest where the water requirement coefficient of corn was 0.98 m³/kg, a value close to Level #2 in the experiment scheme.

表 114　　玉米需水系数分析表

	年份	1989 年	1990 年	1991 年	平均
1	产量(kg/亩)	219.0	350.30	260.30	276.53
1	需水量(m³/亩)	404.44	221.40	293.10	306.17
1	需水系数(m³/kg)	1.84	0.63	1.13	1.20
2	产量(kg/亩)	152.00	242.60	224.10	206.23
2	需水量(m³/亩)	345.70	240.70	310.40	298.93
2	需水系数(m³/kg)	2.27	0.98	1.39	1.55
3	产量(kg/亩)	133.30	171.20	156.70	153.73
3	需水量(m³/亩)	315.00	226.80	228.30	256.70
3	需水系数(m³/kg)	2.36	1.32	1.47	1.72

Table 114 Analysis of Water Requirement Coefficient of Corn

	Year	1989	1990	1991	Average
1	Yield (kg/mu)	219.0	350.30	260.30	276.53
	Water Requirement (m^3/mu)	404.44	221.40	293.10	306.17
	Water Requirement Coefficient (m^3/kg)	1.84	0.63	1.13	1.20
2	Yield (kg/mu)	152.00	242.60	224.10	206.23
	Water Requirement (m^3/mu)	345.70	240.70	310.40	298.93
	Water Requirement Coefficient (m^3/kg)	2.27	0.98	1.39	1.55
3	Yield (kg/mu)	133.30	171.20	156.70	153.73
	Water Requirement (m^3/mu)	315.00	226.80	228.30	256.70
	Water Requirement Coefficient (m^3/kg)	2.36	1.32	1.47	1.72

(四) 甜菜需水系数分析
2.6.3.4 Analysis of Water Requirement Coefficient of Beet

比较本次试验3个不同水平的甜菜产量可见,1号水平的甜菜平均产量最高,达1 757 kg/亩,需水系数为0.21 m^3/kg;3号水平的产量和需水系数最低,分别为1 650.9 kg/亩和0.17 m^3/kg;2号水平的产量和需水系数介于1号水平和3号水平之间。这其中1990年的1号水平产量最高,为1 964 kg/亩,需水系数为0.20 m^3/亩,次为1991年2号水平,产量为1 896.5 kg/亩,需水系数为0.18 m^3/kg。结合甜菜产量和需水量之间关系曲线分析可见,需水量为391.17 m^3/亩(586.8 mm)时,产量达最高峰,需水系数为0.22 m^3/kg,接近试验设计方案中的1号水平(见表115)。

Comparing beet yield in three different levels in the experiment, it was apparent that beet in Level #1 registered the highest average production as 1 757 kg/mu with a water requirement coefficient of 0.21 m^3/kg; while beets in Level #3 showed the lowest yield and water requirement coefficient of 1 650.9 kg/mu and 0.17 m^3/kg respectively and the yield and the water requirement coefficient of beet in Level #2 fell in between Level #1 and 3. Among them, the highest yield, 1 964 kg/mu, came from beet in Level #1 in 1990 with the water requirement coefficient of 0.20 m³/mu, the second was 1 896.5 kg/mu for Level #2 in 1991 with the water requirement coefficient of 0.18 m³/kg. According to the analysis of the relation curve of beet yield and water requirement, it can be seen that when the water requirement was 391.17 m³/mu (586.8 mm), the yield reached the peak and the water requirement coefficient was 0.22 m³/kg, which approximated to that of beet in Level #1 in this experiment (see Table 115).

Table 115 Analysis of Water Requirement Coefficient of Beet

	Year	1989	1990	1991	Average
1	Yield (kg/mu)	1793.0	1964.0	1514.0	1757.0
1	Water Requirement (m³/mu)	317.67	385.13	425.0	375.93
1	Water Requirement Coefficient (m³/kg)	0.18	0.20	0.28	0.21
2	Yield (kg/mu)	1658.0	1551.3	1896.5	1701.9
2	Water Requirement (m³/mu)	256.67	359.67	332.0	316.11
2	Water Requirement Coefficient (m³/kg)	0.15	0.23	0.18	0.19
3	Yield (kg/mu)	1 632.7	1 722.0	1 598.0	1 650.9
3	Water Requirement (m³/mu)	214.0	315.67	290.0	273.22
3	Water Requirement Coefficient (m³/kg)	0.13	0.18	0.18	0.17

（五）作物需水系数分析研究小结
2.6.3.5 Summary of Analysis Research of Crop Water Requirement Coefficient

通过对春小麦、苜蓿、玉米及甜菜的需水系数的试验研究，现就主要结果归纳如下：

Through the study on the water requirement coefficient of spring wheat, alfalfa, corn and beet, main results are summarized as follows：

春小麦的最优需水量为 289.14 m^3/亩，需水系数为 0.17 m^3/kg，选择试验方案为 2 号水平。

When choosing Level #2, the optimal water requirement of spring wheat was 289.14 m^3/mu and its water requirement coefficient was 0.17 m^3/kg.

苜蓿的最优需水量为 540 m^3/亩，需水系数为 0.804 m^3/kg，选择试验方案为 2 号水平。

The optimal water requirement of alfalfa was 540 m^3/mu and its water requirement coefficient was 0.804 m^3/kg, so the scheme of Level #2 was selected.

玉米的最优需水量为 273 m^3/亩，需水系数为 0.98 m^3/kg，接近试验方案的 2 号水平。

The optimal water requirement of corn was 273 m^3/mu and its water requirement coefficient was 0.98 m^3/kg, so the scheme of Level #2 was selected.

甜菜的最优需水量为 391.17 m^3/亩，需水系数为 0.22 m^3/kg，选择试验方案为 1 号水平（见表 116）。

The optimal water requirement of beet was 391.17 m^3/mu and its water requirement coefficient was 0.22 m^3/kg, so the scheme of Level #1 was selected (see Table 116).

表 116　　　　　　　　作物需水系数分析统计表

作物		春小麦	苜蓿	玉米	甜菜
需水量	m^3/亩	289.14	540.0	273	391.17
	mm	433.71	810.0	409.3	586.8
需水系数(m^3/kg)		0.17	0.804	0.98	0.22
选择水平		2	2	2	1

Table 116 Statistics of Crop Water Requirement Coefficient

Crop		Spring Wheat	Alfalfa	Corn	Beet
Water Requirement	m³/mu	289.14	540.0	273	391.17
	mm	433.71	810.0	409.3	586.8
Water Requirement Coefficient (m³/kg)		0.17	0.804	0.98	0.22
Selected Level		2	2	2	1

四、作物产量的方差分析
2.6.4 Analysis of Variance of Crop Yield

方差分析是对两个以上样本间的差异程度进行检验的一种统计方法。一组试验资料方差的大小表明变异的程度。我们通过对作物产量的方差分析,进一步探讨作物不同水平与重复间的差异。

The analysis of variance is a statistical method to test the degree of difference between two or more samples. The value of the variance of a set of experimental data indicates the degree of variation. Through the analysis of variance on crop yield, we come to further explore the differences between crops' different levels and repetitions.

(一)方差分析原理
2.6.4.1 Principle of Variance Analysis

一个样本的方差就是平方和除以自由度。进行方差分析首先要将方差分解成几个部分,而分解方差就需分解总平方和与自由度。

The variance of a sample is the sum of squares divided by degree of freedom. The variance analysis firstly divides the variance into several parts, while dividing the variance requires the division of the total sum of squares and degree of freedom.

总平方和:
Total sum of squares:

$$SS_T = \sum_1^{nR} (X_{Rn} - \overline{X})^2 = \sum_1^{nR} X^2 - C$$

矫正数(C):

$$C = \frac{(\sum X)^2}{nR} = \frac{T^2}{nR}$$

Correction(C):
$$C = \frac{(\sum X)^2}{nR} = \frac{T^2}{nR}$$

总的变异是 nR 个试验数的变异,所以,
The total variation is the variation of nR test numbers, so

自由度(U):
Degree of freedom (U):

$$U = R - 1$$

总平方和(SS_T)可分为处理间平方和(SS_t)和组间(重复)平方和(SS_e)。
The total sum of squares (SS_T) is composed of the sum of square within treatments (SS_T) and groups (repetitions) (SS_E).

处理(水平)间平方和(SS_t):
The sum of squares within treatments (levels) (SS_T):

$$SS_t = n \sum_1^R (\overline{X}_R - \overline{X})^2 = \frac{\sum_1^R T_R^2 - C}{n}$$

处理间的变异是每重复平均数 R 的变异,故自由度 $V = R - 1$。
The variation within treatments is also the variation of the average number R in repetitions, so the degree of freedom $V = R - 1$.

组间(重复)平方和 SS_e:
The sum of squares within groups (repetitions) SS_E:

$$SS_E = \sum_1^R [\sum_1^n (X_{Rn} - \overline{X}_R)^2 = SS_T - SS_t]$$

组间(重复)的变异是由每组内几个观测值引起的,每组自由度 $n-1$,但是共有 R 组,所以组间(重复)自由度为 $(n-1)R$。
The variation between groups (repetitions) came from several observed values in each group, and the degree of freedom of each group was $n - 1$ and there were totally R groups. Therefore the degree of freedom within groups (repetitions) turned out to be $(n-1)R$.

总自由度为处理(水平)间自由度与组间(重复)自由度之和:
And the total degree of freedom equaled the sum of degree of freedom within treatments (levels) and groups (repetitions):

$$(nR - l) = (R - l) + R(n - l)$$

总平方和为处理(水平)间平方和与组间(重复)平方和之和：
The total sum of squares equaled the sum of sums of square within treatments (levels) and sums of square within groups (repetitions):

$$\sum_1^{nR} (X_{Rn} - \overline{X})^2 = n \sum_1^R (\overline{X}_R + \overline{X})^2 + \sum_1^R \left[\sum_1^n (X_{Rn} - \overline{X}_R)^2 \right]$$

总均方(S_T^2):
(S_T^2) Total mean square:

$$S_T^2 = \frac{\sum (X - \overline{X})^2}{nR - 1}$$

处理(水平)间均方(S_t^2):
(S_t^2) Mean square within treatments (levels):

$$S_t^2 = \frac{n \sum (\overline{X}_R - \overline{X})^2}{R - 1}$$

组内(重复)均为(S_e^2):
(S_e^2) Mean square within groups (repetitions):

$$S_e^2 = \frac{\sum (X - \overline{X}_R)^2}{R(n - 1)}$$

F 为两个样本的均方 S_1^2 与 S_2^2 之比:
F is the ratio of mean squares S_1^2 and S_2^2 of two samples:

$$F = \frac{S_1^2}{S_2^2}$$

F 受两个样本均方 S_1^2 和 S_2^2 的两个自由度 U_1 和 U_2 的影响。在 F 假设测验中，S_1^2 是平方差大者，S_2^2 是平方差小者(一般为误差均方)。所以计算的 F 值都大于1；当 F 小于1时，即为否定假设。

第二章 作物需水量与需水规律
2 Water Requirement of Crops and Its Law

F is under the influence of S_1^2 and S_2^2 two degrees of freedom U_1 and U_2. In hypothesis testing about F, S_1^2 refers to the larger square error while S_2^2 is the smaller one (usually error mean square). So the calculated F value is greater than 1. When F is less than 1, it is a negative assumption.

依据上述原理,现将春小麦、苜蓿、玉米及甜菜的产量方差分析及 F 检验结果分述如下。

In terms of the principle above, the analysis of the variance of yield and the experiment results of F with spring wheat, alfalfa, corn and beet are described as follows.

(二)小麦产量的方差分析及 F 检验
2.6.4.2 Variance Analysis of Wheat Yield and its F Value Checks

小麦是根据 3 个水平、3 个重复设置,其产量方差分析结果见表 117。

On the basis of 3 levels and 3 repetitions, the analysis result of wheat yield variance is demonstrated in Table 117.

表 117　　春小麦产量方差分析计算表

重复	区组			T_r
	I	II	III	
1	274.0	315.8	188.3	778.1
2	336.0	423.8	229.3	989.2
3	144.9	149.6	270.1	564.6
T_t	754.9	889.3	687.7	2331.9

Table 117　　Calculation for Variance Analysis of Spring Wheat Yield

Repetition	Zone			T_r
	I	II	III	
1	274.0	315.8	188.3	778.1
2	336.0	423.8	229.3	989.2
3	144.9	149.6	270.1	564.6
T_t	754.9	889.3	687.7	2331.9

总自由度:$f_{总} = 2 + 3 \times (3 - 1) = 8$

Total degree of freedom: $f_{total} = 2 + 3 \times (3 - 1) = 8$

处理间自由度：$f_{处} = R - 1 = 3 - 1 = 2$

Degree of freedom within treatments: $f_{treatments} = R - 1 = 3 - 1 = 2$

组内自由度：$f_{组} = n - 1 = 3 - 1 = 2$

Degree of freedom in the group: $f_{group} = n - 1 = 3 - 1 = 2$

矫正系数：$C = \dfrac{T^2}{nR} = \dfrac{2\,331.9^2}{3 \times 3} = 604\,195.3$

Compensation coefficient: $C = \dfrac{T^2}{nR} = \dfrac{2\,331.9^2}{3 \times 3} = 604\,195.3$

总平方和：$SS_T = \sum X^2 - C = 67\,563.1$

Total sum of squares: $SS_T = \sum X^2 - C = 67\,563.1$

水平间平方和：$SS_t = \sum T_r^2/R - C = 634\,243.1 - 604\,195.3 = 30\,047.8$

Sum of squares within levels: $SS_t = \sum T_r^2/R - C = 634\,243.1 - 604\,195.3 = 30\,047.8$

重复间平方和：$SS_e = \sum T_t^2/n - C = 7\,024.63$

Sum of squares within repetitions: $SS_e = \sum T_t^2/n - C = 7\,024.63$

$$SS_{误} = SS_T - SS_t - SS_e = 30\,490.67$$

水平间均方：$SS_t^2 = \dfrac{30\,047.8}{3 - 1} = 15\,023.9$

Mean square within levels: $SS_t^2 = \dfrac{30\,047.8}{3 - 1} = 15\,023.9$

重复间均方：$SS_e^2 = \dfrac{7\,024.63}{3 \times (3 - 1)} = 1\,170.77$

Mean square within repetitions: $SS_e^2 = \dfrac{7\,024.63}{3 \times (3 - 1)} = 1\,170.77$

表118 列出了小麦产量方差分析结果，由表118 可见，$F < F_{0.05}$，说明水平、重复间对试验结果均无显著差异。

Table 118 outlines the variance analysis results of wheat yield. As is shown in Table 118, $F < F_{0.05}$, indicating that the sum of squares within levels or repetitions has no significant influence on experimental results.

表 118　　　　　　　　春小麦产量方差分析特征值表

变异源	自由变 DF	平方和 SS	均方 Ms	F	$F_{0.05}$	$F_{0.05}$
重复间 SS_e	2	7 024.63	3 512.32	0.46	6.94	18.0
水平间 SS_t	2	30 047.8	15 023.9	1.97	6.94	18.0
误差	2	30 490.67	7 622.7			
总变异	8	67 563.1				

Table 118　　　Eigenvalues for Variance Analysis of Spring Wheat Yield

Variance Rources	Free Variables DF	Sum of Squares SS	Mean Squares Ms	F	$F_{0.05}$	$F_{0.05}$
Within Repetitions S_e	2	7 024.63	3 512.32	0.46	6.94	18.0
Within Levels SS_t	2	30 047.8	15 023.9	1.97	6.94	18.0
Error	2	30 490.67	7 622.7			
Total Variance	8	67 563.1				

(二)苜蓿产量的方差分析及 F 检验
2.6.4.2 Variance Analysis of Alfalfa Yield and its F Value Checks

苜蓿产量的方差分析结果及 F 检验见表 119 和表 120。

Variance analysis results of alfalfa yield and its F value checks are outlined in Table 119 and Table 120.

总自由度：$f_{总} = 3 \times 3 - 1 = 8$

Total degree of freedom $f_{total} = 3 \times 3 - 1 = 8$

处理间自由度：$f_{处} = R - 1 = 3 - 1 = 2$

Degree of freedom within treatments $f_{treatments} = R - 1 = 3 - 1 = 2$

组间自由度：$f_{组} = n - 1 = 3 - 1 = 2$

Degree of freedom within groups $f_{groups} = n - 1 = 3 - 1 = 2$

矫正系数：$C = \dfrac{\sum T^2}{nR} = \dfrac{5\,748.5^2}{3 \times 3} = 3\,671\,694.7$

Compensation coefficient：$C = \dfrac{\sum T^2}{nR} = \dfrac{5\,748.5^2}{3 \times 3} = 3\,671\,694.7$

总平方和：$S_{总} = \sum_{1}^{nR} X^2 - C = 3\,713\,851.75 - 3\,671\,694 = 42\,157.75$

Total sum of squares: $S_{total} = \sum_{1}^{nR} X^2 - C = 3\,713\,851.75 - 3\,671\,694 = 42\,157.75$

表119　　　　　　　　　　　　苜蓿产量方差分析表

重复	区组			T_t	T_r
	I	II	III		
1	727.8	727.8	687.3	2 142.9	714.30
2	672.9	571.3	650.2	1 894.4	631.47
3	616.8	567.8	526.6	1 711.2	570.4
T_r	2 017.5	1 866.9	1 864.1	5 748.5	
$\overline{X}_r$	672.5	622.3	621.37		638.72

Table 119　　　　　　　　Variance Analysis of Alfalfa Yield

Repetition	Zone			T_t	T_r
	I	II	III		
1	727.8	727.8	687.3	2 142.9	714.30
2	672.9	571.3	650.2	1 894.4	631.47
3	616.8	567.8	526.6	1 711.2	570.4
T_r	2 017.5	1 866.9	1 864.1	5 748.5	
$\overline{X}_r$	672.5	622.3	621.37		638.72

处理间平方和：$S_{处} = \dfrac{\sum T^2}{K} - C = 3\,702\,992.4 - 3\,671\,694 = 31\,298.4$

Sum of squares within treatments: $S_{treatments} = \dfrac{\sum T^2}{K} - C = 3\,702\,992.4 - 3\,671\,694 = 31\,298.4$

重复间平方和：$S_{重} = \dfrac{\sum T\gamma^2}{n} - C = 3\,676\,830.22 - 3\,671\,694 = 5\,136.22$

Sum of squares within repetitions: $S_{repetitions} = \dfrac{\sum T\gamma^2}{n} - C = 3\,676\,830.22 - 3\,671\,694 = 5\,136.22$

误差平方和：$S_{误} = S_{总} - S_{处} - S_{重} = 5\ 723.13$

Error's sum of squares：$S_{error} = S_{total} - S_{treatments} - S_{repetitions} = 5\ 723.13$

表120　　　　　　　　　　苜蓿产量方差分析特征值表

变异源	自由度 f	平方和 SS	均方 Ms	F	$F_{0.051\ 1}$	$F_{0.01}$
水平	2	31 298.4	15 649.2	1.53	6.94	18.00
重复	2	5 136.22	2 568.11	1.79	6.94	18.00
误差	4	5 723.13	1 430.78			
总和	8	42 157.75	5 269.27			

Table 120　　　　　　　　Eigenvalues for Variance Analysis of Alfalfa Yield

Variance sources	Degree of freedom f	Sum of Squares SS	Mean Squares Ms	F	$F_{0.0511}$	$F_{0.01}$
Level	2	31 298.4	15 649.2	1.53	6.94	18.00
Repetition	2	5 136.22	2 568.11	1.79	6.94	18.00
Error	4	5 723.13	1 430.78			
Total	8	42 157.75	5 269.27			

由方差分析得出结论：不同的需水量对苜蓿产量无明显影响。

Conclusion can be drawn from the variance analysis that different water requirements have no significant effect on alfalfa yield.

但有个明显规律：设计的土壤水分下限值越高，需水量越大，产量也越高，产量和需水量的曲线呈直线相关。

However, there are obvious laws that the higher the lower limit value of soil water, the higher the water requirement and the yield, and the yield and water requirement curve had a linear correlation.

（三）玉米产量分析及 F 检验
2.6.4.3 Variance Analysis of Corn Yield and its F Value Checks

玉米产量方差分析计算及 F 检验分析结果见表 121 和表 122。
Variance analysis results of corn yield and its F value checks are outlined in Table 121 and Table 122.

表 121　　　　　　　　　　玉米产量方差分析表

重复	I	II	III	T_t
1	720.2	581.4	459.8	1 761.4
2	549.2	285.5	510.0	1 344.7
3	518.3	313.8	108.3	940.4
T_r	1 787.7	1 180.7	1 078.1	4 046.5

Table 121　　　　　　　Variance Analysis of Corn Yield

Repetition	I	II	III	T_t
1	720.2	581.4	459.8	1 761.4
2	549.2	285.5	510.0	1 344.7
3	518.3	313.8	108.3	940.4
T_r	1 787.7	1 180.7	1 078.1	4 046.5

表 122　　　　　　　　玉米产量方差分析特征值表

方差来源	平方和 SS	自由度 f	均方 Ms	F	$F_{0.05}$	$F_{0.01}$
区组间	98 056.4	2	49 028.2	3.24	6.94	18.00
水平间	112 348.7	2	56 174.4	3.72		
误差	60 438.7	4	15 109.7			
总和	271 843.8	8				

Table 122　　　　　Eigenvalues for Variance Analysis of Corn Yield

Variance Sources	Sum of Squares SS	Degree of Freedom f	Mean Squares Ms	F	$F_{0.05}$	$F_{0.01}$
Within Zones	98 056.4	2	49 028.2	3.24	6.94	18.00
Within Levels	112 348.7	2	56 174.4	3.72		
Error	60 438.7	4	15 109.7			
Total	271 843.8	8				

自由度分解：总自由度：$f = 2 + 3 \times (3 - 1) = 8$
区组自由度：$f = n - 1 = 3 - 1 = 2$
处理间自由度：$f = R - l = 2$
误差自由度：$f = 4$

Division of degree of freedom: Total degree of freedom: $f = 2 + 3 \times (3 - 1) = 8$
Degree of freedom in zones: $f = n - 1 = 3 - 1 = 2$
Degree of freedom within treatments: $f = R - 1 = 2$
Error's degree of freedom: $f = 4$

平方和分解：矫正数：$C = \dfrac{T^2}{nR} = \dfrac{4\,046.5^2}{9} = 1\,819\,351.4$
总平方和：$S_{总} = 2\,090\,195.2 - 1\,819\,351.4 = 270\,843.8$
区组平方和：$S_\gamma = 98\,056.4$
水平平方和：$S_t = 112\,348.7$
误差平方和：$S_{误} = S_{总} - (S\gamma + St) = 60\,438.7$

Division of sum of squares: Correction: $C = \dfrac{T^2}{nR} = \dfrac{4\,046.5^2}{9} = 1\,819\,351.4$
Total sum of squares: $S_{\text{total}} = 2\,090\,195.2 - 1\,819\,351.4 = 270\,843.8$
Sum squares in zones: $S_\gamma = 98\,056.4$
Sum of squares within levels: $S_t = 112\,348.7$
Error's sum of squares: $S_{\text{error}} = S_{\text{total}} - (S_\gamma + S_t) = 60\,438.7$

表 123 列出了玉米产量方差分析及 F 检验结果。$f_水 < F_{0.05}$，并且 $f_重 < F_{0.05}$，表明水平和重复间都没有显著差异。

Table 123 outlines the variance analysis results of corn yield. As $F_{\text{levels}} < F_{0.05}$ and $F_{\text{repetitions}} < F_{0.05}$, it indicates that the sum of squares between levels or repetitions have no significant influence on experiment results.

表 123　　　　　　　　　甜菜产量方差分析计算表

重复	重复			T_r
	I	II	III	
1	2 787	2 500	1 113	6 400
2	2 680	1 343	2 450	6 473
3	1 915	1 600	1 280	4 795
T_t	7 382	1 443	4 843	17 668

Table 123　　　　**Calculation of Variance Analysis of Beet Yield**

Repetition	Repetition			T_r
	I	II	III	
1	2 787	2 500	1 113	6 400
2	2 680	1 343	2 450	6 473
3	1 915	1 600	1 280	4 795
T_t	7 382	1 443	4 843	17 668

（四）甜菜产量的方差分析及 F 检验

2.6.4.4 Variance Analysis of Beet Yield and its F Value Checks

在对甜菜进行方差分析及 F 检验时，我们仅对甜菜产量做了分析及检验，而未对甜菜含糖量进行分析和检验。

When conducting variance analysis of beet and its F value checks, we only made analysis and checks on beet yield but its sugar content.

由计算结果表 123 和表 124 可见，甜菜水平及重复向均无显著差异。

As shown by the calculated results in Table 123 and Table 124, the beet's sum of squares between levels or repetitions has no significant difference on experimental results.

自由度分解：总自由度：$f_{总} = nK - 1 = 3 \times 3 - 1 = 8$

区组自由度：$f_{区} = 2$

处理间自由度：$f_{处} = 2$

误差自由度：$f = 8 - 2 - 2 = 4$

Division of degree of freedom: Total degree of freedom: $f_{total} = nK - 1 = 3 \times 3 - 1 = 8$

Degree of freedom in zones: $f_{zones} = 2$

Degree of freedom within treatments: $f_{treatments} = 2$

Error's degree of freedom: $f = 8 - 2 - 2 = 4$

平方和的分解：矫正数：$C = \dfrac{T^2}{nR} = \dfrac{17\ 668^2}{9} = 34\ 684\ 247.11$

总平方和：$S_{总} = \sum X^2 - C = 38\ 114\ 143 - 34\ 684\ 247.11 = 3\ 429\ 896$

水平间平方和：$S_{总} = \sum T_\gamma^2/\gamma - C = 599\ 671$

重复间平方和：$S_{重} = \sum T_t^2/t - C = 1\ 174\ 027$

$S_{误} = 1\ 656\ 198$

Division of sum of squares: Correction $C = \dfrac{T^2}{nR} = \dfrac{17\ 668^2}{9} = 34\ 684\ 247.11$

Total sum of squares: $S_{\text{total}} = \sum X^2 - C = 38\ 114\ 143 - 34\ 684\ 247.11 = 3\ 429\ 896$

Sum of squares within levels: $S_{\text{total}} = \sum T_\gamma^2/\gamma - C = 599\ 671$

Sum of squares within repetitions: $S_{\text{repetitions}} = \sum T_t^2/t - C = 1\ 174\ 027$

$S_{\text{error}} = 1\ 656\ 198$

表 124　　　　甜菜方差分析及 F 检验特征值表

变异源	平方和 SS	自由度 f	均方 Ms	F	$F_{0.05}$	$F_{0.01}$
重复间	1 174 027	2	587 013.5	1.42	6.94	18.00
水平间	599 671	2	299 835.5	0.72	19.25	
误差	1 656 198	4	414 049.5			
总和	3 429 896					

Table 124　　　Eigenvalues of Variance Analysis of Beet Yield and F Value Checks

Variance sources	Sum of Squares SS	Degree of Freedom f	Mean Squares Ms	F	$F_{0.05}$	$F_{0.01}$
Within Repetitions	1 174 027	2	587 013.5	1.42	6.94	18.00
Within Levels	599 671	2	299 835.5	0.72	19.25	
Error	1 656 198	4	414 049.5			
Total	3 429 896					

通过对春小麦、苜蓿、玉米及甜菜产量的方差分析和 F 检验，其结果表明，在本次试验中，各类作物的水平及重复间均无显著差异。

Through variance analysis and F value checks on spring wheat, alfalfa, corn and beet, the results showed that regardless of crops, the sum of squares within levels or repetitions had no significant difference on experimental results in this experiment.

第三章 作物灌溉制度研究

3 Study on Irrigation System of Crops

第一节 春小麦灌溉制度与灌溉效益分摊系数研究

3.1 Irrigation System of Spring Wheat and Its Irrigation Benefit

一、试验区概况
3.1 Overview of Experimental Site

(一) 试验条件
3.1.1.1 Experiment Condition

小麦试验区面积 4.5 亩,加保护区近 5 亩。位于整个试验区的西南面,灌水条件较好。试验选种 3 个品种,即阿春 3 号、新春 2 号和昌春 3 号,都是阿勒泰地区主要的小麦种植品种。

The Wheat Plot covered an area of 4.5 mu and it was nearly 5 mu including protection area. The plot was located in the southwest of the whole experimental site with good irrigation conditions. The three selected varieties, namely Achun #3, Xinchun #2 and Changchun #3, were all major wheat varieties in Altay Prefecture.

(二) 田间基本参数
3.1.1.2 Field Basic Parameter

试验小区土质为沙壤土,无地下水补给。土壤容重 1.58 g/cm³,比重 2.67 g/cm³,田间持水率 10.491%,有机质含量 0.46%,含氮 0.037%,速效磷 2.55 mg/kg,pH = 8.3,土壤偏碱性,缺磷少氮。

The soil in the Plots featured sandy loam without groundwater recharge. With soil bulk density of 1.58 g/cm³, soil specific gravity of 2.67 g/cm³, field capacity of 10.491%, organic content of 0.46%, nitrogen content of 0.037%, available P of 2.55 mg/kg, pH = 8.3, the alkaline soil was lack of phosphorus and nitrogen.

二、试验方案和方法
3.1.2 Experiment Scheme and Methods

该试验方案设计的主导思想是以控制土壤含水率水平为手段,并设计3个水平的施肥条件和种子品种,通过正交试验研究春小麦的优化灌溉制度和灌溉效益分摊系数的研究。

The main idea of the experiment scheme was to control the soil water content level, arrange fertilization conditions in three levels and seed varieties and to study the spring wheat's optimal irrigation scheduling and the sharing coefficient of irrigation benefit by orthogonal experiment.

(一)试验因素和水平
3.1.2.1 Experimental Factors and Levels

试验课题期限为3年,即1989—1990年,采用3因素3水平的正交试验。3因素为:土壤含水率、施肥和作物品种;3水平即施肥量(底肥、羊粪、春耕前施,提高土壤肥力),2 000 kg/亩,3 000 kg/亩,4 000 kg/亩;土壤含水率。控制水平(占田间持水率的百分比,50%,60%,70%);3品种:新春2号、阿春3号和昌春3号。

The duration of the experimental subject was 3 years, namely from 1989 to 1990, using orthogonal experiment with 3 factors and 3 levels. The 3 factors were soil water content, fertilization and crop varieties. The 3 levels were the amount of fertilizer (base fertilizer and sheep manure used before spring plowing to improve soil fertility) of 2 000 kg/mu, 3 000 kg/mu and 4 000 kg/mu; soil water content (controlled at 50%, 60% and 70% of the field capacity); 3 varieties, that is, Xinchun #2, Achun #3 and Changchun #3.

(二)试验方案
3.1.2.2 Experiment Scheme

由于该试验因素和水平都达到3个,因此试验方案采用正交试验即采用 $L_k(m^j)$ 正交试验表。其中 K 代表处理数, m 代表水平, j 代表因素,在本试验中具体表示为 $L_p(3^3)$,即 P 处理,3水平,3因素,其排列见表125。

Since experiment factors and levels were up to 3, hence orthogonal test was adopted, namely $L_k(m^j)$ orthogonal test table. Among them, K represented item number, m for levels and j for factors, which were specified as $L_p(3^3)$ in this experiment, namely P for treatments, 3 levels and 3 factors. Table 125 outlines their arrangement.

表 125　　　　　　　　　　　1989 年—1991 年春小麦正交试验表

因素	A 品种		B 施肥量(kg/亩)		C 土壤含水量(%)	
1	1	阿春 3 号	1	2 000	1	50
2	1	阿春 3 号	2	3 000	2	60
3	1	阿春 3 号	3	4 000	3	70
4	2	新春 2 号	1	2 000	2	60
5	2	新春 2 号	2	3 000	3	70
6	2	新春 2 号	3	4 000	1	50
7	3	昌春 3 号	1	2 000	3	70
8	3	昌春 3 号	2	3 000	1	50
9	3	昌春 3 号	3	4 000	2	60

Table 125　　　　　**Orthogonal Test of Spring Wheat** (1989 – 1991)

Factor	A Varieties		B Amount of Fertilizer (kg/mu)		C Soil Water Content (%)	
1	1	Achun #3	1	2 000	1	50
2	1	Achun #3	2	3 000	2	60
3	1	Achun #3	3	4 000	3	70
4	2	Xinchun #2	1	2 000	2	60
5	2	Xinchun #2	2	3 000	3	70
6	2	Xinchun #2	3	4 000	1	50
7	3	Changchun #3	1	2 000	3	70
8	3	Changchun #3	2	3 000	1	50
9	3	Changchun #3	3	4 000	2	60

（三）试验方法

3.1.2.3　Experimental Methods

将 3 个因素,3 个水平加 3 个对照共 30 个试验小区,分成 3 个重复,每个重复有 9 个处理。每个试验小区面积控制在 0.15 亩,处理号是随机排列,处理排布是西北走向,保持

第三章 作物灌溉制度研究
3 Study on Irrigation System of Crops

同一重复间土壤的土质一样,并排设置3个重复,在试验小区种植了必要的保护区。每一重复都有单独的引水设施,并有量水堰计水量,秒表计时,再根据量水堰的流量公式计算并控制每次的每一处理的灌水量。灌水方式采用畦灌,根据取土样测得的当时土壤含水率值与试验设计要求的田间持水率相比较确定是否灌水和灌水量。根据试验规范,在播前和秋收后,做一个100 cm的取土,中间每隔5 d取土一次,取土深度根据小麦的生育阶段来确定。

According to 3 factors, 3 levels and 3 contrasts, these 30 experimental plots are divided into 3 repetitions with 9 treatments each. Each experimental plot's area was limited in 0.15 mu. The treatment number was random and arranged in north-west trend. The soil in a group remained the same and every 3 repetitions were laid side by side. And necessary protection area had been made available in the experimental plots. Every repetition was equipped with separate water diversion facilities, measuring weir and stopwatch, calculating and controlling the amount of irrigation for each group according to the flow formula of the measuring weir. Using border irrigation; compared the soil water content measured from sample soil then and the field capacity required by this experimental design to identify whether irrigation was necessary and irrigation amount. According to the experimental specification, a 100 cm × 100 cm is worked out soil sample plot respectively before sowing and after autumn harvest, sampled the soil every five days during this period and the depth of sample soil was determined by the wheat growth and development period.

三、耗水量分析
3.1.3 Analysis of Water Consumption

(一)春小麦生育期各旬耗水量
3.1.3.1 Water Consumption of Spring Wheat in Every Ten Days of the Growth and Development Period

春小麦生育期各旬耗水量大体呈现两头小中间大的趋势,这与小麦的生长特征密切相关,6月中旬,春小麦正处于灌浆期,这是小麦生长的关键阶段,需要大量的水进行有机物的合成,表现出在这个阶段的耗水量较大,5月中下旬处于拔节—抽穗阶段,需水量也较大,到4月中下旬,耗水量较低,所以,春小麦生育期各旬的耗水量呈波峰状,6月中、下旬达到最大值(见表126)。

In general, the water consumption of wheat in every ten days of the growth and development period took on an oval-shaped tendency, which started and ended with a small amount but swelled in the middle, showing a close correlation with the wheat growth characteristics. In the middle of June, spring wheat was in the filling period which was a key stage for wheat growth and needed plentiful water to carry on the synthesis of organics, thus requiring a large amount of water at this stage. The wheat was in the jointing-heading stage in the middle of May, the water requirement was still high while in the last twenty days of April, it remained quite low. Therefore, the water consumption of wheat in every ten days of the growth and development period was a wave-crest-like figure which peaked at mid-to-late of June (see Table 126).

表 126　　　　　　　　春小麦生育期各旬耗水量统计表　　　　　　　　单位：m³/亩

时段	阿春3号			新春2号			昌春3号		
	50%	60%	70%	50%	60%	70%	50%	60%	70%
10/4 – 20/4	26.87	22.50	21.34	18.23	25.47	19.40	26.52	21.14	21.86
21/4 – 30/4	25.70	17.34	26.51	16.43	20.06	21.60	20.73	21.16	23.99
1/5 – 10/5	23.51	21.34	30.21	15.78	18.44	16.15	13.51	13.32	10.99
11/5 – 20/5	21.55	25.36	15.48	19.34	18.52	15.36	14.94	22.04	25.47
21/5 – 31/5	18.09	19.20	17.01	19.70	17.86	14.76	15.11	17.38	21.07
1/6 – 10/6	38.98	48.15	23.81	31.68	21.67	17.90	37.97	32.74	47.96
11/6 – 20/6	36.29	47.26	43.33	36.00	29.94	38.58	27.06	32.01	25.23
21/6 – 30/6	44.83	44.61	20.46	38.80	47.05	40.98	42.67	22.87	38.20
1/7 – 10/7		35.96	21.07	21.07	16.60	27.44	18.34	13.24	20.78

Table 126　　Statistics of Water Consumption of Spring Wheat in Every Ten Days of the Growth and Development Period　　Unit: m³/mu

Period	Achun #3			Xinchun #2			Changchun #3		
	50%	60%	70%	50%	60%	70%	50%	60%	70%
10/4 – 20/4	26.87	22.50	21.34	18.23	25.47	19.40	26.52	21.14	21.86
21/4 – 30/4	25.70	17.34	26.51	16.43	20.06	21.60	20.73	21.16	23.99
1/5 – 10/5	23.51	21.34	30.21	15.78	18.44	16.15	13.51	13.32	10.99
11/5 – 20/5	21.55	25.36	15.48	19.34	18.52	15.36	14.94	22.04	25.47
21/5 – 31/5	18.09	19.20	17.01	19.70	17.86	14.76	15.11	17.38	21.07
1/6 – 10/6	38.98	48.15	23.81	31.68	21.67	17.90	37.97	32.74	47.96
11/6 – 20/6	36.29	47.26	43.33	36.00	29.94	38.58	27.06	32.01	25.23
21/6 – 30/6	44.83	44.61	20.46	38.80	47.05	40.98	42.67	22.87	38.20
1/7 – 10/7		35.96	21.07	21.07	16.60	27.44	18.34	13.24	20.78

（二）春小麦各生育阶段耗水量

3.2　Water Consumption of Spring Wheat in Every Growth and Development Period

从表 127 可看出，前期播前灌水量较大，这是因为，本地区年降水量小，春季多风少

3 Study on Irrigation System of Crops

雨,试验小区属沙性土质,持水性差,播前墒度差。为有利于出苗,进行播前灌而且灌水量较大,以利于小麦顺利出苗。纵观整个生育阶段,耗水量峰值出现在孕穗—抽穗—灌浆期,这是小麦生长的关键阶段。

As shown in Table 127, the irrigation amount was quite large in the early period for annual precipitation was low in this region, with windy and rainless weather in spring as well as the sandy soil mass in experimental plot, which bound water poorly and showed low soil water content. Before sowing, abundant water is irrigated to help the wheat sprout smoothly. Throughout the growth and development period, the water consumption of wheat peaked at booting-heading-filling period which was a critical period for wheat growth.

表 127　　　　　　　春小麦各生育阶段耗水量统计表　　　　　单位:m³/亩

生育阶段	阿春3号			新春2号			昌春3号		
	50%	60%	70%	50%	60%	70%	50%	60%	70%
播前灌	43.76	31.36	36.21	27.83	33.78	32.88	40.88	35.37	37.01
播种—出苗	37.94	25.95	35.35	23.06	24.59	29.13	26.55	28.71	24.06
出苗—分蘖	49.62	65.58	45.73	38.28	34.52	38.80	34.74	46.73	41.19
分蘖—拔节	13.74	13.10	12.18	17.86	16.99	14.32	14.11	16.95	16.78
拔节—孕穗	36.12	39.84	25.24	24.06	16.58	16.06	26.30	27.93	31.64
孕穗—抽穗	61.44	70.45	57.56	52.55	34.53	48.40	52.00	46.17	52.40
抽穗—灌浆	65.93	61.68	35.99	63.83	62.12	73.47	65.65	42.07	47.61
灌浆—成熟		40.91	26.09	27.26	17.00	38.05	24.47	18.45	23.58

Table 127　　Statistics of Water Consumption of Spring Wheat in Every Growth and Development Period　　Unit: m³/mu

Growth and Development Period	Achun #3			Xinchun #2			Changchun #3		
	50%	60%	70%	50%	60%	70%	50%	60%	70%
Irrigation Before Sowing	43.76	31.36	36.21	27.83	33.78	32.88	40.88	35.37	37.01
Seeding – Seedling	37.94	25.95	35.35	23.06	24.59	29.13	26.55	28.71	24.06
Seedling – Tillering	49.62	65.58	45.73	38.28	34.52	38.80	34.74	46.73	41.19
Tillering – Jointing	13.74	13.10	12.18	17.86	16.99	14.32	14.11	16.95	16.78
Jointing – Booting	36.12	39.84	25.24	24.06	16.58	16.06	26.30	27.93	31.64
Booting – Heading	61.44	70.45	57.56	52.55	34.53	48.40	52.00	46.17	52.40
Heading – Filling	65.93	61.68	35.99	63.83	62.12	73.47	65.65	42.07	47.61
Filling – Maturing		40.91	26.09	27.26	17.00	38.05	24.47	18.45	23.58

(三) 春小麦耗水规律
3.1.3.3 Water Consumption Pattern of Spring Wheat

从表 127 中可以看出,春小麦在整个生育阶段中,前期和后期耗水量较少,中期较大,呈现一个慢慢递增逐渐达到一个最大值,然后又减少的波状趋势。耗水的高峰出现在 6 月中旬左右,也就是小麦的抽穗—灌浆阶段。因为,这个生育阶段是小麦生长的旺盛时期,需要大量的光、水、热进行光合作用,将无机质合成有机质,小麦籽粒逐渐成形并日趋饱满,就要求有充分的灌水保障,因此,在整个生育阶段中耗水高峰值出现在这个阶段。随着籽粒的逐渐饱满,小麦逐渐成熟,小麦的生长也渐渐平缓,需水强度变小,耗水量减少,这样,小麦的耗水情况在整个生育阶段表现出的规律就是前期和后期较小,中间大。

From Table 127, it can be seen that spring wheat consumed less water in the early and late stages and more water in the middle during the whole growth and development period, taking on a wavy form which increased slowly to reach a peak value and then declined. Water consumption peaked at the middle of June when the wheat grew into heading-filling period. This stage marked the prime of wheat growth, requiring lots of light, water and heat to conduct photosynthesis to transform inorganic substances into organics. As wheat grains gradually took shape and became full, sufficient irrigation was required. Hence, the peak value of water consumption occurred in this stage. With grains growing full, the wheat was ripe, its growth slowed down, water requirements dropped as well as water consumption. This way, the water consumption of wheat along the whole growth and development period remained minor in the early and late stages but quite high in the middle.

四、方差分析
3.1.4 Variance Analysis

(一) 极差分析
3.1.4.1 Range Analysis

从表 128 中可以看出,品种对小麦产量的影响最大,灌溉制度对产量的影响次之,底肥对产量的影响相对最小。从极差值 R 可看出,试验结果 3 因素的最优组合是 A2B3C3,而正交试验方案设计表 128 中没有这个处理,与该处理最接近的组合是 A2B2C3,则试验结果最优给合是 A2B2C3,即:品种新春 2 号,底肥 3 000 kg/亩,土壤含水率控制水平 70%(占田间持水率)。

It can be seen from Table 128 that the variety had the greatest impact on wheat yield, followed by irrigation scheduling, and the base fertilizer had the least impact on yield. As it can be seen from the range R, the optimal combination of the 3 factors in the experiment is A2B3C3, while this treatment doesn't exist in Table 128 of orthogonal test design. The closest combination to this treatment is A2B2C3, so the optimal combination of results is A2B2C3, namely the combination of Xinchun #2, base fertilizer of 3 000 kg/mu and soil water content controlled at 70% of field capacity.

自由度分解：总自由度：$f = nk - 1 = 26$

区组自由度：$f = n - 1 = 2$

处理间自由度：$f = k - l = 8$

Division of degree of freedom: Total degree of freedom: $f = nk - 1 = 26$

Degree of freedom in zones: $f = n - 1 = 2$

Degree of freedom within treatments: $f = k - 1 = 8$

表 128　　　　　　春小麦正交试验结果分析表

因素	A 品种	B 施肥量	C 灌溉制度	产量（kg/亩）
1	1	1	1	140.05
2	1	2	2	167.29
3	1	3	3	218.95
4	2	1	2	190.11
5	2	2	3	212.01
6	2	3	1	178.66
7	3	1	3	153.92
8	3	2	1	155.78
9	3	3	2	156.15
K_1	526.29	484.08	474.49	
K_2	580.78	535.08	513.55	
K_3	465.85	553.76	584.88	
$\overline{K_1}$	175.43	161.36	158.16	
$\overline{K_2}$	193.59	178.36	171.18	
$\overline{K_3}$	38.31	184.59	194.96	
R	38.31	23.20	36.80	

Table 128 Results Analysis of Spring Wheat Orthogonal Test

Factor	A Varieties	B Amount of Fertilizer	C Irrigation Scheduling	Output (kg/mu)
1	1	1	1	140.05
2	1	2	2	167.29
3	1	3	3	218.95
4	2	1	2	190.11
5	2	2	3	212.01
6	2	3	1	178.66
7	3	1	3	153.92
8	3	2	1	155.78
9	3	3	2	156.15
K_1	526.29	484.08	474.49	
K_2	580.78	535.08	513.55	
K_3	465.85	553.76	584.88	
$\overline{K}_1$	175.43	161.36	158.16	
$\overline{K}_2$	193.59	178.36	171.18	
$\overline{K}_3$	38.31	184.59	194.96	
R	38.31	23.20	36.80	

(二)方差分析

3.1.4.2 Variance Analysis

春小麦产量方差分析见表129。

The variance analysis of spring wheat yield is shown in Table 129.

表 129　　　　　　　　春小麦产量方差分析表　　　　　　　　单位：kg/亩

处理	区组			T_t	X_t
	I	II	III		
1	157.97	126.84	135.34	420.15	140.05
2	207.79	172.94	121.18	501.88	167.29
3	237.35	224.99	192.58	654.92	218.31
4	182.34	201.68	186.01	570.03	190.01
5	256.18	198.34	181.51	636.03	212.01
6	173.15	190.17	172.67	535.99	178.66
7	194.01	173.68	94.07	461.76	153.92
8	160.27	173.09	134.01	467.37	155.78
9	185.84	152.38	130.24	468.46	156.15
T_γ	1 754.9	1 614.11	1 347.61	4 716.62	
X_γ	194.99	179.35	149.73		174.69

Table 129　　　　　Variance Analysis Table of Spring Wheat Yield　　　　　Unit：kg/mu

Treatment	Zone			T_t	X_t
	I	II	III		
1	157.97	126.84	135.34	420.15	140.05
2	207.79	172.94	121.18	501.88	167.29
3	237.35	224.99	192.58	654.92	218.31
4	182.34	201.68	186.01	570.03	190.01
5	256.18	198.34	181.51	636.03	212.01
6	173.15	190.17	172.67	535.99	178.66
7	194.01	173.68	94.07	461.76	153.92
8	160.27	173.09	134.01	467.37	155.78
9	185.84	152.38	130.24	468.46	156.15
T_γ	1 754.9	1 614.11	1 347.61	4 716.62	
X_γ	194.99	179.35	149.73		174.69

误差自由度：$f = 26 - 2 - 8 = 16$

Error's degree of freedom：$f = 26 - 2 - 8 = 16$

平方和分解：$c = T^2/nk = 4\ 716^2/3 \times 9 = 823\ 944.61$

Division of sum of squares：$c = T^2/nk = 4\ 716^2/3 \times 9 = 823\ 944.61$

总平方和：$S_\text{总} = \sum_1^n X^2 - C = (157.97^2 + 207.19^2 + \cdots + 130.24^2) - 823\,944.61$
$= 34\,600.75$

Total sum of squares：$S_\text{total} = \sum_1^n X^2 - C = (157.97^2 + 207.19^2 + \cdots + 130.24^2) - 823\,944.61 = 34\,600.75$

$$S_\text{区} = \sum T_r^2/k - c = 9\,508.40$$

$$S_\text{zones} = \sum T_r^2/k - c = 9\,508.40$$

$$S_\text{处} = \sum T_t^2/n = 17\,787.12$$

$$S_\text{groups} = \sum T_t^2/n = 17\,787.12$$

$$S_\text{误} = S_\text{总} - S_\text{区} - S_\text{处} = 7\,304.48$$

$$S_\text{error} = S_\text{total} - S_\text{zones} - S_\text{groups} = 7\,304.48$$

分析表 130 可得出结论：$F_\text{区} = 10.41 > F_{0.01} = 6.23$，表示区组间差异极显著 $F_\text{处} = 4.87 > F_{0.01} = 3.89$，处理间差异极显著。这表明本试验的设计方案较好，具有较好的代表性。

By analyzing Table 130, we can draw the conclusion：when $F_\text{zones} = 10.41 > F_{0.01} = 6.23$, the differences within zones were significant; when $F_\text{treatments} = 4.87 > F_{0.01} = 3.89$, the differences within treatments were notable. This shows that the designing scheme of this experiment is good and representative.

表 130　　春小麦产量 F 检验表

方差来源	平方和 SS	自由度 f	均方 Ms	F	$F_{0.05}$	$F_{0.01}$
区组间	9 508.4	2	4 754.2	10.41	3.63	6.23
处理间	17 787.12	8	2 223.39	4.87	2.59	3.89
误差	7 304.48	16	456.53			
总和	34 600.75	26				

Table 130　　Check List for F Value of Spring Wheat Yield

Variance Sources	Sum of Squares SS	Degree of Freedom f	Mean Squares Ms	F	$F_{0.05}$	$F_{0.01}$
Within Zones	9 508.4	2	4 754.2	10.41	3.63	6.23
Within Treatments	17 787.12	8	2 223.39	4.87	2.59	3.89
Error	7 304.48	16	456.53			
Total	34 600.75	26				

五、边际分析
3.1.5 Marginal Analysis

（一）生产函数
3.1.5.1 Production Function

任何一种产品的生产过程，都是指投入资源的生产过程。这种表示产品数量和投入资源的关系可表示为生产函数关系，其表达式为：

The production process of any product refers to the production process of resources input. This relation between quantity of products and resources input can be expressed as a production function, which can be represented as:

$$Y = f(X)$$

式中　Y——产品数量；
　　　X——资源投入量；$X = X_1 \cdot X_2 \cdots X_n$，为生产 Y 产品时所需的各种生产资源的投入量。

Where　Y = product quantity
　　　X = the quantity of resources input; $X = X_1 \cdot X_2 \cdots X_n$ means the total input of various resources needed to produce Y.

在众多的资源中，有人工可以控制的和有限量的称之为变动资源；还有不可控制或难以控制的和无限量的称之为固定资源。本文仅以变动资源中的水为研究对象，其他均视为不变资源，即只研究水分对春小麦生产的影响，则水分生产函数表示为：

Among the numerous resources, there are controlled and limited resources, which are called variable resources. Those that are uncontrollable, unmanageable or unlimited resources are called fixed resources. In this book, only water in variable resources is taken as a research object, while other resources are regarded as invariable resources, that is, we only study water's impact on spring wheat production and describe the water production function as:

$$Y = f(\theta)$$

式中　θ——单位面积上的水分投入量；
　　　Y——产量。

Where　θ = the amount of water applied in per unit area
　　　Y = yield.

投入量的价值：$T_c = P_X \cdot X$

Value of water applied: $T_c = P_X \cdot X$

所得产量的价值: $T_P = P_Y \cdot Y$
Production value: $T_P = P_Y \cdot Y$

净效益: $Z = P_Y \cdot Y - (K + Px \cdot X)$
Net benefits: $Z = P_Y \cdot Y (K + Px \cdot X)$

式中　T_c——投入水量的价值；
　　　T_p——所得产品的价值；
　　　P_X——水的单价；
　　　P_Y——产品的单价；
　　　X——投入的总水量；
　　　Y——所得总产量；
　　　Z——净效益；
　　　K——固定生产要素的总价值为常数。

Where　T_c = the value of water applied
　　　　T_p = product value
　　　　P_X = unit price of water
　　　　P_Y = unit price of product
　　　　X = total amount of water applied
　　　　Y = total production
　　　　Z = net benefit
　　　　K = the constant that refers to total value of fixed production factors.

固定生产要素中，肥料：33.36 元/亩，种子费：14.07 元/亩，机耕费：7 元/亩，收割、运输费：8 元/亩。

In fixed production factors, fertilizer was 33.36 yuan/mu, seeds 14.07 yuan/mu, tractor-ploughing fees 7 yuan/mu, and costs for harvest and transportation 8 yuan/mu.

$$K = 33.36 + 14.07 + 7 + 8 = 62.43 \text{ 元/亩}$$
$$K = 33.36 + 14.07 + 7 + 8 = 62.43 \text{ yuan/mu}$$

净效益最大时：
When the net benefit reached the maximum value：

$$\frac{d_z}{d_x} = 0 \text{ 即} \frac{d_z}{d_x} = \frac{d_Y}{d_x} \cdot P_Y - P_X = 0$$

$$\frac{d_z}{d_x} = 0, \text{ namely } \frac{d_z}{d_x} = \frac{d_Y}{d_x} \cdot P_Y - P_X = 0$$

则
Then
$$P_Y \cdot d_Y = P_X \cdot d_X$$
$$P_Y \cdot d_Y = P_X \cdot d_X$$

故

$$\frac{d_Y}{d_X} = \frac{P_X}{P_Y}$$

Thus

$$\frac{d_Y}{d_X} = \frac{P_X}{P_Y}$$

（二）边际产量
3.1.5.2 Marginal Yield

边际产量（MPP）是指每增加一个单位的变动资源时所能增加总产量的数量。即在某一水平的投入下，由于增加一个单位的投入而较上一投入水平所增加的产品数量。其公式如下：

Marginal yield (MPP) is the total product increased by each additional unit of variable resources. That is to say, under a certain level of input, the number of products increased due to the increase of one unit of investment compared with the previous level. The function is written as follows：

$$M = \frac{\Delta Y}{\Delta X} \quad 或 \quad M = \frac{d_Y}{d_X} = \frac{d_f}{d_X} = f'$$

$$M = \frac{\Delta Y}{\Delta X} \quad or \quad M = \frac{d_Y}{d_X} = \frac{d_f}{d_X} = f'$$

M 表示边际产量或边际生产力，ΔX 表示变动资源的增加量，引入价格概念：$P_Y \cdot \Delta Y$ 称为边际产值，$P_X \cdot \Delta X$，称为边际成本，当边际产值大于边际成本时，就产生了纯利润，否则，生产过程的利润为负值。

M represents MP (marginal product) or marginal productivity, while ΔX is the increment of variable resources. When the concept of price is introduced in：$P_Y \cdot \Delta Y$ is marginal production value and $P_X \cdot \Delta X$ is marginal cost. When the marginal production value overtakes the marginal cost, the net profit is gained; otherwise, the profit in the production process is negative.

（三）总产量、平均产量、边际产量和生产弹性
3.1.5.3 Total Production, Average Product, Marginal Yield and Production Flexibility

小麦生产过程中，对增加的每单位的作用及产生的影响程度作边际分析，为此引出几个基本概念：

A marginal analysis of the role and impact increased per unit is made in the wheat production process. To this end, we introduce several basic concepts：

总产量——投入一定量的变动资源所引起的产出总额（Y）。

Total production refers to gross production gained from inputting a certain amount of variable resources (Y).

平均产量——每单位变动资源可平均引起的产量(A)。
Average product refers to production gained from per unit variable resources (A).

边际产量——连续追加的每单位变动资源所引起的总产量增加额(M),即:
Marginal yield refers to the increment of total production arising from the continuous increase of per unit variable resources (M), then:

$$A = \frac{Y}{X} \quad M = \frac{\Delta Y}{\Delta X}$$

根据导数的几何意义:函数 $Y = f(X)$ 在点 X 处的导数 $f'(X)$ 等于函数 $Y = (X)$ 曲线上在 X 点切线的斜率,即:
According to the geometrical significance of the derivative: the derivative of function $Y = f(X)$ at X is equal to the slope of tangent line at X on the curve of function $Y = (X)$, namely:

$$M = d_Y / d_X = f'(X)$$

生产弹性—反映产量增长对于投入资源的敏感程度,也就是上面提到的产量增加的幅度,和资源增加幅度的比例关系,即:
Production flexibility reflects the sensitivity of production growth to the resources input, namely the proportional relation between the increment of resources and production mentioned above, that is:

$$E_P = \Delta Y / Y = \frac{\Delta Y}{\Delta X} / \frac{Y}{X} = M/A$$

根据春小麦的试验资料,计算春小麦总产量,平均产量,边际产量,生产弹性,将结果列入表131中,并据此表绘制春小麦水分生产函数曲线,见图35。
Based on experimental data of spring wheat, the total production, average product, marginal yield and production flexibility of spring wheat are calculated, listing results in Table 131 and drawing out the water production function curve of spring wheat accordingly. See figure 35.

从图中可以看出,总产量(Y)随着耗水量的增加,Y 的增加幅度由慢变快,以后又逐渐变缓,当耗水量为 296 m³/亩时,总产量达到最大值,即图中曲线顶点,$Y = 220$ kg/亩。这以后,再继续增加水源的投入,不但不能获得产量的增加反而导致产量开始下降。在图中产量曲线的顶点,这点的斜率值为0,即这时的边际产量(M)等于0。
It can be seen from the figure that the total production (Y) raises as the water

3 Study on Irrigation System of Crops

consumption increases, the rate of increase is slow at first and then grows fast, and later it gradually slows down. When the water consumption is 296 m³/mu, the total production reaches the maximum, namely the peak of the curve at this point $Y = 220$ kg/mu. After that, continue adding water, the production doesn't increase, instead it begins to decline. At the top of the production curve in the figure, the slope at this point is zero, which means that the marginal yield (M) is zero.

表131　　　　　　　　　　春小麦试验结果表

处理	耗水量 X (m³/亩)	总产量 Y (kg/亩)	平均产量 A (kg/亩)	水量增量 ΔX(m³/亩)	产量增量 ΔY(kg/亩)	边际产量 M	生产弹性 E_P
1	233.80	140.05	0.60				
2	260.61	167.29	0.64	26.81	27.24	1.02	1.59
3	219.29	218.95	1.00	-41.32	51.66	-1.25	-1.25
4	186.96	190.01	1.02	-32.33	-28.94	0.90	0.88
5	290.58	212.01	0.73	103.62	22.00	0.21	0.29
6	220.95	178.66	0.81	-69.63	-33.35	0.48	0.59
7	251.83	153.92	0.61	30.88	-24.74	0.80	1.31
8	216.86	155.78	0.72	-34.97	1.86	-0.05	-0.07
9	189.10	156.15	0.83	-27.76	0.37	-0.01	-0.01

Table 131　　　　　　　　**Experimental Results of Spring Wheat**

Treatment	Water Consumption X(m³/mu)	Total Production Y (kg/mu)	Average Product A (kg/mu)	Increment of Water Amount ΔX(m³/mu)	Production increment ΔY (kg/mu)	Marginal yield M	Production flexibility E_P
1	233.80	140.05	0.60				
2	260.61	167.29	0.64	26.81	27.24	1.02	1.59
3	219.29	218.95	1.00	-41.32	51.66	-1.25	-1.25
4	186.96	190.01	1.02	-32.33	-28.94	0.90	0.88
5	290.58	212.01	0.73	103.62	22.00	0.21	0.29
6	220.95	178.66	0.81	-69.63	-33.35	0.48	0.59
7	251.83	153.92	0.61	30.88	-24.74	0.80	1.31
8	216.86	155.78	0.72	-34.97	1.86	-0.05	-0.07
9	189.10	156.15	0.83	-27.76	0.37	-0.01	-0.01

平均产量(A)的曲线,随着耗水量的增加,也有一个增减过程,但最大值的出现与总产量的最大值点不一致。

As water consumption increases, average product (A) also demonstrates a changing process, but the point of maximum average product does not go in line with that of total production.

平均产量(A)的曲线,随着耗水量的增加,也有一个增减过程,但最大值的出现与总

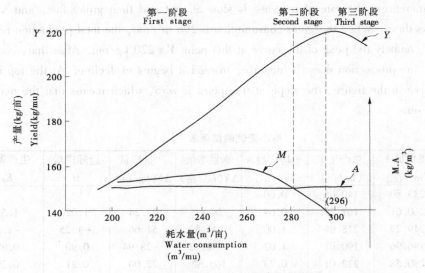

图 35　春小麦水分生产函数曲线
Figure 35　Water Production Function Curve of Spring Wheat

产量的最大值点不一致。

As water consumption increases, average product (A) also demonstrates a changing process, but the point of maximum average product does not go in line with that of total production.

综观图中 3 条曲线,有两个重要的转折点:一是边际产量曲线或与平均产量曲线的交点。在该点处,边际产量与平均产量相等。二是总产量曲线的顶点,边际产量在这个转折点上为零,过此点后转为负值。以这两点为界限,连续投入资源的生产过程,可以划分为 3 个阶段;从原点到平均产量和边际产量的交点处为第一阶段。这个阶段的特点是边际产量高于平均产量,因而引起平均产量逐渐提高,总产量的增加幅度大于资源的增加幅度。因此,只要资源条件允许,就不应该在这个阶段停止投入资源。第二个阶段是从平均产量和边际产量的交点到总产量最高点之间,这一阶段的特点是边际产量小于平均产量,因而引起平均产量的逐渐降低,总产量增加的幅度小于资源增加的幅度,即出现报酬递减。而总产量曲线的顶点是在一定技术条件下可能达到的一种界限。在这一阶段,总产量按报酬递减的形式继续增长,可以获得较高的产量,在一定限度内,继续投入资源也可获得收益。第三阶段是在总产量达到最大值点以后,也就是边际产量转为负值以后。在这一阶段,增加资源的投入反而引起了总产量的降低,把这一阶段称为生产的绝对不合理阶段。

Looking at the 3 curves, there are two important turning points: one is the intersection of the marginal yield curve and the average product curve. At this point, marginal yield equals average product. The other is the peak of total production curve, where marginal yield here is zero and turns negative after this point. Within the two points, the production process with continuous resources investment can be divided into 3 stages; the first stage starts from the origin and ends at the intersection of the marginal yield and the average product. This stage features that marginal yield overtops average product, thus average product gradually raises and the increment of total production exceeds that of resources. So, as long as resource condition permits, resources input should not be stopped at this stage. The second stage starts from the intersection of the marginal yield and the average product and ends at the peak of total production. This stage features that average product overtops marginal yield, causing that average product gradually drops and the increment of total production is under that of resources. In other words, returns diminish. However, the peak of the total production curve is a limit which may be reached under certain technical conditions. At this stage, total production continues to grow in the form of diminishing returns, resulting in higher yield and, within a limit, profits can be obtained from continuous resource investment. The third stage begins from the maximum of total production, that is to say, the point where marginal yield turns negative. As this stage, additional resource input leads to a reduction in total production. Thus this stage is called the absolutely unreasonable stage of production.

(四) 价值生产函数及最大纯收益
3.1.5.4 Value Production Function and Maximum Net Return

任何物品的生产必须以经济效益为中心。生产的目的,在一般情况下,并不只是为了取得最高产量或产值,主要还是要争取获得较多的盈利,在考虑如何获得最高盈利时,只需分析投入多少变动资源最有利,这就要涉及到变动资源与产品两者的价格。这当中有一条重要准则:当边际产值($P_X \cdot \Delta Y$)等于边际资源成本($P_X \cdot \Delta X$)时,也就是当边际产值与边际成本的比值等于1时,盈利最大。可以得出这样的结论:当边际产值大于边际资源成本时,可以增加盈利,因而可以继续投入变动资源,以获得更多的盈利。当边际产值小于边际成本时,再增加投入变动资源,不但不能增加盈利,反而会减少盈利或扩大亏损。因而,盈利最大的资源投入量是当边际产值等于边际资源的成本时。用公式表示:

The production of any article must center on economic benefit. The purpose of production, in general, is not merely to achieve maximum production or output, but mainly to strive for more profits. While considering how to obtain the highest profit, just analyze how much variable resources are most economical, which involves the prices of variable resources and products. There is an important principle: when the value of marginal production ($P_X \cdot \Delta Y$) equals the cost of marginal resources ($P_X \cdot \Delta X$), namely when the ratio of the value of marginal production to the cost of marginal resources is equal to 1, profit maximizes. It can be concluded that when the marginal production value is greater than the marginal resource cost, profit increases, so additional variable resources can be invested to obtain more profits. When the marginal cost overtakes the marginal production value, adding variable resources will not increase the profit, instead, it reduces the profit or enlarges the loss. As a result, the profit is the highest when the value of marginal production equals the cost of marginal resources. This formula is represented as:

$$P_Y \cdot \Delta Y = P_X \cdot \Delta X$$

把上式改写成:$\Delta Y/\Delta X = P_X/P_Y$,即:

Rewrite the formula as:$\Delta Y/\Delta X = P_X/P_Y$, that is:

$$M = P_X/P_Y$$

据边际分析原理将价值生产函数分成 3 个阶段:第一阶段为开始到平均产值最高点。在此阶段随水分投入量的不断增加,平均产值不断增加,边际产值大于平均产值最后相等。因此,在此阶段水分的投入量至少应施到第一阶段的终点;第二阶段为自第一阶段终点到产值最高点。此阶段平均产值和边际产值均随水分投入量的增多而递减直到边际产值为零;第三阶段自产值最高点以后。此阶段随水分投入量增加而产值减少,边际产值为负(见表 132、图 36)。

According to the principle of marginal analysis, the value production function falls into 3 stages: the first stage ranges from the origin to the zenith of average production value. Here the average production value climbs up as water amount increases, marginal production value firstly exceeds average production value and then the two are equal. Therefore, water amount applied for production should achieve the end point of the first stage; the second stage is from the endpoint of the first stage to the peak of production value. Here both average and marginal production values diminish as water amount increases until marginal production value becomes zero. The third stage begins from the peak of production value. In this stage additional water results in the loss of production value and marginal production value becomes negative (see Table 132 and Figure 36).

由价值生产函数的 3 个阶段说明:分析投入量不论停留在第一阶段和进入第三阶段均属不合理。在第一阶段时,水分投入量的平均报酬在不断增大,资源得不到合理发挥。在第三阶段时,由于水分投入量的增加产值反而减少,显然为不合理的。只有处在第二阶段才能使水分及其他资源的投入得到合理发挥。

As illustrated by three stages of the value production function: it is unreasonable to focus on the first or the third stage when analyzing resource investment. At the first stage, the average return of water is on the increase, not putting the resource in good use. At the third stage, it is obviously improper when the amount of water increases, production value cuts down. Only at the second stage water and other resources are well used.

获得最大纯效益的充分必要条件为边际产值等于边际成本。据此条件由图 36 中查得最优水分投入量 $Q^x = 289$ m³/亩,该点恰好落在第二阶段,对应的单位面积纯收益最大

值 $Z^* = 58.50$ 元/亩。

The sufficient and necessary condition for obtaining the maximum net return is that the marginal production value equals the marginal cost. Based on this, we get the optimal water amount in Figure 36, namely $Q^x = 289$ m³/mu. This point falls into the second stage exactly, and the corresponding maximum net return per unit area Z^* is 58.50 yuan/mu.

表 132　　春小麦价值生产函数计算表 ($P_X = 0.04$　$P_Y = 0.58$)

处理号	耗水量 X (m³/亩)	产量 Y (kg/亩)	平均产值 $\Delta P = \Delta P_Y$ (元/m³)	边际产值 $M_P = M \cdot P_Y$ (元/m³)	耗水成本 $T_c = P_X \cdot X$ (元/亩)	边际成本 $M_c = P_X$ (元/m³)	资源投入 K (元/亩)	产值 $T_p = Y \cdot P_Y$ (元/亩)	纯收益 $Z = T_p - T_c - K$ (元/亩)
1	233.80	140.05	0.348		9.352	0.04	62.43	81.23	9.45
2	260.61	167.29	0.371	0.592	10.42	0.04	62.43	97.02	24.17
3	219.29	218.95	0.580	-0.725	8.77	0.04	62.43	126.99	55.79
4	186.96	190.01	0.592	0.522	7.48	0.04	62.43	110.21	40.3
5	290.58	212.01	0.423	0.122	11.62	0.04	62.43	122.97	48.92
6	220.95	178.66	0.470	0.278	8.84	0.04	62.43	103.62	32.35
7	251.83	153.92	0.354	0.464	10.07	0.04	62.43	89.27	16.77
8	216.86	155.78	0.418	-0.029	8.67	0.04	62.43	90.35	19.25
9	189.10	156.15	0.481	-0.006	7.56	0.04	62.43	90.57	20.58

Table 132　　Computation Sheet for Value Production Function of Spring Wheat ($P_X = 0.04$　$P_Y = 0.58$)

Treatment No.	Water Consumption X (m³/mu)	Output Y (kg/mu)	Average Production Value $\Delta P = \Delta \cdot P_Y$ (yuan/m³)	Marginal Production Value $M_P = M \cdot P_Y$ (yuan/m³)	Cost of Water Consumption $T_c = P_X \cdot X$ (yuan/mu)	Marginal Cost $M_c = P_X$ (yuan/m³)	Resources Input K (yuan/mu)	Production Value $T_p = Y \cdot P_Y$ (yuan/mu)	Net Return $Z = T_p - T_c - K$ (yuan/mu)
1	233.80	140.05	0.348		9.352	0.04	62.43	81.23	9.45
2	260.61	167.29	0.371	0.592	10.42	0.04	62.43	97.02	24.17
3	219.29	218.95	0.580	-0.725	8.77	0.04	62.43	126.99	55.79
4	186.96	190.01	0.592	0.522	7.48	0.04	62.43	110.21	40.3
5	290.58	212.01	0.423	0.122	11.62	0.04	62.43	122.97	48.92
6	220.95	178.66	0.470	0.278	8.84	0.04	62.43	103.62	32.35
7	251.83	153.92	0.354	0.464	10.07	0.04	62.43	89.27	16.77
8	216.86	155.78	0.418	-0.029	8.67	0.04	62.43	90.35	19.25
9	189.10	156.15	0.481	-0.006	7.56	0.04	62.43	90.57	20.58

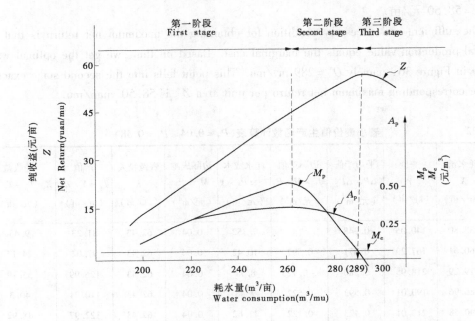

图36 春小麦价值生产函数曲线图
Figure 36 Value Production Function Curve of Spring Wheat

六、春小麦水分生产函数
3.1.6 Water Production Function of Spring Wheat

表133列出了春小麦各生育阶段耗水量、产量试验成果。由表134可见,作物产量受水分的制约,产量与水分形成一定的函数关系。作物的生长是有阶段性的,不同生产阶段缺水,对产量有不同的影响,并且作物生长发育和产量形成一个连续过程。作物在某一阶段缺水不仅直接影响该生长期,且对下一生长期也产生影响。无论哪一阶段严重缺水,都会导致作物减产甚至绝收。因此,作物与各生长期耗水量具有连乘关系(见表133、表134)。

Table 133 outlines the water consumption and experimental results of yield of spring wheat in each growth and development period. From Table 134, it is clear that crop yield is under the influence of water with a functional relation between them. The growth of crops is of different stages. The effects on yield vary according to water strain in different production stages. And it's a continuous process that crops' growth and development is followed by crops yield. Water shortage in a stage affects not only the current period but also the next one. And severe water shortage, regardless of which stage, gives rise to crop failure even total crop failure. Therefore, there is a serial correlation between crops yield and water consumption in each growth period (see Table 133 and Table 134).

第三章 作物灌溉制度研究
3 Study on Irrigation System of Crops

表 133 春小麦各生育阶段耗水量、产量统计表 单位：m³/亩

生育阶段	1	2	3	4	5	6	7
播种—分蘖	51.28	47.50	51.85	47.99	48.85	51.36	47.75
分蘖—拔节	23.98	32.75	13.34	6.36	21.32	30.15	22.88
拔节—抽穗	31.25	30.63	35.17	31.54	30.86	21.58	21.75
抽穗—灌浆	69.93	69.81	74.19	51.90	57.35	65.59	55.41
灌浆—成熟	77.52	79.92	59.22	69.17	62.57	48.68	41.31
全生育期(m³/亩)	253.96	260.61	233.80	186.96	220.95	216.86	189.10
产量(kg/亩)	297.56	162.57	138.72	173.28	195.38	157.49	160.50

Table 133 Statistics of Water Consumption and Yield of Spring Wheat in Every Growth and Development Period   Unit: m³/mu

Growth and development period	1	2	3	4	5	6	7
Sowing – Tillering	51.28	47.50	51.85	47.99	48.85	51.36	47.75
Tillering – Jointing	23.98	32.75	13.34	6.36	21.32	30.15	22.88
Jointing – Heading	31.25	30.63	35.17	31.54	30.86	21.58	21.75
Heading – Filling	69.93	69.81	74.19	51.90	57.35	65.59	55.41
Filling – Maturing	77.52	79.92	59.22	69.17	62.57	48.68	41.31
Whole Growth and Development Period (m³/mu)	253.96	260.61	233.80	186.96	220.95	216.86	189.10
Output (kg/mu)	297.56	162.57	138.72	173.28	195.38	157.49	160.50

这种水量和产量的连乘关系用数学表达式表现出来就是作物水分生产函数相乘模型，可用詹森(Jehsen)1968年提出的下述模式来表达。

The serial correlation between water and crops yield is of multiple model of crops water production function when written in mathematical expression. The following pattern proposed by Jehsen in 1968 can be used to describe this.

· 254 ·   干旱瘠薄土地作物需水量与灌溉制度研究（以新疆阿勒泰地区为例）
A Case Study on Crop Water Requirement and Irrigation
Scheduling in Arid and Barren Land（Altay Prefecture, Xinjiang）

表 134　　　　　　　春小麦生育阶段相对耗水量和相对产量统计表　　　　　　单位：m³/亩

生育阶段	播种—分蘖	分蘖—拔节	拔节—抽穗	抽穗—灌浆	灌浆—成熟	全生育期	产量（kg/亩）
1	1.000 0	1.000 0	1.000 0	1.000 0	1.000 0	1.000 0	1.000 0
2	0.926 3	1.365 7	0.980 2	0.998 3	1.005 2	1.026 2	0.546 3
3	1.011 7	0.556 3	1.125 4	1.060 9	0.763 9	0.926 6	0.466 2
4	0.935 8	0.265 2	1.009 3	0.742 2	0.892 3	0.736 2	0.582 3
5	0.952 6	0.889 1	0.987 5	0.820 1	0.807 1	0.870 0	0.656 6
6	1.001 6	1.257 3	0.690 3	0.937 2	0.628 0	0.853 9	0.529 3
7	0.931 2	0.954 1	0.696 0	0.792 4	0.532 9	0.744 6	0.539 3

Table 134　　Statistics of Corresponding Water Consumption and Yield
of Spring Wheat in Every Growth and Development Period　　Unit: m³/mu

Growth and development period	Sowing-Tillering	Tillering-Jointing	Jointing-Heading	Heading-Filling	Filling-Maturing	Whole Growth and Development Period	Output (kg/mu)
1	1.000 0	1.000 0	1.000 0	1.000 0	1.000 0	1.000 0	1.000 0
2	0.926 3	1.365 7	0.980 2	0.998 3	1.005 2	1.026 2	0.546 3
3	1.011 7	0.556 3	1.125 4	1.060 9	0.763 9	0.926 6	0.466 2
4	0.935 8	0.265 2	1.009 3	0.742 2	0.892 3	0.736 2	0.582 3
5	0.952 6	0.889 1	0.987 5	0.820 1	0.807 1	0.870 0	0.656 6
6	1.001 6	1.257 3	0.690 3	0.937 2	0.628 0	0.853 9	0.529 3
7	0.931 2	0.954 1	0.696 0	0.792 4	0.532 9	0.744 6	0.539 3

$$\frac{Y}{Y_M} = \sum_{i=1}^{n} \left(\frac{W_i}{W_{Mi}}\right) \lambda_i \tag{1}$$

式中　n——作物生长划分的阶段数；

　　　W_{Mi}——作物在阶段 i 供水充足时的最大耗水量，m³/亩；

　　　W_i——作物在阶段 i 供水不足时的实际耗水量，m³/亩；

　　　Y_M——作物各阶段耗水量达到最大值$\{W_{M1}, W_{M2}, \cdots, W_{MN}\}$时的最高产量，kg/亩；

　　　Y——作物生长阶段由于供水不足，只达到$\{W_1, W_2, \cdots, W_n\}$时的产量，kg/亩；

　　　λ_i——阶段 i 的敏感指标。

Where　n = the number of crops growth stages

　　　W_{Mi} = the maximum water consumption in stage i when there is sufficient water supply, m³/mu

　　　W_i = the real water consumption in stage i when there is water shortage, m³/mu

　　　Y_M = the production peak when the water consumption of crops $\{W_{M1}, W_{M2}, \cdots, W_{MN}\}$ peaks in every period, kg/mu

Y = crops yield due to water strain $\{W_1, W_2, \cdots, W_n\}$ in growth stages, kg/mu

λ_i = sensitive indicator in stage i.

将(1)两边取自然对数,化为下式:
Draw the natural log of both sides of (1) and turn it into the following formula:

$$\ln\left(\frac{Y}{Y_M}\right) = \sum_{i=1}^{n} \lambda_i \ln\left(\frac{W_i}{W_{Mi}}\right) \tag{2}$$

令 $\ln\left(\frac{Y}{Y_M}\right) = Z$ $\ln\left(\frac{W_i}{W_{Mi}}\right) = X$

Let $\ln\left(\frac{Y}{Y_M}\right) = Z$ $\ln\left(\frac{W_i}{W_{Mi}}\right) = X$

则(2)式化简成
Then, the formula (2) gets simplified as

$$Z = \sum_{i=1}^{n} \lambda_i X_i \tag{3}$$

成为一个多元线性函数:将表 134 中的数据代入(3)中采用消元法求得春小麦各生育阶段水分敏感指数:

A multivariate linear function: plug the data in Table 134 into formula (3) and by elimination we get the sensitive index of water in every growth and development period of spring wheat:

播种—分蘖	分蘖—拔节	拔节—抽穗	抽穗—灌浆	灌浆—成熟
0.221 2	0.021 6	0.136 2	0.418 2	0.231 4
Sowing – Tillering	Tillering – Jointing	Jointing – Heading	Heading – Filling	Filling – Maturing
0.221 2	0.021 6	0.136 2	0.418 2	0.231 4

将上述数据代入公式(1):
Put the data above into formula (1):

$$Y/Y_M = (W_1/W_{M1})^{0.221\,2}(W_2/W_{M2})^{0.021\,6}(W_3/W_{M3})^{0.136\,2}(W_4/W_{M4})^{0.418\,2}(W_5/W_{M5})^{0.231\,4}$$

化简后得到:

After simplification, we get:

$$Y = 253.96(0.0195W_1)^{0.2212}(0.0417W_2)^{0.0216}(0.032W_3)^{0.1362}(0.0143W_4)^{0.4182}$$
$$(0.0129W_5)^{0.1362}$$

从水分敏感指数可以看出,春小麦各生育阶段缺水的敏感程度,对春小麦产量产生不同程度的影响。各生育阶段对水分敏感度反应大小顺序为:④>⑤>①>③>②。

It can be seen from water sensitivity index that the effect on spring wheat yield varies according to the level of sensitivity on water shortage in every growth and development period of spring wheat. In every growth and development period, the order of water sensitivity is: ④ > ⑤ > ① > ③ > ②.

这说明春小麦在抽穗—灌浆阶段是对水分反应最敏感的阶段,小麦籽粒逐渐饱满,光合作用最强烈,是小麦在整个生育期间需水的高峰期。

This indicates that spring wheat is most sensitive to water in heading – filling period when the wheat grain is gradually growing full with the strongest photosynthesis, which is a peak of water consumption in wheat's growth and development.

七、春小麦灌溉制度的优化设计
3.1.7 Optimized Scheme for Irrigation Scheduling of Spring Wheat

用动态规划的方法,按时间顺序,将春小麦整个生育阶段划分为5个阶段。每个阶段根据土壤含水量和降水量做出决策,即灌水时间、灌水次数和灌水定额。各阶段决策所组成的最优决策就是春小麦的最优灌溉制度。

By means of dynamic planning, the whole growth and development period of spring wheat is divided into five stages in chronological order. At every stage, decisions are made according to the time, frequency and irrigation water quota on the basis of soil water content and precipitation. The optimal decision made up by decision at each stage is the optimal irrigation scheduling of spring wheat.

(一)建立动态规划数学模型
3.1.7.1 Establish a Mathematical Model of Dynamic Planning

1. 变量设置

Ⅰ Variables set

阶段变量 i:根据作物生育阶段过程划分为 i 个生育阶段;
Stage i: divide the growth and development period of crops into i stages;

状态变量 θ_i：各生育阶段土壤含水量；
Condition θ_i: Soil water content at every period;

决策变量 X_i：各生育阶段灌水量。
Decision X_i: irrigation amount at every period.

2. 约束条件

II Constraint condition

作物耗水量约束：
Constraints on crops water consumption：

$$W_{mini} \leqslant W_i \leqslant W_{maxi}$$

式中 W_{mini}——各生育阶段适宜下限耗水量；
         W_{maxi}——各生育阶段供水充足条件下最大作物耗水量。
Where W_{mini} = the proper minimum water consumption in every growth and development period
        W_{maxi} = the maximum water consumption in every growth and development period when water is sufficient.

土壤含水量约束：

Constraints on soil water content：

$$\theta_{min} \leqslant \theta_i \leqslant \theta_{max}$$

式中 θ_{min}——土壤适宜含水量下限；
        θ_{max}——田间持水率。
Where θ_{min} = the proper minimum soil water content
        θ_{max} = field capacity.

灌溉定额约束：
Constraints on irrigation quota：

$$\sum_{i=1}^{n} X_i = X^*$$

式中 X_i——各生育阶段灌水量；
        X^*——生育阶段优化灌溉定额。
Where X_i = irrigation amount in every period
        X^* = optimized irrigation quota in every period.

3. 状态转移方程
III State transfer equation

$$M_i = M_{i-1} X_i - W_i$$

式中 M_{i-1}、M_i——阶段始末可供作物吸收的水量。
Where M_{i-1}, M_i = water for crops uptaking at the beginning and end of each stage.

$$M_i = 667 H_\gamma (\theta_i - \theta_{\min})$$

式中 H——土壤计划湿润层深度；
γ——计划湿润层内平均土壤容重；
θ_i——土壤含水率状态；
$\theta_{\min}$——凋萎含水量。

Where H = the depth of the soil wetting layer
γ = average soil bulk density in the soil wetting layer
θ_i = the state of soil water content
$\theta_{\min}$ = soil water content when crops wither.

4. 目标函数
IV Objective function

采用 Jehsen, 水量与产量模型：
Use Jehsen water and yield model：

$$\frac{Y}{Y_{\max}} = \sum_{i=1}^{n} \left(\frac{W_i}{W_{\max i}}\right) \lambda_i$$

式中 Y——作物供水不足时达到的产量；
$Y_{\max}$——作物供水充足时达到的产量；
n——作物生育阶段数；
W_i——作物供水不足时在阶段 i 的实际耗水量；
λ_i——各生育阶段水分敏感指数。

Where Y = yield when water is insufficient
$Y_{\max}$ = maximum yield when water is sufficient
n = number of crops growth and development periods
W_i = the real water consumption in stage i when there is water shortage
λ_i = water sensitive index in every growth and development period.

根据价值生产函数分析得出优化灌溉定额 $X^* = 289$ m³/亩，灌水次数 10 次，灌水定额为 28.9 m³/亩（取整数 29 m³/亩）。
Based on value production function, we analyze and get the optimized irrigation quota of

$X^* = 289$ m³/mu, while the frequency of irrigation is 10 times and irrigation water quota, 28.9 m³/mu (round down to 29 m³/mu).

5. 效益函数
V Benefit function

$$f_i = r_i(M_i, I_i, d_i) = (W_i/W_{mi})\lambda_i$$

式中　$W_i = \min\{W_{\text{max}i}, M_i, +X_i\}$;
　　　M_i——土壤可供作物吸收的水量; $M_i = 667\gamma H(\theta_{qi} - \theta_{凋})$
　　　θ_{qi}——阶段初始的土壤含水量;
　　　$H = 0.6$ m, $\theta_{凋} = 4.0\%$;
　　　I_i——i 阶段和 i 阶段以后可供灌水量其值为 29 的倍数即:29、58、87、116、145、174、203、232、261、289 m³/亩;
　　　d_i——第 i 阶段灌水次数 $0 \leq d_i \leq I_i/29$。

Where　$W_i = \min\{W_{\text{max}i}, M_i, +X_i\}$
　　　M_i = water uptaken by crops from the soil; $M_i = 667\gamma H(\theta_{qi} - \theta_{\text{wilt}})$
　　　θ_{qi} = soil water content at the beginning of the stage
　　　$H = 0.6$ m, $\theta_{\text{wilt}} = 4.0\%$
　　　I_i = irrigation amount available at and after stage i, the value is a multiple of 29, namely 29, 58, 87, 116, 145, 174, 203, 232, 261, 289 m³/mu
　　　d_i = irrigation frequency at stage i and $0 \leq d_i \leq I_i/29$.

递推计算过程,采用逆时序递推。即从第五阶段开始,依次递推至第一阶段,求出各阶段开始的土壤含水量 θ_{gi}(及相应的土壤供水量 M_i)和可供该阶段及该阶段后各阶段使用的水量 I_i,以及各种组合情况下的最优决策 d_i^*, f_i^*,计算结果见表 135、136、137。

The reverse sequential recursion is used to work out the calculation process. That is, from the fifth stage to the first stage, the soil water content is worked out at the start of the stage θ_{gi} (and corresponding soil water supply M_i), the water available at this stage and stages later on I_i, and the optimal decision d_i^* and f_i^* in each combination. The calculation results are shown in Table 135, 136 and 137.

有了表 135、136、137 就可以得出在各种初始条件下的最优灌溉制度。为此,我们将由 1 阶段进行正向递推可得出灌水决策过程。递推方法如下:

With Table 135, 136 and 137, we can figure out the optimal irrigation scheduling in various starting conditions. Therefore, we will carry out forward recursion to obtain the decision-making process of irrigation by the first stage. Recursive method is as follows:

根据初始土壤含水率在表 136 中查得与之对应的 d_1^* 和 f_1^* 及 M_1,由递推方程分别求出第二阶段、第六阶段、第四阶段和第五阶段的 M 和 I 值,再查表 136 和表 137 可得出最优的灌水决策 d_1^*、d_2^*、d_3^*、d_4^* 和 d_5^*。至此,优化灌溉制度就确定了。表 138 是给出初始土壤含水率为9%和70%的两种情况下的最优灌水决策。

First, according to the initial soil water content, the corresponding d_1^*, f_1^* and M_1 are found in Table 136. Then with recursive equation we figure out the value of M and I at the second, third, fourth and fifth stages respectively, at last refer to Table 136 and 137 to obtain the optimal irrigation decision d_1^*、d_2^*、d_3^*、d_4^* and d_5^*. So far, optimized irrigation scheduling has been settled. Table 138 provides different irrigation decisions on the conditions that initial soil water content are 9% and 7%.

当初始含水率为9%时,相应5个阶段在水量上的分配为:58、29、29、58、116(m³/亩),相应最优产量为:

When initial soil water content is 9%, the water distribution in the corresponding five stages: 58, 29, 29, 58 and 116 (m³/mu), and corresponding maximum yield:

$$Y^* = f_1^* \cdot Y_m = 1.00 \times 220 = 220 \text{ kg/亩}$$

当初始含水率这7%时,相应5个阶段在水量上的分配为:58、29、29、58、116(m³/亩),相应最优产量为:

When initial soil water content is 7%, the distribution of water in the corresponding five stages: 58, 29, 29, 58 and 116 (m³/mu), and corresponding maximum yield:

$$Y^* = f_1^* \cdot Y_m = 0.996 \times 220 = 219.12 \text{ kg/亩}$$

表135　　　　　　　　　　　　　优化灌溉制度递推表

土壤水分状况		第五阶段递推计算 $W_{ms}=77.52$ m³/亩, $\lambda_5=0.2314$ $f_5^* = d_5^{max}\{f_5\}$ $= d_5^{max}\{W_5/77.52\}^{0.2314}$ 其中: $W_5 = \min\{77.52, M_5 + 29d_5\}$ $0 \le d_5 \le I_5/29$			第四阶段递推计算 $Wm_4=69.93$ m³/亩; $\lambda_4=0.4182$ $f_4^* = d_4^{max}\{(\frac{W_4}{69.93})^{0.4182}\} \cdot f_5^*(M_5, I_5)$ 其中: $W_4 = \min(69.93, M_4+29d_4)$ $0 \le d_4 \le I_4/29$ $M_5 = M_4+29d_4-W_4$　$I_5=I_4-29d_4$				
含水率 $Q_{5.4}$	供水量 $M_{5.4}$	d_5^*/f_5^*			d_4^*/f_4^*				
		$I_5=29$	$I_5=58$	$I_5=87$	$I_4=58$	$I_4=87$	$I_4=116$	$I_4=145$	$I_4=174$
11.0	31.38	1/0.944	2/1	2/1	0/0.669	0/0.715	1/0.940	2/1	2/1
10.0	26.89	1/0.927	21/	21/	0/0.626	0/0.671	1/0.910	2/1	2/1
9.0	22.41	1/0.909	2'/1	2/1	0/0.581	0/0.621	1/0.879	2/1	2/1
8.0	17.93	1/0.8904	2/0.995	3/1	0/0.529	0/0.566	1/0.846	2/1	2/1
7.0	13.45	1/0.870	2/0.981	3/1			1/0.775	2/1	2/1
6.0	8.96	1/0.848	2/0.967	3/1				2/0.982	3/1
5.0	4.48	1/0.823	2/0.951	3/1				2/0.954	3/1
4.0	0	1/0.797	2/0.935	3/1					

Table 135   **Recursion Table for Optimized Irrigation Scheduling**

Condition of Soil Water		Recursive calculation for the fifth stage $W_{ms} = 77.52 \ \text{m}^3/\text{mu}, \ \lambda_5 = 0.2314$ $f_5^* = d_5^{\max}\{f_5\}$ $= d_5^{\max}\{W_5/77.52\}^{0.2314}$ Where $W_5 = \min\{77.52, M_5 + 29d_5\} \ 0 \leq d_5 \leq I_5/29$			Recursive calculation for the fourth stage $Wm_4 = 69.93 \ \text{m}^3/\text{mu}; \ \lambda_4 = 0.4182$ $f_4^* = d_4^{\max}\{(\frac{W_4}{69.93})^{0.4182}\}f_5^*(M_5, I_5)$ Where: $W_4 = \min(69.93, M_4 + 29d_4)$ $0 \leq d_4 \leq I_4/29$ $M_5 = M_4 + 29d_4 - W_4 \quad I_5 = I_4 - 29d_4$				
Water Content $Q_{5.4}$	Water Supply $M_{5.4}$	\multicolumn{3}{c}{d_5^* / f_5^*}			\multicolumn{5}{c}{d_4^* / f_4^*}				
		$I_5 = 29$	$I_5 = 58$	$I_5 = 87$	$I_4 = 58$	$I_4 = 87$	$I_4 = 116$	$I_4 = 145$	$I_4 = 174$
11.0	31.38	1/0.944	2/1	2/1	0/0.669	0/0.715	1/0.940	2/1	2/1
10.0	26.89	1/0.927	21/	21/	0/0.626	0/0.671	1/0.910	2/1	2/1
9.0	22.41	1/0.909	2′/1	2/1	0/0.581	0/0.621	1/0.879	2/1	2/1
8.0	17.93	1/0.8904	2/0.995	3/1	0/0.529	0/0.566	1/0.846	2/1	2/1
7.0	13.45	1/0.870	2/0.981	3/1			1/0.775	2/1	2/1
6.0	8.96	1/0.848	2/0.967	3/1				2/0.982	3/1
5.0	4.48	1/0.823	2/0.951	3/1				2/0.954	3/1
4.0	0	1/0.797	2/0.935	3/1					

表 136   优化灌溉制度递推表

土壤水分状况		第三阶段递推计算 $W_{M3} = 31.25 \ \text{m}^3/亩 \quad \lambda_3 = 0.1362$ $f_3^* = d_3^{\max}\{(\frac{W_3}{W_{M3}})^{0.1362}\} \cdot f_4^*\{M_4, I_4\}$ 其中:$W_3 = \min\{31.25, M_3 + 29d_3\}$ $0 \leq d_3 \leq I_3/29$ $M_4 = M_3 + 29d_3 - W_3 \quad I_4 = I_3 - 29d_3$					
含水率 Q_3	供水量 M_3	\multicolumn{6}{c}{d_3^* / f_3^*}					
		$I_3 = 58$	$I = 87$	$I = 116$	$I = 145$	$I = 174$	$I = 203$
11.0	31.38	0/0.552	0/0.737	0/0.866	0/0.926	0/1	0/1
10.0	26.89	0/0.540	0/0.721	0/0.860	0/0.906	0/0.980	1/1
9.0	22.41	0/0.527	0/0.704	0/0.839	0/0.884	0/0.956	1/1
8.0	17.93		0/0.683	0/0.810	0/0.857	0/0.927	1/1
7.0	13.45		0/0.657	0/0.783	0/0.824	0/0.892	1/1
6.0	8.96		0/0.621	0/0.741	0/0.780	0/0.844	1/1
5.0	4.48		0/0.565	0/0.675	0/0.710	0/0.768	1/1
4.0							

Table 136　　　　　Recursion Table for Optimized Irrigation Scheduling

condition of soil water		Recursive calculation for the third stage					
		$W_{M3} = 31.25 \text{ m}^3/\text{mu} \quad \lambda_3 = 0.1362$					
		$f_3^* = d_3^{\max}\{(\dfrac{W_3}{W_{M3}})^{0.1362}\} \cdot f_4^*\{M_4, I_4\}$					
		Where：$W_3 = \min\{31.25, M_3 + 29d_3\}$					
		$0 \leqslant d_3 \leqslant I_3/29$					
		$M_4 = M_3 + 29d_3 - W_3 \quad I_4 = I_3 - 29d_3$					
Water Content	Water Supply	d_3^*/f_3^*					
Q_3	M_3	$I_3 = 58$	$I = 87$	$I = 116$	$I = 145$	$I = 174$	$I = 203$
11.0	31.38	0/0.552	0/0.737	0/0.866	0/0.926	0/1	0/1
10.0	26.89	0/0.540	0/0.721	0/0.860	0/0.906	0/0.980	1/1
9.0	22.41	0/0.527	0/0.704	0/0.839	0/0.884	0/0.956	1/1
8.0	17.93		0/0.683	0/0.810	0/0.857	0/0.927	1/1
7.0	13.45		0/0.657	0/0.783	0/0.824	0/0.892	1/1
6.0	8.96		0/0.621	0/0.741	0/0.780	0/0.844	1/1
5.0	4.48		0/0.565	0/0.675	0/0.710	0/0.768	1/1
4.0							

表 137　　　　　优化灌溉制度递推表

土壤水分状况		第二阶段递推计算			第一阶段递推计算		
		$Wm_2 = 23.98 \text{ m}^3/\text{亩}, \lambda_2 = 0.0216$			$Wm_1 = 51.28 \text{ m}^3/\text{亩}; \lambda_1 = 0.2213$		
		$f_2^* = d_2^{\max}\{(W_2/23.98)^{0.0216}\} \cdot f_3^*\{M_3, I_3\}$			$f_1^* = d_1^{\max} \cdot \{(W_1/51.28)^{0.2213} \cdot$		
		其中：$W_2 = \min\{23.98, M_2 + 29d_2\}$			$f_2^*\{M_2, I_2\}\}$		
		$0 \leqslant d_2 \leqslant I_2/29$			其中：$W_1 = \min(51.28, M_1 + 29d_1)$		
		$M_3 = M_2 + 29d_2 - W_{42}$			$0 \leqslant d_1 \leqslant I_1/29$		
		$I_3 = I_2 - 29d_2$			$M_2 = M_1 + 29d_1 - W_1$		
					$I_2 = I_1 - 29d_1$		
含水率	供水量	d_2^*/f_2^*			d_1^*/f_1^*	
$Q_{2.1}$	$M_{2.1}$	$I_2 = 174$	$I_2 = 203$	$I_2 = 232$	$I_1 = 261$	$I_1 = 290$
11.0	31.38	0/0.958	0/1	0/1	0/0.897	1/1
10.0	26.89	0/0.929	0/1	0/1	0/0.867	2/1
9.0	22.41	0/0.915	0/0.988	1/1	0/0.833	2/1
8.0	17.93	0/0.910	0/0.984	1/1	0/0.793	2/1
7.0	13.45	0/0.904	0/0.978	1/1	0/0.744	0/0.996
6.0	8.96	0/0.896	0/0.969	1/1	0/0.680	2/0.991
5.0	4.48	0/0.883	0/0.954	1/1	0/0.583	2/0.984
4.0	0					

第三章 作物灌溉制度研究
3 Study on Irrigation System of Crops

Table 137 Recursion Table for Optimized Irrigation Scheduling

condition of soil water		Recursive calculation for the second stage			Recursive calculation for the first stage	
		$Wm_2 = 23.98 \text{ m}^3/\text{mu}, \lambda_2 = 0.0216$ $f_2^* = d_2^{\max}\{(W_2/23.98)^{0.0216}\} \cdot f_3^* = \{M_3, I_3\}$ Where: $W_2 = \min\{23.98, M_2 + 29d_2\}$ $0 \leq d_2 \leq I_2/29$ $M_3 = M_2 + 29d_2 - W_2$ $I_3 = I_2 - 29d_2$			$Wm_1 = 51.28 \text{ m}^3/\text{mu}, \lambda_1 = 0.2213$ $f_1^* = d_1^{\max}\{(W_1/51.28)^{0.2213}\} \cdot f_2^* = \{M_2, I_2\}$ Where: $W_1 = \min(51.28, M_1 + 29d_1)$ $0 \leq d_1 \leq I_1/29$ $M_2 = M_1 + 29d_1 - W_1$ $I_2 = I_1 - 29d_1$	
Water Content	Water Supply	d_2^* / f_2^*			d_1^* / f_1^*	
$Q_{2,1}$	$M_{2,1}$	$I_2 = 174$	$I_2 = 203$	$I_2 = 232$	$I_1 = 261$	$I_1 = 290$
11.0	31.38	0/0.958	0/1	0/1	0/0.897	1/1
10.0	26.89	0/0.929	0/1	0/1	0/0.867	2/1
9.0	22.41	0/0.915	0/0.988	1/1	0/0.833	2/1
8.0	17.93	0/0.910	0/0.984	1/1	0/0.793	2/1
7.0	13.45	0/0.904	0/0.978	1/1	0/0.744	0/0.996
6.0	8.96	0/0.896	0/0.969	1/1	0/0.680	2/0.991
5.0	4.48	0/0.883	0/0.954	1/1	0/0.583	2/0.984
4.0	0					

表 138 不同初始含水率最优决策

初始土壤含水率	阶段	M_i (m³/亩)	I_i (m³/亩)	d^*	f^*	θ_i
9%	1	22.41	290	2	1.00	9.0
	2	29.13	232	1		10.50
	3	34.15	203	1		14.62
	4	31.9	174	2		14.12
	5	19.97	116	4	11.46	
7%	1	13.45	290	2	0.996	7.0
	2	20.17	232	1		11.50
	3	25.19	203	1		12.62
	4	22.94	174	2		12.12
	5	11.01	116	4		9.46

Table 138　　**Optimal Decision on Different Initial Water Contents**

Initial Soil Water Content	Stage	M_i (m^3/mu)	I_i (m^3/mu)	d^*	f^*	θ_i
9%	1	22.41	290	2	1.00	9.0
	2	29.13	232	1		10.50
	3	34.15	203	1		14.62
	4	31.9	174	2		14.12
	5	19.97	116	4	11.46	
7%	1	13.45	290	2	0.996	7.0
	2	20.17	232	1		11.50
	3	25.19	203	1		12.62
	4	22.94	174	2		12.12
	5	11.01	116	4		9.46

(二) 灌水时间 T 的确定

3.1.7.2　Irrigation Time T Definition

从农田水量平衡可知：

Based on the water balance of farmland, it can be seen that:

$$ET = P_e + G + \Delta W \text{ 或 } e_t \cdot T = P_e + g \cdot T + \Delta W$$
$$ET = P_e + G + \Delta W \text{ or } e_t \cdot T = P_e + g \cdot T + \Delta W$$

$$T = \frac{\Delta W + P_e}{e_t - g}$$

式中　T——未来灌水的间隔时间；

　　　ΔW——土壤有效储水量；

　　　P_e——有效降雨量；

　　　e_t——日耗水强度，mm/d，$e_t = K_c \cdot ET$；

　　　g——地下水补给强度，mm/d。

Where　T = the interval of irrigation in the future

　　　ΔW = effective soil water storage

　　　P_e = effective precipitation

　　　e_t = water consumption intensity per day, mm/d, $e_t = K_c \cdot ET$

　　　g = recharge intensity of groundwater, mm/d.

本试验区地下水水位在3 m 以下，所以无地下水补给 $g=0$。

The groundwater level in this experimental area is below 3 m, so there is no groundwater recharge and $g = 0$.

表 139 中 T_1, T_2 分别表示初始土壤含水率为 9%、7% 两种情况的优化灌溉间隔天数。

In Table 139, T_1, T_2 mean the optimized irrigation interval when initial soil water contents are 9% and 7% respectively.

表 139　　　　　　　　　春小麦优化灌溉制度灌水时间表

生育阶段	播种—分蘖	分蘖—拔节	拔节—抽穗	抽穗—灌浆	灌浆—成熟
9% · ΔW_1 (mm)	61.86	72.17	100.49	97.05	78.77
7% · ΔW_2 (mm)	68.73	79.04	79.87	83.30	65.02
P_e (mm)	0	3.55	4.57	5.0	18.90
e_t (mm)	7.17	12.82	12.31	12.74	12.74
T_1 (d)	9	6	9	8	8
T_2 (d)	10	7	7	7	7

Table 139　　Timetable for the Optimized Irrigation Scheduling of Spring Wheat

Growth and Development Period	Sowing – Tillering	Tillering – Jointing	Jointing – Heading	Heading – Filling	Filling – Maturing
9% · ΔW_1 (mm)	61.86	72.17	100.49	97.05	78.77
7% · ΔW_2 (mm)	68.73	79.04	79.87	83.30	65.02
P_e (mm)	0	3.55	4.57	5.0	18.90
e_t (mm)	7.17	12.82	12.31	12.74	12.74
T_1 (d)	9	6	9	8	8
T_2 (d)	10	7	7	7	7

至此，我们就得出春小麦的优化灌溉制度，从而给出初始含水率为 9% 和 7% 两种情况下的优化灌溉制度（见表 140）。

Thus, we get the optimized irrigation scheduling of spring wheat and the two conditions when initial water content are 9% and 7% (see Table 140).

表140　　　　　　　　　　　春小麦灌溉制度分析表

初始土壤含水率	阶段	灌水定额(m³/亩)	灌水次数	灌水间隔天数(d)
9%	1	29	2	1
	2		1	6
	3		1	9
	4		2	8
	5		4	8
7%	1	29	2	10
	2		1	7
	3		1	7
	4		2	7
	5		4	7

Table 140　　　　　Analysis of Irrigation Scheduling of Spring Wheat

Initial Soil Water Content	Stage	Irrigation Water Quota (m³/mu)	Irrigation Frequency	Irrigation Interval (days)
9%	1	29	2	1
	2		1	6
	3		1	9
	4		2	8
	5		4	8
7%	1	29	2	10
	2		1	7
	3		1	7
	4		2	7
	5		4	7

八、灌溉效益分摊系数
3.1.8 Sharing Coefficient of Irrigation Benefit

农作物灌溉效益分摊系数,是分析计算水利经济的重要参数。农业增产是灌溉工程设施和农业技术措施综合因素作用的结果。因此,灌溉效益分摊实验,是农业措施和灌溉因素相结合的复因子试验(见表141)。

Crops's haring coefficient of irrigation benefit is a major parameter for analyzing and calculating the economy of irrigation works. Agricultural yield increase is the joint result of comprehensive factors, such as irrigation facilities and agricultural technical measures. Hence, the experiment of irrigation benefit share is a multi-factor experiment combining agricultural measures and irrigation factors (see Table 141).

表 141　　　　　　　　　　春小麦分摊系数试验成果表

项目	Y_1	Y_2	Y_3	Y_4
	（农低+低水）	（农低+高水）	（农高+低水）	（农高+高水）
产量(kg/亩)	140.05	153.92	178.66	218.95
灌水量(m³/亩)	233.80	251.83	220.95	219.29

Table 141　　　　　Experimental Results of Spring Wheat Sharing Coefficient

Item	Y_1	Y_2	Y_3	Y_4
	(low agrotechnique + low irrigation amount)	(low agrotechnique + high irrigation amount)	(high agrotechnique + low irrigation amount)	(high agrotechnique + high irrigation amount)
Yield (kg/mu)	140.05	153.92	178.66	218.95
Irrigation Amount (m³/mu)	233.80	251.83	220.95	219.29

（一）一般农业技术水平措施下灌溉增产率(%)

3.1.8.1　Yield-increase Rate with General Agricultural Technique (%)

$$\Delta C_Y l = [(Y_2 - Y_1)/Y_1] \times 100 = \frac{153.92 - 140.05}{140.05} \times 100 = 9.9\%$$

（二）高农业技术水平措施下灌溉增产率(%)

3.1.8.2　Yield-increase Rate with High Agricultural Technique (%)

$$\Delta C_Y h = [(Y_4 - Y_3)/Y_3] \times 100 = \frac{218.95 - 178.66}{178.66} \times 100 = 22.55\%$$

（三）效益分摊系数

3.1.8.3　Sharing Coefficient of Irrigation Benefit

$$\sum{}_{水} = [(Y_2 + Y_4) - (Y_1 + Y_3)]/2(Y_4 - Y_1) = 0.343$$

$$\sum{}_{Water} = [(Y_2 + Y_4) - (Y_1 + Y_3)]/2(Y_4 - Y_1) = 0.343$$

$$\sum_{\text{农}} = [(Y_3 + Y_4) - (Y_1 + Y_2)]/2(Y_4 - Y_1) = 0.657$$

$$\sum_{\text{Agrotechnique}} = [(Y_3 + Y_4) - (Y_1 + Y_2)]/2(Y_4 - Y_1) = 0.657$$

九、试验结论
3.1.9 Conclusion

春小麦最优灌溉定额为289 m³/亩,灌水定额为29 m³/亩,灌水次数10次,最大纯收益58.50元/亩。

The optimized irrigation quota of spring wheat is 289 m³/mu and irrigation water quota is 29 m³/mu with 10 times and the maximum net return of 58.50 yuan/mu.

把春小麦整个生育期分为5个生育阶段,建立优化灌溉数学模式,推算出土壤初始条件下的优化灌溉方案,即优化灌水定额的分配,结果见表142。

Through dividing the whole growth and development period of spring wheat into 5 stages and establishing mathematical model for optimized irrigation scheduling, we figure out the optimized irrigation scheme under different initial soil conditions, namely the allocation of better irrigation water quota. The results are shown in Table 142.

表142　　　　　　　　春小麦各生育阶段灌水分配表

生育阶段	播种—分蘖	分蘖—拔节	拔节—抽穗	抽穗—灌浆	灌浆—成熟
灌水次数	2	1	1	2	4
灌水量	58	29	29	58	116

Table 142　Allocation of Irrigation Water Quota for Spring Wheat in Each Growth and Development Period

Growth and Development Period	Sowing – Tillering	Tillering – Jointing	Jointing – Heading	Heading – Filling	Filling – Maturing
Irrigation Frequency	2	1	1	2	4
Irrigation Amount	58	29	29	58	116

春小麦各生育阶段的水分敏感指数为:分蘖:0.2212,拔节:0.0216,抽穗:0.1362,灌浆:0.4182,成熟:0.2314。各生育阶段对水分敏感反映程度的大小顺序为:灌浆>成熟>分蘖>抽穗>拔节。

The water sensitivity index of spring wheat in each growth and development period: tillering: 0.2212, jointing: 0.0216, heading: 0.1362, filling: 0.4182, maturing: 0.2314.

The order of water sensitivity indexes at each stage is filling > maturing > tillering > heading > jointing.

春小麦效益分摊系数为：
The spring wheat's sharing coefficient of irrigation benefit：

一般农业技术水平下灌溉增产率大于90%；
With general agrotechnique, the yield – increase rate tops 90%；

高农业技术水平下灌溉增产率为22.55%；
While the rate remains 22.55% with high agrotechnique；

灌溉效益分摊系数：0.343；
Sharing coefficient of irrigation benefit：0.343；

农业效益分摊系数：0.657。
Sharing coefficient of agricultural benefit：0.657.

通过灌溉效益分摊系数的分析计算，得到的结果与本试验区的实际环境中春小麦的生长情况相符合。即在土质条件较差的种植环境下，农业技术措施对小麦产量具有一定的影响。

Through the analysis and calculation of the sharing coefficient of irrigation benefit, the result is consistent with the growth of spring wheat in the practical condition of this experimental zone. Even in the planting environment with poor soil conditions, agricultural technical measures have an impact on wheat yield.

植物生长发育要求一定的热、光、水和营养，缺一不可，而该试验课题是建立在环境条件较差的基础上，仅从优化灌溉方面对小麦单产及经济效益研究，经过灌溉效益分摊系数的研究表明，在作物生长环境较差的地区，如土壤有机质、钾、磷等含量偏低，土质较差地区，在优化灌溉制度前提下，适当增加农业技术水平措施投入，可以提高春小麦的单产并获得较好的经济效益。

In the crops' growth and development, a certain amount of heat, light, water and nutrition is necessary, while this experimental subject is based on poor environment conditions, merely studying the per unit yield and economic benefits of wheat from irrigation methods. According to the study on sharing coefficient of irrigation benefits, in a poor environment where the soil is of poor quality and lacking of soil organic matter, potassium and phosphorus, proper increase of agricultural technique helps enhance per unit area yield of spring wheat and gain favorable economic benefits on the basis of optimized irrigation scheduling.

第二节 玉米灌溉制度与灌溉效益分摊系数研究

3.2 Irrigation System of Corn and Its Irrigation Benefit

一、试验区概况
3.2.1 Overview of Experimental Site

玉米所采用的品种是阿单一号，试验区面积为0.12亩，种距行距60 cm，株距25 cm，每亩株数4 400株，试验设3次重复。

The variety of corn: Adan #1, the area of experimental zone: 0.12 mu, row spacing: 60 cm, plant spacing: 25 cm, the number of corn plants per mu: 4 400, 3 repetitions.

试验区土壤为砂壤土，田间持水率为11.23%，干容重为1.445 g/cm³，基础肥力不高，有机质含量为0.49%。

The quality of soil in experimental zone: sandy loam, field capacity: 11.23%, dry bulk density: 1.445 g/cm³, basic fertility: low, organic content: 0.49%.

二、试验方案与方法
3.2.2 Experiment Scheme and Methods

（一）试验的目的和意义
3.2.2.1 Goal and Significance of the Experiment

玉米灌溉制度与灌溉效益分摊系数的研究是对玉米在不同农业措施和灌溉水平下进行灌溉制度的分析及农业措施和灌溉水平对玉米产量产生的影响及所占比例。

The irrigation scheduling and sharing coefficient of irrigation benefit for corn is aimed at analyzing the irrigation scheduling under different agricultural measures and irrigation levels and the effect of agricultural measures and irrigation levels.

第三章 作物灌溉制度研究
3　Study on Irrigation System of Crops

（二）试验方案的设计
3.2.2.2　Experiment Scheme Design

试验设计的思想以灌溉水生产函数为主体,结合灌溉效益进行试验研究,通过一次试验,获得多项成果,设计方案见表143。

The experiment was conducted by following the irrigation water production function and combining with irrigation benefits to achieve multiple results by one time experiment. The scheme is outlined in Table 143.

（三）试验方法
3.2.2.3　Experimental Methods

试验采用小区灌溉,小区长19 m,宽4.2 m,面积0.12亩,试验分7个处理,3次重复共21个小区,采用顺序排列,灌溉方式采用畦灌。玉米播种量3 kg/亩,播种带肥磷酸二铵10 kg/亩,生育期农先中耕2次,施尿素2次,肥量20 kg/亩。农中中耕1次,施尿素10 kg/亩。

The irrigation plot in this experiment was 19 m long and 4.2 m wide, covering an area of 0.12 mu. The experiment set 7 treatments and 3 repetitions in total 21 plots in a sorted order and used the way of border irrigation. Seeding rate of corn: 3 kg/mu, and fertilizer of seeding zone: diammonium phosphate 10 kg/mu, in the growth and development period cultivate 2 times before agricultural measures, apply urea 2 times, fertilizing amount: 20 kg/mu. Cultivate one time in the middle of agricultural measures and apply urea 10 kg/mu.

表143　各生育期土壤湿度（占田间持水率%）

处理	农业措施	拔节	抽雄—散粉	吐丝—灌浆
1	农中	60	65	60
2	农先	60	65	60
3	农先	40	65	60
4	农先	60	45	60
5	农先	60	65	45
6	农先	40	65	45
7	农中	40	45	45

Table 143　　　　Soil Humidity in Each Growth and Development Period
(Percentage in field capacity, %)

Treatment	Agricultural Measures	Jointing Period	Tasseling-Pollinating	Silking-Filling
1	In the Middle of Agricultural Measures	60	65	60
2	Before Agricultural Measures	60	65	60
3	Before Agricultural Measures	40	65	60
4	Before Agricultural Measures	60	45	60
5	Before Agricultural Measures	60	65	45
6	Before Agricultural Measures	40	65	45
7	In the Middle of Agricultural Measures	40	45	45

三、玉米相对耗水量分析
3.2.3 Analysis of Relative Water Consumption of Corn

本次试验是将玉米的全生育期划分为苗期、拔节期、抽雄期、灌浆期和成熟期,其生育期各阶段耗水量见表144。

The experiment divided the whole period of corn into seeding, jointing, tasseling, filling and maturing stages, and water consumption in each stage is showed in Table 144.

表144　　　　　　　　　玉米生育期耗水量表　　　　　　　　　单位:m³/亩

处理	苗期	拔节	抽雄	灌浆	成熟	总耗水量
1	72.96	39.59	81.59	52.15	45.18	288.14
2	62.36	42.42	79.99	37.46	47.00	269.22
3	65.87	36.39	71.27	37.08	32.71	243.31
4	68.65	27.98	39.61	15.15	37.42	189.00
5	62.05	23.62	49.24	23.03	27.70	185.13
6	75.90	25.43	53.79	22.95	35.02	213.09
7	66.85	19.07	51.33	37.07	27.60	201.86

第三章 作物灌溉制度研究
3 Study on Irrigation System of Crops

Table 144　　　　　Water Consumption of Corn at Each Stage　　　　Unit: m³/mu

Treatment	Seedling	Jointing	Tasseling	Filling	Maturing	Total Water Consumption
1	72.96	39.59	81.59	52.15	45.18	288.14
2	62.36	42.42	79.99	37.46	47.00	269.22
3	65.87	36.39	71.27	37.08	32.71	243.31
4	68.65	27.98	39.61	15.15	37.42	189.00
5	62.05	23.62	49.24	23.03	27.70	185.13
6	75.90	25.43	53.79	22.95	35.02	213.09
7	66.85	19.07	51.33	37.07	27.60	201.86

四、玉米耗水量与相对产量分析
3.2.4 Analysis of Water Consumption and Relative Yield of Corn

玉米优化灌溉制度试验通过 2 年的试验，通过平均方法来计算具体成果见表 145。

Through 2-year experiment, we get the optimized irrigation scheduling for corn and work out the concrete results in Table 145 with the averaging method.

表 145　　　　　1990—1991 年度玉米灌溉制度成果表　　　　单位：kg/亩、m³/亩

处理	1990 年		1991 年		平均	
	产量	水量	产量	水量	产量	水量
1	416.92	344.02	402.15	233.86	409.54	288.94
2	460.28	329.37	414.10	209.06	437.16	267.22
3	343.77	286.12	407.4	200.50	375.46	243.31
4	290.22	245.22	368.27	132.77	329.25	189.00
5	303.36	244.71	355.00	125.55	329.18	185.13
6	355.17	281.05	356.93	145.12	356.05	213.09
7	281.39	242.05	296.67	161.67	289.03	201.86

Table 145 Corn Irrigation Scheduling Results in 1990–1991   Units: kg/mu, m³/mu

Treatment	1990		1991		Average	
	Output	Water Amount	Output	Water Amount	Output	Water Amount
1	416.92	344.02	402.15	233.86	409.54	288.94
2	460.28	329.37	414.10	209.06	437.16	267.22
3	343.77	286.12	407.4	200.50	375.46	243.31
4	290.22	245.22	368.27	132.77	329.25	189.00
5	303.36	244.71	355.00	125.55	329.18	185.13
6	355.17	281.05	356.93	145.12	356.05	213.09
7	281.39	242.05	296.67	161.67	289.03	201.86

五、边际分析
3.2.5 Marginal Analysis

（一）玉米水分生产函数
3.2.5.1 Water Production Function of Corn

作物产量与光合作用强度有密切关系，而光合作用又受蒸腾失水的影响。同时作物在其生产发育不同阶段的蒸腾失水量及其对缺水的敏感程度是不同的，并且在某一阶段缺水，不仅影响本生育阶段，而且对下一阶段也产生影响，无论哪个阶段缺水都会导致作物不同程度的减产，所以作物产量与各生育阶段的缺水具有连乘的集合关系。

Crop yield has a close relation with photosynthetic intensity, while photosynthesis is affected by water loss in transpiration. At the same time, the water loss in transpiration of the crop and its sensitivity to water shortage vary in different stages. Water shortage in a stage affects not only the current stage but also the next one. No matter which stage suffers from water shortage, it will lead to different degrees of crop failure, therefore showing a multiple aggregation relationship between crop yield and water shortage in each stage.

作物水分生产函数相乘模型，用 M·E·Jensen 1968 年提出的模式表达：

The multiplication model of crop water production function is expressed as the following with what was proposed by M·E·Jensen in 1968:

$$Y/Y_M = \sum_{i=1}^{n} \left(\frac{W_i}{w_{Mi}}\right)^{\lambda_i}$$

式中　　n——作物生产划分的阶段数；

　　　　W_{Mi}——作物在 i 阶段充足供水时的最大耗水量，m³/亩；

3 Study on Irrigation System of Crops

W_i——作物在 i 阶段供水不足时的实际耗水量，m^3/亩；
Y_M——作物各阶段耗水量达到最大值时的最高产量，kg/亩；
Y——作物各生长阶段由于供水不足只能达到的产量，kg/亩；
λ_i——阶段 i 的水分敏感指数。

Where n = the number of crops production stages
W_{Mi} = the highest water consumption in stage i when there is sufficient water supply (m^3/mu)
W_i = the real water consumption in stage i when there is insufficient water supply (m^3/mu)
Y_M = the production peak when the water consumption of crops peaks in each stage (kg/mu)
Y = crops yield due to water strain in each stage (kg/mu)
λ_i = water sensitivity index in stage i.

本次试验处理 1 为最大耗水量，以处理 1 为参照进行水分敏感指数的计算。

In the experiment, treatment 1 consumed the most water, thus taking treatment 1 as reference group to calculate the water sensitivity index.

将 Jensen 公式两边取自然对数：
The natural logarithm of both sides of Jensen's formula is taken as：

$$\ln(Y/Y_M) = \sum \lambda_i \ln(W_i/W_{Mi})$$

令 $\ln(Y/Y_M) = Z$ $\ln(W_i/W_{Mi}) = X_i$
Let $\ln(Y/Y_M) = Z$ $\ln(W_i/W_{Mi}) = X_i$

则为 $Z = \sum \lambda_i \cdot X_i$
Then $Z = \sum \lambda_i \cdot X_i$

成为一个多元线型方程，代入试验的 7 组灌溉数据，列出五元一次方程，利用矩阵高斯消元法，求出各阶段 λ_i 的值，根据表 146 及表 147，计算出玉米各阶段水分敏感指数：

Here we get a multiple linear equation. Substitute the 7 groups' irrigation data in this equation and write down a linear equation in five unknowns. By Gaussian elimination, we figure out the value of λ_i in each stage. According to Table 146 and Table 147, the water sensitivity index of corn is worked out in each stage：

玉米的水分生产函数为：
The water production functions of corn：

$$Y/Y_M = (W_1/W_{M1})^{0.2111}(W_2/W_{M2})^{0.8317}(W_3/W_{M3})^{0.6237}(W_4/W_{M4})^{0.4807}(W_5/W_{M5})^{0.2623}$$

即
$$Y = 409.54(0.0137W_1)^{0.2111}(0.0253W_2)^{0.8317}(0.0123W_3)^{0.6237}$$
$$(0.0192W_4)^{0.4807}(0.0221W_5)^{0.2623}$$

Namely
$$Y = 409.54(0.0137W_1)^{0.2111}(0.0253W_2)^{0.8317}(0.0123W_3)^{0.6237}$$
$$(0.0192W_4)^{0.4807}(0.0221W_5)^{0.2623}$$

苗期	拔节期	抽雄期	灌浆期	成熟期
0.2111	0.8317	0.6237	0.4807	0.2623

Seedling	Jointing	Tasseling	Filling	Maturing
0.2111	0.8317	0.6237	0.4807	0.2623

表 146　玉米耗水量及相对产量计算表　　　　　单位：m³/亩、kg/亩

处理	苗期	拔节	抽雄	灌浆	成熟	总耗水量	产量
1	72.96	39.59	81.59	52.15	45.18	288.94	409.54
2	62.36	42.42	79.99	37.46	47.00	261.42	437.19
3	65.87	36.39	71.27	37.08	32.71	241.31	375.46
4	68.65	27.98	39.61	15.15	37.42	189.00	329.46
5	62.05	23.62	49.24	23.03	27.70	185.13	329.18
6	75.90	25.43	53.79	22.95	35.02	213.07	356.05
7	66.85	19.07	51.33	37.07	27.60	201.86	289.03

Table 146　Computation for Water Consumption and Relative Yield of Corn

Units: m³/mu, kg/mu

Treatment	Seedling	Jointing	Tasseling	Filling	Maturing	Total Water Consumption	Output
1	72.96	39.59	81.59	52.15	45.18	288.94	409.54
2	62.36	42.42	79.99	37.46	47.00	261.42	437.19
3	65.87	36.39	71.27	37.08	32.71	241.31	375.46
4	68.65	27.98	39.61	15.15	37.42	189.00	329.46
5	62.05	23.62	49.24	23.03	27.70	185.13	329.18
6	75.90	25.43	53.79	22.95	35.02	213.07	356.05
7	66.85	19.07	51.33	37.07	27.60	201.86	289.03

表 147 玉米相对耗水量及相对产量计算表

处理	苗期	拔节	抽雄	灌浆	成熟	全生育期	相对产量
1	1.000 0	1.000 0	1.000 0	1.000 0	1.000 0	1.000 0	1.000 0
2	0.854 7	1.071 5	0.980 6	0.718 3	1.040 3	0.931 8	1.067 5
3	0.902 8	0.919 2	0.873 7	0.711 0	0.724 0	0.842 1	0.916 8
4	0.940 9	0.706 7	0.485 6	0.290 5	0.828 2	0.654 1	0.804 5
5	0.850 5	0.596 6	0.603 7	0.441 6	0.613 1	0.640 7	0.803 8
6	1.040 3	0.642 3	0.659 4	0.440 1	0.775 1	0.737 5	0.869 4
7	0.916 3	0.481 7	0.629 3	0.710 8	0.610 9	0.698 6	0.705 7

Table 147 Computation for Relative Water Consumption and Relative Yield of Corn

Treatment	Seedling	Jointing	Tasseling	Filling	Maturing	Whole Growth and Development Period	Relative Yield
1	1.000 0	1.000 0	1.000 0	1.000 0	1.000 0	1.000 0	1.000 0
2	0.854 7	1.071 5	0.980 6	0.718 3	1.040 3	0.931 8	1.067 5
3	0.902 8	0.919 2	0.873 7	0.711 0	0.724 0	0.842 1	0.916 8
4	0.940 9	0.706 7	0.485 6	0.290 5	0.828 2	0.654 1	0.804 5
5	0.850 5	0.596 6	0.603 7	0.441 6	0.613 1	0.640 7	0.803 8
6	1.040 3	0.642 3	0.659 4	0.440 1	0.775 1	0.737 5	0.869 4
7	0.916 3	0.481 7	0.629 3	0.710 8	0.610 9	0.698 6	0.705 7

由水分敏感指数看出,玉米各阶段缺水率的高低,对玉米产量将产生不同程度的影响,各生育阶段对水分敏感度反应大小为:拔节期＞抽雄期＞灌浆期＞成熟期＞苗期。

With the water sensitivity index, it is shown that corn yield varies according to the degree of water strain in each stage. And the order of water sensitivity indexes of each growth stage is: jointing period ＞ tasseling period ＞ filling period ＞ maturing period ＞ seedling period.

这说明玉米在拔节期对水分的反应最敏感,是玉米需水的临界期,必须保证水分的充足供应;苗期水分敏感度最小,土壤水分只需满足其顺利出苗,度过蹲苗期即可。

It demonstrates that corn is most sensitive to water in jointing period, thus it is considered as the critical period of corn water requirement, in which sufficient water supply shall be guaranteed. Being least sensitive to water in seedling period indicates that the soil water content shall only ensure the corn seedling emerges from the ground smoothly in this period.

（二）玉米产量与边际产量
3.2.5.2 Corn Yield and Marginal Yield

边际产量的概念是指相对意义上的最后一次的"增加或变化率"，所以边际产量就是最后增加或升增加的数量。在农业过程中，年增一单位的资源投入，称为边际投入（ΔX）；与此相对应所增加或减少的产量就是边际产量（ΔY）；二者比率 $\Delta Y/\Delta X$ 表示边际生产力，当引入价格概念后，投入资源单价乘以边际投入（$P_X \cdot \Delta X$）称为边际成本，产品单价乘以边际产量（$P_X \cdot \Delta Y$）称为边际产值。根据经济效果原理，边际成本必须小于边际收益，否则边际利润为负值，就会降低经济效益。

The concept of marginal yield refers to the last "increase or rate of change" on a relative level, so the marginal yield equals to the last increase or increment. In agricultural production, every unit of resources input is called marginal input (ΔX); accordingly, the yield, whether increases or decreases, is marginal yield (ΔY); the ratio of $\Delta Y/\Delta X$ represents marginal productivity. When price is introduced, the unit price of resources input multiplied by marginal input ($P_X \cdot \Delta X$) is called marginal cost, and the unit price multiplied by marginal yield ($P_X \cdot \Delta X$) gets the marginal production value. In terms of the principle of economic effects, marginal cost must be less than marginal benefit, otherwise marginal profit will be negative and economic benefit will wane.

边际分析是指对连续增加的各单位因素的作用，及其产生影响程度的分析，进行边际分析必须计算作物的总产量、平均产量和边际产量。

Marginal analysis refers to the analysis of the effect of continuously enhancing single factor and their degree of impact. Conducting marginal analysis needs to figure out the total production, average product and marginal yield of crops.

总产量（Y）是指施用一定量的变动资源所产生的产品总量。

Total production (Y) is the total amount of products produced by a certain amount of variable resources.

平均产量（A）是指每单位变动资源可平均引起的产量。

Average product (A) refers to the production gained from per unit variable resources.

$$A = Y/X$$

边际产量（M）是指连续追加的各单位变动资源所引起的总产量的增加额：

Marginal yield (M) refers to the increment of total production from the continuous increase of per unit variable resources:

$$M = \Delta Y/\Delta X$$

根据导数的意义:函数 $Y=f(X)$ 在点 X 处的导数 $f'(X)$,等于函数 $Y=f(X)$ 曲线在 X 点切线的斜率,即:

According to the significance of derivatives: the derivative of function $Y=f(X)$ at X point is equal to the slope of tangent line at X point on the curve of function $Y=(X)$, namely:

$$M = dY/dX = f'(x)$$

生产弹性(E_P)是指反映产量增加对于投入资源的敏感程度,也就是产量增加幅度与资源增加幅度的比例关系。

Production flexibility (E_p) reflects the sensitivity of production growth to the resources input, namely the proportional relation between the increment of resources and production.

$$E_P = (\Delta Y/Y/\Delta X/X) = (\Delta Y/\Delta X) \cdot (X/Y) = M/A$$

根据1990—1991年玉米灌溉试验的资料,分别计算玉米的总产量、平均产量、边际产量和生产弹性列于表148,并绘制玉米物质生产函数曲线见图37。

Total yield, average yield, marginal yield and production elasticity (see Table 148) are all obtained by the calculation based on the corn irrigation experiment data from 1990 to 1991, and a production function curve is drawn (see Figure 37).

由图37看出,总产量(Y)曲线随水量增加而总产量的增加,由慢变快,以后又逐渐减慢,当耗水量为270 m³/亩时,总产量达到顶点427 kg/亩左右时,这时再增加水量,不但不能增产,反而会引起产量下降,由于总产量曲线达到顶点时,这个点斜率为0,这时的边际产量也等于0。

It's easy to find out from the figure that with the increasing water amount, the total yield (Y) increases from slowly to quickly, and then slowly. When the water consumption reaches 270 m³/mu, the total yield peaks at about 427 kg/mu. The yield will stop increasing if the irrigation amount keeps increasing or it will decrease if the irrigation amount is excessive. As total production reaches the peak, where the slope is zero, marginal yield is zero too.

平均产量(A)曲线随耗水量增加而增加,但平均产量的最高点与总产量的最高点并不一致。

The average product (A) curve climbs up as water consumption soars, but the peak of average product doesn't identify with that of total production.

这3条曲线有2个重要的转折点,一是边际曲线与平均产量相交处,边际产量等于平

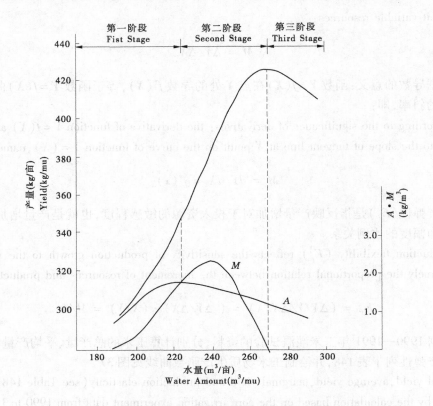

图 37　玉米物质生产函数曲线
Figure 37　Matter Production Function Curve of Corn

均产量为1.76；二是总产量曲线的顶点，边际产量在这个点上为零，过后转为负值，以原点到平均产量的最高点为第一阶段，这个阶段的特点是边际产量高于平均产量，因而引起平均产量逐渐提高，总产量的增加幅度小于资源增加的幅度，出现报酬递减。总产量曲线顶点是在一定技术条件下所可能达到的一种界限，这时的边际产量为零。在这一阶段总产量按照报酬递减的形势增长，可以获得较高的产量，在一定限度内，继续投入资源仍可获得利益，所以是资源投入的最适点，即可以获得最大收益的点，因此称第二阶段为生产合理阶段；第三阶段是在总产量曲线的最高点以后，也就是边际产量转为负值以后的阶段。在这一阶段内越增加投入越减产，显然这是生产绝对不合理的阶段。

These three curves have two significant turning points, one is where marginal curve intersects with average product curve, at which marginal yield is 1.76 kg/m³ and equals average product; the other is where total production peaks while marginal yield becomes zero and turns negative after this point. Taking the origin to the peak of average product as the first stage, this stage features marginal yield overtaking average product. Therefore average product gets boosted, and the increment of total production cannot compare with that of resources input, resulting in diminishing returns. The peak of total production curve is a limit that can be reached on a certain technical condition, and at this point the marginal yield is 0. At this stage, the total production rises with diminishing returns, making high yield possible. Within a certain limit, additional resources input still generates benefits, so it is the optimal point for resources input and the point that can bring the maximum return. Therefore, the second stage is called a reasonable stage of production. The third stage starts from the zenith of total production, namely where marginal yield turns negative. The more the resources, the less the yield. Thus this stage is the absolutely unreasonable stage of production.

表 148 玉米生产函数表统计表

编号	处理号	灌溉定额 X	总产量 Y	平均产量 A	水量增加 A	产量增加 ΔY	边际产量 M	生产弹性 E_p
1	7	201.86	289.03	1.43				
2	6	213.09	356.05	1.67	11.23	67.02	5.97	3.57
3	5	185.13	329.18	1.78	−27.96	−26.87	0.96	0.54
4	4	189.00	329.25	1.74	3.87	0.07	0.02	0.01
5	3	243.31	375.46	1.54	54.31	46.21	0.85	0.55
6	2	269.22	437.19	1.62	25.91	61.73	2.38	1.59
7	1	288.94	409.54	1.42	19.72	27.65	1.40	0.99

Table 148 Statistics of Corn Production Function

No.	Treatment No.	Irrigation Quota X	Total Yield Y	Average Yield A	Increased Water Amount A	Increased Yield ΔY	Marginal Yield M	Production Flexibility E_p
1	7	201.86	289.03	1.43				
2	6	213.09	356.05	1.67	11.23	67.02	5.97	3.57
3	5	185.13	329.18	1.78	−27.96	−26.87	0.96	0.54
4	4	189.00	329.25	1.74	3.87	0.07	0.02	0.01
5	3	243.31	375.46	1.54	54.31	46.21	0.85	0.55
6	2	269.22	437.19	1.62	25.91	61.73	2.38	1.59
7	1	288.94	409.54	1.42	19.72	27.65	1.40	0.99

六、玉米产量的方差分析
3.2.6 Variance Analysis of Corn Yield

方差分析是对两个以上样本间的差异程度进行比较检验的一种统计方法,对玉米产量进行方差分析见表149,以检验试验结果是否显著,对2年玉米试验的产量进行分析。

Analysis of variance is a statistical method to compare the degree of deviation between two or more samples. By analyzing the corn yield in the 2-year-long experiment, analysis of variance for corn yield is shown in Table 149 to test whether the experimental results are remarkable.

表 149　　　　　　　　　　　玉米产量方差分析表

处理	区组		T_t	$\overline{X}_t$
1	416.92	402.15	819.08	409.54
2	460.28	414.10	874.38	437.19
3	343.77	407.14	750.92	375.46
4	290.22	368.27	658.50	329.25
5	303.36	355.00	658.36	329.18
6	355.17	356.13	712.10	356.05
7	281.11	293.67	578.06	289.06
T_r	2 451.11	2 597.26	5 048.38	
$\overline{X}_r$	350.16	371.04	721.20	$\overline{X} = 360.60$

Table 149　　　　　　　　Variance Analysis Table of Corn Yield

Treatment	Zone		T_t	$\overline{X}_t$
1	416.92	402.15	819.08	409.54
2	460.28	414.10	874.38	437.19
3	343.77	407.14	750.92	375.46
4	290.22	368.27	658.50	329.25
5	303.36	355.00	658.36	329.18
6	355.17	356.13	712.10	356.05
7	281.11	293.67	578.06	289.06
T_r	2 451.11	2 597.26	5 048.38	
$\overline{X}_r$	350.16	371.04	721.20	$\overline{X} = 360.60$

自由度分解：总自由度：$f = nk - 1 = 14 - 1 = 13$

　　　　　　区组自由度：$f = 1$

　　　　　　处理间自由度：$f = 7 - 1 = 6$

误差自由度：$f=(n-1)(k-1)=6$

Division of degree of freedom：Total degree of freedom：$f=nk-1=14-1=13$

Degree of freedom in zones：$f=1$

Degree of freedom within treatments：$f=7-1=6$

Error's degree of freedom：$f=(n-1)(k-1)=6$

平方和分解：矫正数：$C=\dfrac{T^2}{nk}=\dfrac{5\ 048.38^2}{14}=1\ 820\ 438.62$

总平方和：$S_{总}=\sum\limits_{1}^{nk}X^2-C=1\ 859\ 536.58-1\ 820\ 438.62=39\ 097.36$

区组平方和：$S_{区}=\sum(\dfrac{Tr^2}{k})-C=1\ 518.49$

处理间平方和：$S_{处}=\sum(\dfrac{Tt^2}{n})-C=33\ 367.9$

误差平方和：$S_{误}=S_{总}-S_{区}-S_{处}=4\ 210.89$

Division of sum of squares：Correction $C=\dfrac{T^2}{nk}=\dfrac{5\ 048.38^2}{14}=1\ 820\ 438.62$

Total sum of squares $S_{\text{total}}=\sum\limits_{1}^{nk}X^2-C=1\ 859\ 536.58-1\ 820\ 438.62=39\ 097.36$

Sum squares in zones $S_{\text{zones}}=\sum(\dfrac{Tr^2}{k})-C=1\ 518.49$

Sum of squares within treatments $S_{\text{treatments}}=\sum(\dfrac{Tt^2}{n})-C=33\ 367.9$

Error's sum of squares $S_{\text{error}}=S_{\text{total}}-S_{\text{zones}}-S_{\text{treatments}}=4\ 210.89$

将上述数据列入方差分析表（见表150）中进行 F 检验。

Put the data above into the following variance analysis table (see Table 150) and check the value of F.

表 150　　　　　　　　　　玉米方差分析结果表

方差来源	平方和(SS)	自由度(f)	均方(MS)	F	$F_{0.05}$	$F_{0.01}$
区组间	1 518.49	1	1 518.49	2.16	4.67	9.07
处理间	33 367.98	6	5 561.33	7.92	2.92	4.62
误差	4 210.89	6	701.82			
总和	39 097.36	13				

Table 150　　　　　　　Variance Analysis Result of Corn

Variance Sources	Sum of Squares (SS)	Degree of Freedom (f)	Mean Squares (MS)	F	$F_{0.05}$	$F_{0.01}$
Between Zones	1 518.49	1	1 518.49	2.16	4.67	9.07
Within Treatments	33 367.98	6	5 561.33	7.92	2.92	4.62
Error	4 210.89	6	701.82			
Total	39 097.36	13				

由上表可知,区组间不显著($F=2.16 \leqslant F_{0.05}=4.67$),而处理间差异极显著($F=7.92 > F_{0.01}=4.62$),说明各处理所采取的不同的农业水平和灌溉水平对作物的产量有极为显著的影响。

From the table above, there is not so much differences within zones ($F=2.16 \leqslant F_{0.05}=4.67$), but the difference within treatments is quite outstanding ($F=7.92 > F_{0.01}=4.62$), indicating that agricultural technique levels and irrigation levels applied in each treatment do exert a huge influence on crop yield.

七、价值生产函数及最大纯收益
3.2.7　The Value Production Function and Maximum Net Benefit

在生产过程中,由于投入的生产资源数量不同,所得出的产品数量也就不同,这种物质数量间的生产关系,以函数的形式来反映就称为生产函数。In the production process, the amount of production varies due to the quantity of productive resources. The productive relation between matter and quantity, in the shape of function, is called production function.

$$Y = f(X) \quad X = (X_1 、 X_2 \cdots X_n)$$

式中　Y——单位面积的产量;
　　　X——资源投入的集合;
　　　$X_1, \cdots, X_n$——各项资源投入量;
　　　$1、2、3, \cdots, n$——各分项资源序号。

Where　Y = yield per unit area
　　　　X = the aggregation of input resources
　　　　$X_1, \cdots, X_n$ = different kinds of input resources
　　　　$1, 2, 3, \cdots, n$ = serial number of resources.

在众多资源中,有人工可以控制的和有限量的,称为变动资源,有不可控制或难以控制的无限量,称之为固定资源。本次课题只研究可变动资源中水分对玉米生产的影响,即只研究投入的水量与所得的产量两种物质的数量关系,即:$Y=f(X)$,两边乘以物质产品的单价,即得投入水量的价值和所得产品的价值,两种价值的关系为价值函数。

Among the numerous resources, there are controlled and limited resources, which are called variable resources. Those that are uncontrollable, unmanageable or unlimited resources are called fixed resources. This project only studies the effect of water in variable resources on corn production, namely the quantitative relation between the water amount input and the yield gained or $Y=f(X)$. Multiply both sides by the unit price of the product and get the price of the water input and the product. The relation between the two prices takes the form of cost function.

投入水量的价值为:$T_C = P_X \cdot X$
Cost of water input:$T_C = P_X \cdot X$

所得产量的价值为:$T_P = P_Y \cdot Y$
Value of production gained:$T_P = P_Y \cdot Y$

2 年效益为:$Z = P_r \cdot Y - (K + P_X \cdot X)$
Benefit of the 2 years:$Z = P_r \cdot Y - (K + P_X \cdot X)$

式中　T_C——投入水量的价值;
　　　T_P——所得产量的价值;
　　　P_X——水的单价,0.04 元/m³;
　　　P_Y——产品的单价,0.42 元/kg;
　　　X——投入的总水量;
　　　Y——总产品;
　　　Z——净效益;
　　　K——固定生产要素的投入,包括种子、肥料、机构管理、收割费用等的总价值。

Where　T_C = the value of water applied
　　　T_P = the value of crop yield
　　　P_X = unit price of water,0.04 yuan/m³
　　　P_Y = unit price of the product,0.42 yuan/kg
　　　X = total amount of water applied
　　　Y = gross product
　　　Z = net benefit
　　　K = input in fixed production factors, including the total cost of seeds, fertilizer, institutional management, harvest fee and etc.

固定投入中：化肥，单价 0.75 元/kg；农先投 20 kg/亩；农中投 15 kg/亩。

In fixed input resources: fertilizer, 0.75 yuan/kg; before agricultural measures, apply fertilizer with the degree of 20 kg/mu; while in the middle of agricultural measures, 15 kg/mu.

机耕费单价：16 元/亩，种子单价：1.2 元/kg。

Unit price of tractor-ploughing: 16 yuan/mu, unit price of seeds: 1.2 yuan/kg.

管理费：农先水平 20 元/亩，农中水平 15 元/亩。

Management fee: before agricultural measures, 20 yuan/mu, in the middle of agricultural measures, 15 yuan/mu.

$$K_{农先} = 20 \times 0.75 + 16 + 1.2 \times 3 + 20 = 54.6 \text{ 元/亩}$$
$$K_{\text{before agricultural measures}} = 20 \times 0.75 + 16 + 1.2 \times 3 + 20 = 54.6 \text{ yuan/mu}$$

$$K_{农中} = 12 \times 0.75 + 16 + 1.2 \times 3 + 15 = 43.60 \text{ 元/亩}$$
$$K_{\text{in the middle of agricultural measures}} = 12 \times 0.75 + 16 + 1.2 \times 3 + 15 = 43.60 \text{ yuan/mu}$$

计算价值生产函数的目的在于分析投入多少变动资源才能获得最大的经济效益。从边际分析上得知，最高产量点是生产行为处于生产函数的第二阶段的末端，但从生产价值来看，最高产量点不一定是获得最大利润值的点，还应考虑产量及其变动价格资源，才能找到资源经济效益的最佳点。生产单位获得利润的最适资源施用量，应该是变动资源价格比率等于资源边际产量，即：

To calculate the value production function is to analyze how much variable resources should be input to help obtain the maximum economic benefits. From the marginal analysis, it's clear that the maximum yield appears at the end of the second stage of production function. But seen from productive value, the point where the maximum yield appears doesn't identify with the point where the maximum profit is gained. Additional consideration needs to go to yield and the price of variable resources to find out the optimal point where resources are given full play to generate economic benefits. For every unit benefit produced, the optimum resources application amount should come out where the ratio of the price of variable resources equals the marginal yield of resources, that is:

$$M = P_Y/P_X = \lim(\Delta Y/\Delta X) = d_Y/d_X$$
$$P_X \cdot \Delta X = P_Y \cdot \Delta Y$$

将生产函数相对应的各种价值、生产函数及纯收益绘图表示（见表 151、图 38），可看出边际产值线与边际成本线交于一点，该点表示边际产值等于边际成本，此点在第二生产阶段内，属于合理生产行为，满足经济效益的最大充分条件，故为最大纯收益点。它对应总产值线上，并不是总产值最大点，总产值最大点产量为 427 kg/亩，产值是 117.75 元/亩，最大纯收益点产量是 425 kg/亩，耗水量是 425 kg/亩，产值是 118 元/亩，耗水量是 264 m³/亩。

Draw out the cost or price in production function, production function and net return (see Table 151 and Figure 38). It can be seen that marginal production value intersects with marginal cost at a point, which means that marginal production value is equal to marginal cost. And this point falls into the second stage, which belongs to reasonable production behaviors, being the point of maximum net return for it meets the highest sufficient condition of economic benefit. Correspondingly, it doesn't mean the point of maximum gross output. Where gross output reaches the highest point, the yield is 427 kg/mu, productive value 117.75 yuan/mu, while for the point of maximum net return, the yield is 425 kg/mu, water consumption reaches 425 kg/mu or 264 m³/mu, and productive value is 118 yuan/mu.

表 151　　边际分析与生产价值函数（$P_X = 0.04, P_r = 0.42$）

处理号	耗水量 X (m³/亩)	产量 r (kg/亩)	平均产量 $A_p = A \cdot P_r$ (元/m³)	边际产量 $M_p = M \cdot P_r$ (元/m³)	耗水成本 $T_c = P_x \cdot X$ (元/亩)	边际成本 $M_c = P_x$ (元/m³)	资源投入 K (元/亩)	产值 $T_p = Y \cdot P_x$ (元/亩)	纯收量 $Z = T_p \cdot T_c \cdot K$ (元/亩)
7	201.86	289.03	0.600 6		8.07	0.04	43.6	121.39	69.72
6	213.09	356.05	0.701 4	2.507 4	8.52	0.04	54.6	149.54	86.42
5	185.13	329.18	0.747 6	0.403 2	7.41	0.04	54.6	138.26	76.25
4	189.02	329.05	0.730 8	0.008 4	7.56	0.04	54.6	138.37	76.21
3	243.21	375.46	0.646 8	0.357 0	9.73	0.04	54.6	157.69	93.36
2	269.32	437.16	0.680 4	0.999 6	10.77	0.04	54.6	183.62	118.25
1	288.94	409.54	0.596 4	0.588 8	11.56	0.04	43.60	172.62	116.85

Table 151　Marginal Analysis and Production Value Function ($P_X = 0.04, P_r = 0.42$)

Treatment No.	Water Consumption X (m³/mu)	Yield r (kg/mu)	Average Product $A_p = A \cdot P_r$ (yuan/m³)	Marginal Yield $M_p = M \cdot P_r$ (yuan/m³)	Water Consumption Cost $T_c = P_x \cdot X$ (yuan/mu)	Marginal Cost $M_c = P_x$ (yuan/m³)	Resource Input K (yuan/mu)	Production Value $T_p = Y \cdot P_x$ (yuan/mu)	Net Return $Z = T_p \cdot T_c \cdot K$ (yuan/mu)
7	201.86	289.03	0.600 6		8.07	0.04	43.6	121.39	69.72
6	213.09	356.05	0.701 4	2.507 4	8.52	0.04	54.6	149.54	86.42
5	185.13	329.18	0.747 6	0.403 2	7.41	0.04	54.6	138.26	76.25
4	189.02	329.05	0.730 8	0.008 4	7.56	0.04	54.6	138.37	76.21
3	243.21	375.46	0.646 8	0.357 0	9.73	0.04	54.6	157.69	93.36
2	269.32	437.16	0.680 4	0.999 6	10.77	0.04	54.6	183.62	118.25
1	288.94	409.54	0.596 4	0.588 8	11.56	0.04	43.60	172.62	116.85

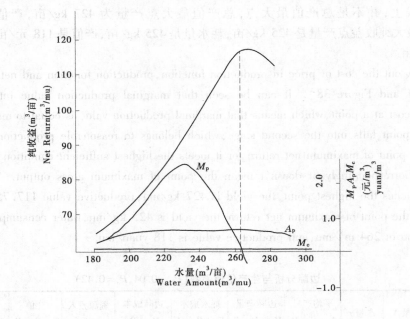

图 38 玉米生产函数及纯收益关系图
Table 38　Relation between Production Function and Net Return of Corn

八、玉米灌溉制度的优化设计
3.2.8　Optimized Scheme for Irrigation Scheduling of Corn

动态规划原理是求解作物优化灌溉制度数学模型的重要方法。用动态规划方法，按时间顺序将作物整个生育期划分为若干个生育阶段，每个阶段根据土壤含水量的初始值作出决策，即作出灌水时间、灌水次数和灌水定额，各阶段的决策组成的最优策略就是作物的最优灌溉制度。

The principle of dynamic planning is an important method of working out the mathematical model of optimized irrigation scheduling for crops. With dynamic planning, the growth and development period of crops can be divided into several stages in chronological order. At each stage, based on the initial water content decisions on irrigation time, irrigation frequency and irrigation water quota are made. The best strategy made up by decisions at each stage is the optimum irrigation scheduling of crops.

下面为动态规化数学模型：
The mathematical model of dynamic planning is as followed：

（一）变量设置
3.2.8.1 Variables Setting

阶段变量 i：将作物生育期划分 i 个生育阶段；
Stage variable i: the growth and development period of crops is divided into i stages;

状态变量 Q_i：各生育期阶段土壤含水量 Q_i；
Condition variable Q_i: soil water content Q_i at every stage;

决策变量 X_i：各生育阶段的灌水量 X_i。
Decision variable X_i: irrigation amount at every stage X_i.

（二）约束条件
3.2.8.2 Constraint Condition

作物耗水量约束：
Constraints on crops water consumption：

$$W_{\mathrm{min}i} \leqslant W_i \leqslant W_{\mathrm{max}i}$$

式中　$W_{\mathrm{min}i}$——各生育阶段适宜下限耗水量；
　　　W_i——各生育阶段耗水量；
　　　$W_{\mathrm{max}i}$——各生育阶段供水充足条件下最大作物耗水量。
Where　$W_{\mathrm{min}i}$ = the proper minimum water consumption in every growth and development period
　　　W_i = water consumption at each growth and development period
　　　$W_{\mathrm{max}i}$ = the maximum water consumption in every growth and development period when water is sufficient.

土壤含水率约束：
Constraints on soil water content：

$$Q_{\mathrm{min}i} \leqslant Q_i \leqslant Q_{\mathrm{max}i}$$

式中　$Q_{\mathrm{min}i}$——土壤水分适宜下限；
　　　$Q_{\mathrm{max}i}$——田间持水率，为 11.23%。
Where　$Q_{\mathrm{min}i}$ = the proper minimum soil water content
　　　$Q_{\mathrm{max}i}$ = field capacity (11.23%).

灌溉定额约束：
Constraints on irrigation quota：

$$X^* = \sum X_i$$

式中　X_i——各生育期阶段灌水量；
　　　X^*——生育期优化灌溉定额。
Where　X_i = irrigation amount in every period
　　　X^* = optimized irrigation quota in every period.

（三）目标函数
3.2.8.3　Objective Function

采用 M・E・Jensen 的水量与产量关系模型：
Apply the relation model of water content and yield by M・E・Jensen：

$$\frac{Y}{Y_{max}} = \max \sum_i^n \left(\frac{W_i}{W_{maxi}}\right)^{\lambda_i}$$

式中　Y——作物供水不足时达到的产量；
　　　Y_{max}——作物供水充足时达到的产量；
　　　W_i——作物供水不足时在阶段 i 的实际耗水量；
　　　λ_i——各生育阶段水分敏感指数。
Where　Y = yield when water is insufficient
　　　Y_{max} = yield when water is sufficient
　　　W_i = the real water consumption in stage i when there is water shortage
　　　λ_i = water sensitive index in every growth and development period.

根据以上的价值生产函数和最大纯收益分析，得出的最优灌溉定额 $X^* = 264$ m³/亩，亩次净灌水定额为 33 m³/亩，共灌水 8 次。

By analyzing the value production function and the maximum net return, we get the optimum irrigation quota $X^* = 264$ m³/mu, with net irrigation water quota is 33 m³/mu for 8 times.

（四）效益函数
3.2.8.4　Benefit Function

$$f_i = r_i(M_i, I_i, d_i) - (W_i/W_{Mi})^\lambda$$

式中　$W_i = \min\{W_{\text{maxi}}, M_i + X_i\}$ ；

M_i——土壤可供作物吸收的水量，$M_i = 667\gamma H(Q_{qi} - Q_{凋})$ ；

Q_{qi}——阶段初始时的土壤含水率；

r——容量，为 1.445 g/cm³ ；

H——计划湿润层为 0.6 m ；

$Q_{凋}$——作物的凋萎系数为 4.0% ；

I_i——第 i 阶段及 i 阶段以后可供灌水量，其值是 33 的整倍数，即 33、66、99、132、165、198、231、264，单位：m³/亩 ；

d_i——第 i 阶段的灌水次数 $0 \leq d_i \leq I_i/33$。

Where　$W_i = \min\{W_{\text{maxi}}, M_i + X_i\}$

M_i = water for crops uptaking in the soil, $M_i = 667\gamma H(Q_{qi} - Q_{\text{wilt}})$

Q_{qi} = soil water content at the beginning of the stage

r = soil bulk density (1.445 g/cm³)

H = the soil wetting layer (0.6 m)

Q_{wilt} = crops wilting coefficient of 4.0%

I_i = irrigation amount available at and after stage i which is a multiple of 33, namely 33, 66, 99, 132, 165, 198, 231, 264, Unit: m³/mu

d_i = irrigation frequency at stage i and $0 \leq d_i \leq I_i/33$.

（五）状态转移方程

3.2.8.5　State Transfer Equation

$$I_i + 1 = I_i - 33d_i$$
$$M_i + 1 = M_i + X_i - W_i$$

（六）递推方程

3.2.8.6　Recursive Equation

$$f(M_i, I_i) = \max d_i\{(M_i, I_i, d_i) \times f_{itl}(M_{i+1}, I_{i+1})\}$$

式中　i——生育阶段，其值是 1、2、3、4、5。

Where　i = the crop growth and development period, the value of which can be 1, 2, 3, 4 and 5.

递推计算过程，采用逆向递推，从第五阶段开始依次递推到第一阶段，利用递推公式，求出各阶段开始的土壤含水量 Q_{qi} 及相应的土壤供水量 M 和可供该阶段及该阶段以后各阶段使用的水量 I_i 以及各种组合下的最优决策 d_i^*、f_i^*。计算过程见表 152～表 154，土壤含水率查得对应的 d_1^*、d_2^*、d_3^*、d_4^*、d_5^* 即为最优灌溉制度，列出初始含水率为 7% 和 9% 两种初始含水率的情况见表 155。

Recursive calculation process: The reverse recursive method from the fifth stage to the first stage is used to work out the soil water content at the start of stage Q_{qi} and corresponding soil water supply M, the water available at this stage and stages later on I_i, and the optimal decision d_i^* and f_i^* in each combination. Table 152 to Table 154 outline the process of calculation. The corresponding value of d_1^*, d_2^*, d_3^*, d_4^* and d_5^* in the list of soil water content composes the optimal irrigation scheduling. The two conditions of initial water content, 7% and 9%, are stated in Table 155.

表 152 优化灌溉制度递推表

土壤水分状况		第五阶段递推公式 $Wm = 45.18$ m³/亩, $\lambda_5 = 0.2623$ $f_5^4 = d_5^{max}(f_5) = d_5^{max}\{W_5/45.18\}^{0.2629}\}$ 其中:$W_5 = \min\{45.18, M_5 + 33d\}$ $0 \leq d_5 \leq I_5/33$			第四阶段递推公式 $Wm_4 = 52.15$ m³/亩; $\lambda_4 = 0.4607$ $f_4^1 = d_4^{max}\{(W_4/52.15)^{0.0007}\} \cdot f_5(M_5, I_5)$ 其中:$W_4 = \min(52.15, M_4 + 33d_4)$ $0 \leq d_4 \leq I_4/33$ $M_5 = M_4 + 33d_4 - W_4$ $I_5 = I_4 - 33d_4$			
含水量	供水量	d_5^1/f_5^1			d_5^1/f_5^1			
$Q_{5.4}$	$M_{5.4}$	$I_5 = 0$	$I_5 = 33$	$I_5 = 66$	$I_4 = 33$	$I_4 = 66$	$I_4 = 99$	$I_4 = 132$
11.0	40.43	0/0.972	1/1	1/1	0/0.315	0/0.885	1/1	1/1
10.0	34.70	0/0.933	1/1	1/1	0/0.751	0/0.322	1/1	1/1
9.0	28.91	0/0.889	1/1	1/1	0/0.693	0/0.753	1/1	1/1
8.0	23.13	0/0.839	1/1	1/1	0/0.623	0/0.677	1/1	1/1
7.0	17.35	0/0.778	1/1	1/1	0/0.542	0/0.589	1/0.983	2/1
6.0	11.57	0/0.700	1/0.996	2/1	0/0.447	0/0.485	1/0.927	2/1
5.0	5.78	0/0.583	1/0.961	2/1			1/0.867	2/1
4.0	0.00	0/0	1/0.921	2/1				

Table 152 Recursion Table for Optimized Irrigation Scheduling

Condition of Soil Water	for the fifth stage $Wm = 45.18$ m³/mu, $\lambda_5 = 0.2623$ $f_5^4 = d_5^{max} d_5^{max}(f_5) = \{W_5/45.18\}^{0.2629}\}$ Where: $W_5 = \min\{45.18, M_5 + 33d\}$ $0 \leq d_5 \leq I_5/33$	for the fourth stage $Wm_4 = 52.15$ m³/mu; $\lambda_4 = 0.4607$ $f_4^1 = d_4^{max}\{(W_4/52.15)^{0.0007}\} \cdot f_5(M_5, I_5)$ Where: $W_4 = \min(52.15, M_4 + 33d_4)$ $0 \leq d_4 \leq I_4/33$ $M_5 = M_4 + 33d_4 - W_4$ $I_5 = I_4 - 33d_4$

第三章 作物灌溉制度研究
3 Study on Irrigation System of Crops

Continued Table 152

Water Content $Q_{5.4}$	Water Supply $M_{5.4}$	d_5^1/f_5^1 $I_5=0$	$I_5=33$	$I_5=66$	d_5^1/f_5^1 $I_4=33$	$I_4=66$	$I_4=99$	$I_4=132$
11.0	40.43	0/0.972	1/1	1/1	0/0.315	0/0.885	1/1	1/1
10.0	34.70	0/0.933	1/1	1/1	0/0.751	0/0.322	1/1	1/1
9.0	28.91	0/0.889	1/1	1/1	0/0.693	0/0.753	1/1	1/1
8.0	23.13	0/0.839	1/1	1/1	0/0.623	0/0.677	1/1	1/1
7.0	17.35	0/0.778	1/1	1/1	0/0.542	0/0.589	1/0.983	2/1
6.0	11.57	0/0.700	1/0.996	2/1	0/0.447	0/0.485	1/0.927	2/1
5.0	5.78	0/0.583	1/0.961	2/1			1/0.867	2/1
4.0	0.00	0/0	1/0.921	2/1				

表 153 优化灌溉制度递推表

土壤水分状况	第二阶段递推公式	第一阶段递推公式
	$Wm_2 = 39.59 \text{ m}^3/亩; \lambda_2 = 0.8317$ $f_2^1 = d_2^{max}\{(W_2/39.59)^{0.3917}\} \cdot f^4\{M_5, I_5\}$ 其中:$W_5 = \min\{39.59, M_2 + 33d_2\}$ $0 < d_2 < I_5/33$ $M_2 = M_2 + 33d_2 - W_2$ $I_2 = I_2 - 233d_2$	$Wm_1 = 72.96 \text{ m}^3/亩; \lambda_1 = 0.2111$ $f_1^1 = d_1^{max}\{(W_1/72.96)^{0.2111}\} \cdot f_2^1\{M_2, I_2\}$ 其中:$W_1 = \min(72.96, M_1 + 33d_1)$ $0 < d_1 < I_1/33$ $M_2 = M_1 + 33d_1 - W_1$ $I_2 = I_1 - 33d_1$

含水量 $Q_{2.1}$	供水量 $M_{2.1}$	d_2^1/f_2^1 $I_2=198$	$I_2=231$	$I_2=264$	d_1^1/f_1^2 $I_2=264$
11.0	40.48	0/1	1/1	1/1	0/0.883
10.0	34.70	0/0.896	1/1	1/1	0/0.855
9.0	28.91	0/0.770	1/1	1/1	0/0.822
8.0	23.13	0/0.660	1/1	1/1	0/0.813
7.0	17.35	0/0.504	1/1	1/1	0/0.795
6.0	11.57	0/0.348	1/1	1/1	0/0.775
5.0	5.78	0/0.187	1/0.969	1/1	0/0.775
4.0	0.00	0/0	0/0	1/0.907	1/0.903

Table 153   **Recursion Table for Optimized Irrigation Scheduling**

Condition of Soil Water		for the second stage $Wm_2 = 39.59 \text{ m}^3/\text{mu}; \lambda_2 = 0.8317$ $f_2^1 = d_2^{\max}\{(W_2/39.59)^{0.3917}\} \cdot f^4\{M_5, I_5\}$ Where: $W_5 = \min\{39.59, M_2 + 33d_2\}$ $0 < d_2 < I_5/33$ $M_2 = M_2 + 33d_2 - W_2$ $I_2 = I_2 - 233d_2$			for the first stage $Wm_1 = 72.96 \text{ m}^3/\text{mu}; \lambda_1 = 0.2111$ $f_1^1 = d_1^{\max}\{(W_1/72.96)^{0.2111}\} \cdot f_2^1\{M_2, I_2\}$ Where: $W_1 = \min(72.96, M_1 + 33d_1)$ $0 < d_1 < I_1/33$ $M_2 = M_1 + 33d_1 - W_1$ $I_2 = I_1 - 33d_1$
Water Content	Water Supply	d_2^1/f_2^1			d_1^1/f_1^2
$Q_{2,1}$	$M_{2,1}$	$I_2 = 198$	$I_2 = 231$	$I_2 = 264$	$I_2 = 264$
11.0	40.48	0/1	1/1	1/1	0/0.883
10.0	34.70	0/0.896	1/1	1/1	0/0.855
9.0	28.91	0/0.770	1/1	1/1	0/0.822
8.0	23.13	0/0.660	1/1	1/1	0/0.813
7.0	17.35	0/0.504	1/1	1/1	0/0.795
6.0	11.57	0/0.348	1/1	1/1	0/0.775
5.0	5.78	0/0.187	1/0.969	1/1	0/0.775
4.0	0.00		0/0	1/0.907	1/0.903

表 154   优化灌溉制度递推表

土壤水分状况		第三阶段递推公式 $Wm = 81.57 \text{ m}^3/亩; \lambda_5 = 0.6237$ $f_5^4 = d_5^{\max}\{(W_5/81.57)^{0.2237}\} \cdot f_4^1\{M_4, I_5\}$ 其中: $W_5 = \min\{31.57, M_5 + 33d\}$ $0 < d_5 < I_5/33$ $M_4 = M_1 + 33d_5 - W_5$ $I_4 = I_5 - 33d$						
含水量	供水量	d_5^1/f_5^1						
$Q_{5,4}$	M_5	$I_5 = 33$	$I_5 = 66$	$I_5 = 99$	$I_5 = 132$	$I_5 = 165$	$I_5 = 198$	$I_5 = 23$
11.0	40.48	0/0.518	0/0.646	0/0.937	0/0.961	1/1	1/1	1/1
10.0	34.70	0/0.471	0/0.587	0/0.890	0/0.921	1/1	1/1	1/1
9.0	28.91	0/0.420	0/0.534	0/0.842	0/0.870	1/0.945	1/1	1/1
8.0	23.13		0/0.456	0/0.792	0/0.816	1/0.886	1/1	1/1
7.0	17.35		0/0.381	0/0.740	0/0.758	1/0.823	1/1	1/1
6.0	11.57		0/0.296	0/0.686	0/0.716	1/0.778	1/0.969	2/1
5.0	5.78		0/0.192	0/0.629	0/0.680	1/0.741	1/0.923	2/1
4.0	0.00							

Table 154 Recursion Table for Optimized Irrigation Scheduling

Condition of Soil Water	for the third stage
	$Wm = 81.57 \text{ m}^3/\text{mu};\ \lambda_5 = 0.6237$ $f_5^4 = d_5^{\max}\{(W_5/81.57)^{0.2237}\} \cdot f_4^1\{M_4, I_5\}$ Where: $W_5 = \min\{31.57, M_5 + 33d\}$ $0 < d_5 < I_5/33$ $M_4 = M_1 + 33d_5 - W_5$ $I_4 = I_5 - 33d$

Water Content $Q_{5.4}$	Water Supply M_5	d_5^1/f_5^1						
		$I_5 = 33$	$I_5 = 66$	$I_5 = 99$	$I_5 = 132$	$I_5 = 165$	$I_5 = 198$	$I_5 = 231$
11.0	40.48	0/0.518	0/0.646	0/0.937	0/0.961	1/1	1/1	1/1
10.0	34.70	0/0.471	0/0.587	0/0.890	0/0.921	1/1	1/1	1/1
9.0	28.91	0/0.420	0/0.534	0/0.842	0/0.870	1/0.945	1/1	1/1
8.0	23.13		0/0.456	0/0.792	0/0.816	1/0.886	1/1	1/1
7.0	17.35		0/0.381	0/0.740	0/0.758	1/0.823	1/1	1/1
6.0	11.57		0/0.296	0/0.686	0/0.716	1/0.778	1/0.969	2/1
5.0	5.78		0/0.192	0/0.629	0/0.680	1/0.741	1/0.923	2/1
4.0	0.00							

当土壤初始含水率为9%时,相对5个阶段在水量上的分配是:66、33、99、33、33。即苗期灌水2次,拔节期灌1次,抽雄期灌3次,灌浆期灌水1次,成熟期灌水1次,相对应灌溉制度的最优产量是:

When the initial water content is 9%, the water distribution at the five stages is as follows: 66,33,99,33 and 33. In other words, the frequency of irrigation for seeding period is 2 times, jointing period 1 time, tasseling period 3 times, filling period 1 time and maturing period 1 time. And the corresponding maximum yield of this irrigation scheduling is:

$$Y^* = f_1^* \cdot Y_m = 0.822 \times 427 = 350.99 \text{ kg/亩}$$

$$Y^* = f_1^* \cdot Y_m = 0.822 \times 427 = 350.99 \text{ kg/mu}$$

当初始含水率为7%时,相对于5个生育阶段在水量上的分配是:66、99、33、33、33,即苗期灌水2次,拔节期灌水3次,抽雄期灌水1次,灌浆期灌水1次,成熟期灌水1次。相对应这种灌溉制度的最优产量为:

When the initial water content is 7%, the water distribution at the five stages is as follows: 66, 99, 33, 33 and 33. In other words, the frequency of irrigation for seeding period is 2 times, jointing period 3 times, tasseling period 1 time, filling period 1 time and maturing period 1 time. And the corresponding maximum yield of this irrigation scheduling is:

$$Y^* = f_1^* \cdot Y_m = 0.795 \times 427 = 339 \text{ kg/亩}$$
$$Y^* = f_1^* \cdot Y_m = 0.795 \times 427 = 339 \text{ kg/mu}$$

表 155　　　　　　　　　　不同初始土壤含水率最优决策表

初始土壤含水率	阶段	M_i(m³/亩)	I_i(m³/亩)	d^*(次)	f^*	Q_i(%)
9%	1	28.91	264	2		9.0
	2	21.95	198	1		7.80
	3	15.36	165	3	0.822	6.66
	4	32.51	66	1		9.62
	5	13.36	33	1		6.31
7%	1	17.35	264	2		7.0
	2	10.39	198	3		5.80
	3	3.80	165	1	0.795	4.66
	4	21.23	66	1		7.67
	5	2.08	33	1		4.36

Table 155　　　　Optimal Decision on Different Initial Soil Water Content

Initial Soil Water Content	Stage	M_i (m³/mu)	I_i (m³/mu)	d^* (times)	f^*	Q_i (%)
9%	1	28.91	264	2		9.0
	2	21.95	198	1		7.80
	3	15.36	165	3	0.822	6.66
	4	32.51	66	1		9.62
	5	13.36	33	1		6.31
7%	1	17.35	264	2		7.0
	2	10.39	198	3		5.80
	3	3.80	165	1	0.795	4.66
	4	21.23	66	1		7.67
	5	2.08	33	1		4.36

九、灌溉效益分摊系数
3.2.9 Sharing Coefficient of Irrigation Benefit

在试验方案中,用于灌溉效益分摊系数计算的有4个处理,其试验结果见表156。

In this experiment, there are 4 combination treatments for figuring out the sharing coefficient of irrigation benefit. The results are listed in Table 156.

表156 分摊系数试验成果表

水平	Y_1	Y_2	Y_3	Y_4
	农低+水低	农低+水高	农高+水低	农高+水高
产量	289.03	409.54	356.05	437.19
灌水量	201.86	288.54	213.09	469.22

Table 156 Experimental Results of Sharing Coefficient

Level	Y_1	Y_2	Y_3	Y_4
	low agrotechnique + low irrigation amount	low agrotechnique + high irrigation amount	high agrotechnique + low irrigation amount	high agrotechnique + high irrigation amount
Output	289.03	409.54	356.05	437.19
Irrigation Amount	201.86	288.54	213.09	469.22

不同农业技术水平下的灌溉增产率及灌溉效益与农业效益的分摊系数按下式计算：

Yield-increase rate and irrigation benefit under different agricultural technique conditions and the sharing coefficient of agricultural benefit are calculated according to the following formula：

(一)一般农业技术水平措施下的灌溉增产率(%)

3.2.9.1 Yield – increase Rate with General Agricultural Technique (%)

$$\Delta C_Y 1 = [(Y_2 - Y_1)/Y_1] \times 100\% = [(409.54 - 289.03)/289.03] \times 100\% = 41.69\%$$

(二)高农业技术水平措施下的灌溉增产率(%)

3.2.9.2 Yield – increase Rate with High Agricultural Technique (%)

$$\Delta C_Y h = [(Y_4 - Y_3)/Y_3] \times 100\% = [(437.19 - 356.05)/356.93] \times 100\% = 22.73\%$$

(三) 效益分摊系数
3.2.9.3 Sharing Coefficient of Irrigation Benefit

$$\sum\nolimits_{水} = [(Y_2 + Y_4) - (Y_1 - Y_3)]/2(Y_4 - Y_1) = 0.6805$$

$$\sum\nolimits_{water} = [(Y_2 + Y_4) - (Y_1 - Y_3)]/2(Y_4 - Y_1) = 0.6805$$

$$\sum\nolimits_{农} = [(Y_3 + Y_4) - (Y_1 + Y_2)]/2(Y_4 - Y_1) = 0.3195$$

$$\sum\nolimits_{agrotechnique} = [(Y_3 + Y_4) - (Y_1 + Y_2)]/2(Y_4 - Y_1) = 0.3195$$

十、试验研究结论
3.2.10　Conclusion of the Experiment

玉米灌溉制度产量最高时，耗水量为 270 m³/亩，产量是 427 kg/亩，纯收益最大时耗水量是 264 m³/亩，产量是 425 kg/亩，最大收益是 118 元/亩，灌水 8 次，灌水定额 33 m³/亩。

In this corn irrigation scheduling, when corn yield peaks, the water consumption amount is 270 m³/mu, the yield is 427 kg/mu; while the net return reaches its maximum, the water consumption is 264 m³/mu, the yield is 425 kg/mu and the maximum net return, 118 yuan/mu, with irrigation water quota of 33 m³/mu for 8 times.

玉米各生产阶段水分敏感指数反映大小顺序为：拔节 > 抽雄 > 灌浆 > 成熟 > 苗期，敏感指数为：苗期 0.2111，拔节期 0.8317，抽雄期 0.6237，灌浆期 0.4807，成熟期 0.2623。

The order of water sensitivity index of corn at each stage is: jointing > tasseling > filling > maturing > seeding; and the sensitivity index are: 0.2111 at seeding, 0.8317 at jointing, 0.6237 at tasseling, 0.4807 at filling and 0.2623 at maturing.

本书用动态规划原理，建立玉米优化灌溉制度数学模型将玉米分成 5 个生育阶段，采用逆向递推法求出各种土壤初始含水率及各种总灌溉定额下的最优灌溉制度方案，确定为玉米全生育期内灌水 8 次，灌溉定额 264 m³/亩，灌水定额 33 m³/亩，灌水时间为：苗期 2 次，拔节期 1 次，抽雄 3 次，灌浆期 1 次，成熟期 1 次。

This book utilizes the principle of dynamic planning by establishing the mathematical model of optimized irrigation scheduling of corn and divided corn's growth and development period into five stages. After working out the initial soil water content and the optimal irrigation scheduling with different irrigation quota by reverse recursive method, we make sure that corn needs 8 times irrigation in the whole growth and development period, irrigation quota is 264 m³/mu, irrigation water quota is 33 m³/mu and the pattern of irrigation is as follows: seeding period 2 times, jointing period 1 time, tasseling period 3 times, filling period 1 time and maturing period 1 time.

玉米灌溉效益分摊系数,随灌溉保证率升降而变化,在一般农业措施下,灌溉增长率为41.69%,在高水平农业措施下,灌溉增长率为22.73%。

The change of sharing coefficient of corn irrigation benefit is consistent with that of probability of irrigation; the irrigation growth rate is 41.69% under general agricultural condition, and 22.73% under high-level agricultural technology condition.

效益分摊系数反映的是灌溉工程措施和农业技术措施综合因素作用结果。阿勒泰地区玉米的效益分摊系数为:水占0.6805,农业占0.3195。

The sharing coefficient of economic benefit is the joint result of comprehensive factors, such as irrigation facilities and agricultural technical measures. The sharing coefficient of economic benefit of corn in Altay Prefecture is that water accounts for 0.680 5 while agricultural technical measures, 0.319 5.

第三节 苜蓿灌溉制度研究
3.3 Irrigation System of Lucerne

一、试验区概况
3.3.1 Overview of Experimental Site

(一)试验区条件
3.3.1.1 Condition of the Experimental Plots

苜蓿灌溉制度试验区占地1.7亩。其中试验田1.4亩,对照区0.3亩。试验区又划分成3个重复,4个处理,共计12块小试验田,由此往南顺序排列。试验田灌水方式为畦灌,灌水控制方式采用三角形量水堰量水。整个试验田场种植紫花苜蓿。

The experimental plots for alfalfa irrigation scheduling covers an area of 1.7 mu, where experimental field is 1.4 mu and control plot 0.3 mu. The experimental plots are divided into 3 repetitions and 4 treatments, totaling 12 plots. Put the 12 plots from north to south in order. The experimental field applies border irrigation and triangle measuring weir to control irrigation amount. The whole experimental field is used to plant alfalfa.

(二) 试验田基本参数
3.3.1.2 Experimental Field Basic Parameter

试验区土壤为沙壤土,地下水位在 2 m 以下,可不考虑地下水补给。土壤容量 1.58 g/cm³,比重 2.67 g/cm³,田间持水率为 11.239%。土壤略偏碱性,缺磷少氮。

The soil in experimental zone is sandy loam and ground water level is below 2 m, so there is no need to take groundwater recharge into consideration. Soil bulk density: 1.58 g/cm³; soil specific gravity: 2.67 g/cm³; field capacity: 11.239%. The soil is slightly alkaline, lacking phosphorus and nitrogen.

二、试验方案和方法
3.3.2 Experiment Scheme and Methods

该试验方案设计的主导思想是,通过控制土壤含水率水平来控制 4 个不同处理水平的灌水量,再对照 4 个处理的产量和最大纯收益,得出苜蓿的最优灌溉制度。

The guiding principle of this experiment is to control the irrigation amount of 4 treatments by controlling soil water content and then analyze the yield and the maximum net return of these treatments to obtain the optimal irrigation scheduling of alfalfa.

(一) 试验因素和水平
3.3.2.1 Experimental Factors and Levels

试验采用 1 因素,即水,4 水平即 4 个不同的灌溉水平,单纯研究水对苜蓿生长的影响程度。

The experiment has only one factor, which is water. And there are 4 different irrigation levels used to study effects of water on alfalfa growth.

(二) 试验方案
3.3.2.2 Experiment Scheme

由于本课题的研究手段较单一,为 1 因素试验,所以试验方案设计时,将 1 因素划分为 4 水平,再设置 3 个重复,共 12 块试验小区。设计方案见表 157。

The study method of this project makes the experiment a single factor experiment. When designing the experimental scheme, we made the single factor a 4 - level water consumption variable, multiplied by 3 repetitions and totaled 12 experimental plots. Experiment designing scheme is shown in Table 157.

第三章 作物灌溉制度研究
3 Study on Irrigation System of Crops

表 157 　　　　紫花苜蓿灌溉制度设计方案（占田间持水率百分比）

处理号	幼苗期	蔓枝伸长期	开花成熟期
1	65	70	65
2	45	70	65
3	65	45	45
4	65	70	40

Table 157　　Designing Scheme for the Irrigation Scheduling of Alfalfa
（Percentage in Field Capacity）

Treatment No.	Seedling	Branch Extension	Flowering & Maturing
1	65	70	65
2	45	70	65
3	65	45	45
4	65	70	40

　　苜蓿品种选用单一的紫花苜蓿。试验田形状为长方形，长 19.0 m，宽 4.2 m，面积 0.12亩，灌溉方式为畦灌。

　　All alfalfa belongs to the same variety. The shape of experimental field: rectangle; Length: 19.0 m; Width: 4.2 m; Area: 0.12 mu; Irrigation method: border irrigation.

三、苜蓿耗水量分析
3.3.3　Analysis of Alfalfa Water Consumption

（一）苜蓿生育期各旬耗水量
3.3.3.1　Water Consumption of Alfalfa in Every Ten Days of the Growth and Development Period

　　由表 158 可以看出，苜蓿 6 月耗水量最大，4 月最小。6 月是苜蓿第一茬和第二茬的衔接阶段，这时气温较高，苜蓿的生长最旺盛，需要大量的水进行光合作用，枝叶生长也最快，是整个生育阶段的旺盛期。

　　From Table 158, we can see that alfalfa consumes the most water in June, while in April, the least. June links up the first cropping and the second one. At this time alfalfa grows prosperously because of high temperature and needs plenty of water for photosynthesis. As it is a vigorous stage in the growth and development period of alfalfa, branches and leaves grow fastest.

表 158　　　　　苜蓿全生育期各旬耗水量统计表

时段		灌水量	旬灌水模数
月份	旬	（m³/亩）	
4	上	9.17	1.62
	中	9.17	1.62
	下	15.99	2.82
5	上	39.32	6.94
	中	33.35	5.88
	下	31.72	5.60
6	上	54.26	9.57
	中	69.49	12.26
	下	54.62	9.24
7	上	38.64	6.82
	中	33.55	5.92
	下	41.04	7.24
8	上	42.71	7.54
	中	34.75	6.13
	下	52.53	9.72

Table 158　Statistics of Water Consumption of Alfalfa in Every Ten Days of the Growth and Development Period

Period		Irrigation Amount (m³/mu)	Modulus of Irrigation Water for Every 10 Days
Month	Every 10 Days		
4	Early	9.17	1.62
	Middle	9.17	1.62
	Late	15.99	2.82
5	Early	39.32	6.94
	Middle	33.35	5.88
	Late	31.72	5.60
6	Early	54.26	9.57
	Middle	69.49	12.26
	Late	54.62	9.24
7	Early	38.64	6.82
	Middle	33.55	5.92
	Late	41.04	7.24
8	Early	42.71	7.54
	Middle	34.75	6.13
	Late	52.53	9.72

（二）苜蓿各生育阶段耗水量
3.3.3.2 Water Consumption of Alfalfa in Every Growth and Development Period

由表 159 可见，在苜蓿全生育中，苜蓿在第一茬、第二茬、第三茬的生育阶段分别为 68、37、51 d，第一茬生育阶段最长，而第二茬生育阶段最短，第三茬生育阶段介于二者之间。然而第三茬的阶段耗水量则最大，为 197.65 m³/亩，模系数达 34.87，耗水强度达 3.88 m³/(亩·d)。

As shown in Table 159, in the whole growing period of alfalfa, the duration of growing period of the first, second and third croppings lasts 68, 37 and 51 days respectively, where the growth stage of the first cropping lasts the longest days, the second cropping, and the shortest while the third cropping is situated between them. However, the third cropping which registers the largest amount of water consumption refers to 197.65 m³/mu, with modulus of irrigation water up to 34.87 and water consumption intensity up to 3.88 m³/(mu·day).

表 159　　　　　　　　　苜蓿各生育阶段耗水量统计表

	生育期	起止日期 （日/月）	阶段耗水量 （m³/亩）	模系数	耗水强度 （m³/(亩·d)）
第一茬	幼苗期	1/4～20/4	18.34	3.23	0.92
	蔓枝伸长期	21/4～20/5	88.66	15.64	3.06
	开花成熟期	21/5～7/6	69.70	12.3	4.10
	生育阶段	68(d)	176.70	31.18	2.68
第二茬	幼苗期	8/6～17/6	64.92	11.45	6.49
	蔓枝伸长期	18/6～5/7	94.79	16.72	5.27
	开花成熟期	6/7～14/7	32.74	5.78	3.64
	生育阶段	37(d)	192.45	33.95	5.35
第三茬	幼苗期	15/7～25/7	38.79	6.84	3.53
	蔓枝伸长期	26/7～22/8	109.40	19.30	3.91
	开花成熟期	23/8～3/9	49.46	8.73	4.12
	生育阶段	51(d)	197.65	34.87	3.88
全生育期		156(d)	566.80	100	3.70

Table 159　　　　　Statistics of Water Consumption of Alfalfa
in Every Growth and Development Period

Growth and Development Period		Duration (day/month)	Water consumption at each stage (m³/mu)	Modulus	Water consumption intensity (m³/mu · day)
1st Cropping	Seedling	1/4 – 20/4	18.34	3.23	0.92
	Branch Extension	21/4 – 20/5	88.66	15.64	3.06
	Flowering & Maturing	21/5 – 7/6	69.70	12.3	4.10
	Growth and Development	68 (days)	176.70	31.18	2.68
2nd Cropping	Seedling	8/6 – 17/6	64.92	11.45	6.49
	Branch Extension	18/6 – 5/7	94.79	16.72	5.27
	Flowering & Maturing	6/7 – 14/7	32.74	5.78	3.64
	Growth and Development	37 (days)	192.45	33.95	5.35
3rd cropping	Seedling	15/7 – 25/7	38.79	6.84	3.53
	Branch Extension	26/7 – 22/8	109.40	19.30	3.91
	Flowering & Maturing	23/8 – 3/9	49.46	8.73	4.12
	Growth and Development	51 (days)	197.65	34.87	3.88
Total		156 (days)	566.80	100	3.70

根据表 159 中的数据,点绘苜蓿全生育期各生育阶段的耗水量图,并从图 39 中可以看出,苜蓿在各生育阶段中以蔓枝伸长期的耗水量最大。在这一阶段,苜蓿枝叶生长最旺盛,需要大量的水进行光合作用,对苜宿干物质的积累有主要作用,是苜蓿在整个生育阶段的关键时期。

Based on the data in Table 159, we get the figure of alfalfa water consumption at each growth and development period and from Figure 39 it's clear that in the branch extension period alfalfa consumes the most water. At this stage, alfalfa's leaves grow vigorously and require a lot of water for photosynthesis, which helps alfalfa accumulate dry matter. Therefore, it is a critical period for alfalfa in the whole growth and development period.

(三)苜蓿耗水规律

3.3.3.3　Water Consumption Law of Alfalfa

苜蓿是一种密植性植物、根系非常发达,枝叶生长茂盛,植株蒸腾量大,整个生长期其耗水量大。其耗水特点为前期在幼苗期,生长较缓慢,植株蒸发较弱,耗水量较小;中期特别是在蔓枝伸长期植株蒸腾最强,需消耗大量的水分,进行光合作用,合成干物质,在这一

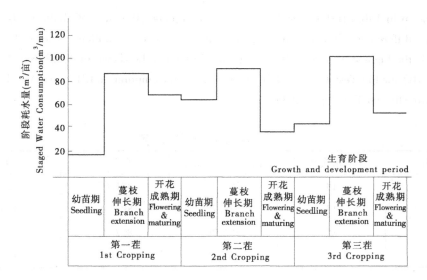

图 39 苜蓿各生育阶段耗水量图
Figure 39 Water Consumption of Alfalfa in Every Growth and Development Period

阶段枝叶生长也最快；后期即开花成熟期其生长速度变缓，蒸腾量也变小，耗水量比上一阶段少。所以，苜蓿在其全生育期内其耗水量呈一抛物线性，中间大，两头小，符合其生理生长特性。

Alfalfa is intensively planted with highly developed root system and flourishing branches and leaves, for its high transpiration. It consumes a mass of water in the whole growth and development period. The characteristics of water consumption of alfalfa are shown as follows: in the early seedling period, alfalfa grows slower, transpires weaker, and consumes less water; in the middle period, especially in the branch extension period, the strong transpiration consumes a large amount of water for photosynthesis and synthesizing dry matter. At this period, the branches and leaves grow fastest; in the later period, namely, the flowering maturity period, compared with the middle period, alfalfa will grow slower, transpire weaker, and consume less water. Therefore, the water consumption of alfalfa in the whole growth and development period is parabolic, that is, high in the middle and low in the ends, which is consistent with its growth characteristics.

四、苜蓿耗水量与相对产量分析
3.3.4 Analysis of Water Consumption and Relative Yield of Alfalfa

由表 160 可见，在不同的 4 种处理下，由于灌水次数、耗水量的不同，苜蓿产量也是不同的，以 3 号处理产量最高，达 897.91 kg/亩，这一点从表 161 还可以详尽得到反映。苜蓿各生育阶段的相对耗水量与相对产量试验成果列入表 162。

As shown in Table 160, under the 4 different treatments, the yield of alfalfa is also different due to the difference of irrigation frequency and water consumption. The yield of No. 3 treatment is the highest, reaching 897.91 kg/mu, which can be shown in detail from Table 161. See Table 162 for the Test Results of Relative Water Consumption and Relative Yield of Alfalfa at Every Growth and Development Period.

表 160　　　　　　　　　　　苜蓿耗水量与相对产量统计表

处理	灌水次数	耗水量(m³/亩)	产量(kg/亩)	需水系数(kg/m³)
1	18	589.23	881.67	1.50
2	17	578.70	883.38	1.54
3	15	566.79	897.91	1.60
4	15	556.10	869.41	1.56

Table 160　　　　Statistics of Alfalfa Water Consumption and Yield

Treatment	Irrigation Frequency	Water Consumption (m³/mu)	Yield (kg/mu)	Water Requirement Coefficient (kg/m³)
1	18	589.23	881.67	1.50
2	17	578.70	883.38	1.54
3	15	566.79	897.91	1.60
4	15	556.10	869.41	1.56

表 161　　　　　　　　苜蓿各生育期阶段耗水量、产量统计表　　　　　　　　单位:m³/亩

生育阶段		幼苗期	蔓枝伸长期	开花成熟期	全生育期	产量(kg/亩)
第一茬	1	41.75	112.61	79.32	231.26	293.60
	2	42.78	57.59	86.27	186.64	303.80
	3	34.33	82.89	67.14	184.36	380.90
	4	28.86	47.92	59.12	135.90	256.10
第二茬	1	18.98	91.51	28.62	139.11	214.50
	2	47.71	82.08	68.21	197.02	220.80
	3	45.50	83.34	62.43	189.27	249.40
	4	22.19	83.13	42.21	147.53	239.70

续表 161

生育阶段		幼苗期	蔓枝伸长期	开花成熟期	全生育期	产量(kg/亩)
第三茬	1	54.00	119.17	45.58	218.86	296.00
	2	33.41	116.75	44.88	195.04	290.70
	3	58.04	112.11	23.05	193.16	264.10
	4	33.42	111.57	34.58	179.57	190.10
平均	1	38.24	107.76	51.17	196.41	268.00
	2	41.30	85.47	66.45	192.90	271.80
	3	45.96	92.11	50.54	189.02	298.10
	4	28.16	80.73	45.30	154.33	228.70

Table 161　Statistics of Alfalfa Water Consumption and Yield in the Whole Growth and Development Period

Unit: m³/mu

Growth and Development Period		Seedling	Branch Extension	Flowering & Maturing	Whole Growth and Development Period	Yield (kg/mu)
1st Cropping	1	41.75	112.61	79.32	231.26	293.60
	2	42.78	57.59	86.27	186.64	303.80
	3	34.33	82.89	67.14	184.36	380.90
	4	28.86	47.92	59.12	135.90	256.10
2nd Cropping	1	18.98	91.51	28.62	139.11	214.50
	2	47.71	82.08	68.21	197.02	220.80
	3	45.50	83.34	62.43	189.27	249.40
	4	22.19	83.13	42.21	147.53	239.70
3rd Cropping	1	54.00	119.17	45.58	218.86	296.00
	2	33.41	116.75	44.88	195.04	290.70
	3	58.04	112.11	23.05	193.16	264.10
	4	33.42	111.57	34.58	179.57	190.10
Average	1	38.24	107.76	51.17	196.41	268.00
	2	41.30	85.47	66.45	192.90	271.80
	3	45.96	92.11	50.54	189.02	298.10
	4	28.16	80.73	45.30	154.33	228.70

苜蓿作物水分生产函数相乘模型,仍采用詹森(Jensen)1968 提出的模式,求得苜蓿各生育阶段的水分敏感指数为:

The multiplicative model of water production function adopts the model proposed by Jensen in 1968 to obtain the water sensitivity index of alfalfa at every growth stage:

幼苗期	蔓枝伸长期	开花成熟期
0.151 9	0.481 0	0.379 2
Seedling	Branch Extension	Flowering & Maturing
0.151 9	0.481 0	0.379 2

表 162　　苜蓿各生育期阶段相对耗水量与相对产量统计表　　单位:m³/亩

生育阶段		幼苗期	蔓枝伸长期	开花成熟期	全生育期	产量(kg/亩)
第一茬	1	1.000 0	1.000 0	1.000 0	1.000 0	1.000 0
	2	1.024 7	0.511 4	1.087 6	0.807 1	1.035 0
	3	0.822 2	0.736 1	0.846 4	0.797 2	1.097 3
	4	0.691 3	0.425 5	0.745 3	0.587 7	0.872 3
第二茬	1	1.000 0	1.000 0	1.000 0	1.000 0	1.000 0
	2	2.525 0	0.897 0	2.383 3	1.415 2	1.030 3
	3	2.408 7	0.888 9	2.181 3	1.359 5	1.163 8
	4	1.174 7	0.908 4	1.474 8	1.572 0	1.119 0
第三茬	1	1.000 0	1.000 0	1.000 0	1.000 0	1.000 0
	2	0.618 7	0.979 7	0.984 6	0.891 2	1.018 0
	3	1.074 8	0.940 8	0.505 8	0.882 6	0.980 9
	4	0.618 9	0.930 2	0.758 7	0.820 5	0.654 2
平均	1	1.000 0	1.000 0	1.000 0	1.000 0	1.000 0
	2	1.109 0	0.793 2	1.298 6	0.981 6	1.014 2
	3	1.234 2	0.854 8	0.987 1	0.962 4	1.112 3
	4	0.756 2	0.749 2	0.885 3	0.785 8	0.853 4

Table 162 Statistics for Water Consumption and Relative Yield of Alfalfa at Every Growth Stage

Unit: m³/mu

Growth and Development Period		Seedling	Branch Extension	Flowering & Maturing	Whole Growth and Development Period	Yield (kg/mu)
1st Cropping	1	1.000 0	1.000 0	1.000 0	1.000 0	1.000 0
	2	1.024 7	0.511 4	1.087 6	0.807 1	1.035 0
	3	0.822 2	0.736 1	0.846 4	0.797 2	1.097 3
	4	0.691 3	0.425 5	0.745 3	0.587 7	0.872 3
2nd Cropping	1	1.000 0	1.000 0	1.000 0	1.000 0	1.000 0
	2	2.525 0	0.897 0	2.383 3	1.415 2	1.030 3
	3	2.408 7	0.888 9	2.181 3	1.359 5	1.163 8
	4	1.174 7	0.908 4	1.474 8	1.572 0	1.119 0
3rd Cropping	1	1.000 0	1.000 0	1.000 0	1.000 0	1.000 0
	2	0.618 7	0.979 7	0.984 6	0.891 2	1.018 0
	3	1.074 8	0.940 8	0.505 8	0.882 6	0.980 9
	4	0.618 9	0.930 2	0.758 7	0.820 5	0.654 2
Average	1	1.000 0	1.000 0	1.000 0	1.000 0	1.000 0
	2	1.109 0	0.793 2	1.298 6	0.981 6	1.014 2
	3	1.234 2	0.854 8	0.987 1	0.962 4	1.112 3
	4	0.756 2	0.749 2	0.885 3	0.785 8	0.853 4

则苜蓿的水分生产函数为：

The water production function of alfalfa is:

$$Y/Y_M = (W_1/WM_1)^{0.151\,9} (W_2/WM_2)^{0.481\,0} (W_3/WM_3)^{0.379\,2}$$

即 $Y = 196.4(0.026\,2W_1)^{0.151\,9}(0.009\,3W_2)^{0.481\,0}(0.019\,5W_3)^{0.379\,2}$

That is $Y = 196.4(0.026\,2W_1)^{0.151\,9}(0.009\,3W_2)^{0.481\,0}(0.019\,5W_3)^{0.379\,2}$

从水分敏感指数可看出，不同生育阶段的缺水对苜蓿生长的影响程度，其敏感度大小顺序是：蔓枝伸长期＞开花成熟期＞幼苗期。

From the water sensitivity index, it can be seen that the effect of water shortage on the growth of alfalfa varies with the growth stage. The order of sensitivity index ranks as follows: branch extension period ＞ flowering and maturing period ＞ seedling period.

五、边际分析
3.3.5 Marginal Analysis

根据苜蓿的试验资料,计算苜蓿总产量、平均产量、边际产量、生产弹性,其结果见表163。并根据表中数据绘制苜蓿水分生产函数曲线,见图40。

According to the experimental data of alfalfa, total production, average product, marginal yield, and production flexibility are calculated out (see Table 163), furthermore, a water production function curve is drawn (see Figure 40).

从图40中可看出,总产量(Y)随耗水量(X)的增加,Y的增加由慢变快,以后又逐渐变缓,当耗水量为578.8 m³/亩,总产量达到最大值为889.9 kg/亩。这以后,再增加耗水量,不但不能增加产量,反而引起产量下降。由于总产量曲线至顶点的斜率为0,则此时对应的边际产量M也为零。平均产量A的曲线,随着耗水量的增加,也有一个增减过程,但最大值的出现与总产量最大值出现不一致。

It can be seen from Figure 40 that the total yield (Y) increases with the increasing water consumption (X) at an accelerating rate, and then gradually slows down. When the water consumption is 578.8 m³/mu, the total production reaches the maximum, 889.9 kg/mu. After that, increasing water consumption only leads to a decline in production, not an increase in production. Since the slope of the apex of the total production curve is 0, the corresponding marginal yield M is zero too. The average product curve A, along with the increase of water consumption, also has a process of increase and decrease, but the occurrence of its maximum value is inconsistent with the maximum value of total production.

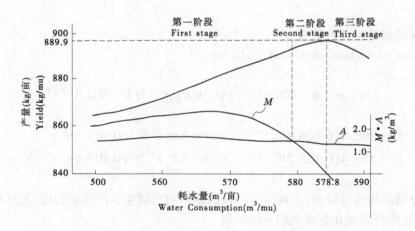

图40 苜蓿水分生产函数曲线
Figure 40　Water Production Function Curve of Alfalfa

第三章 作物灌溉制度研究
3 Study on Irrigation System of Crops

纵观图中3条曲线,有2个重要转折点:一是边际产量曲线与平均产量曲线的交点。在这一点,边际产量与平均产量相等;二是总产量曲线的顶点,边际产量在这一点值为零,过此点后变为负值。以这两点为限,连续投入资源的生产过程可分为3个阶段。从原点到平均产量和边际产量交点处为第一阶段,这个阶段的特点是边际产量高于平均产量,因而引起平均产量的逐渐提高,总产量增加的幅度大于资源增加的幅度,因而只要条件允许就不应在这一阶段停止资源的投入。第二阶段是平均产量和边际产量的交点,这个阶段的特点是边际产量小于平均产量,因而引起平均产量的减少。总产量增加的幅度小于资源增加的幅度,出现报酬递减,而总产量最大值是在一定条件下可达到的,在这一阶段,总产量按报酬递减形式继续增长,可获得较高的产量。第三阶段是总产量达到最大值点以后,也就是边际产量转为负值以后,在这一阶段,增加资源的投入反而引起总产量的降低,则称这一阶段为生产的绝对不合理阶段。

There are 3 curves in the figure with 2 important turning points: one is the crossover point of the marginal yield curve and the average product curve. At this point, marginal yield is equal to the average product; another one is the apex of the total production. At this point, the value of marginal yield is zero, and becomes negative after this point. Based on these two points, the continuous production process can be divided into 3 stages. The first stage is from the base point to the crossover point of average product and marginal yield. As one of the characteristics of this stage, the marginal production is higher than the average yield, which causes a gradual increase in the average yield, and the increase in total production is greater than the increase in resources. Therefore, resources input should not be stopped at this stage as long as conditions permit. The second stage is the crossover point of average production and marginal yield. This stage is characterized by a marginal production that is less than the average yield, thus causing a decrease in average production. If the increase in total production is less than the increase in resources, there will be diminishing returns. The maximum total production is achievable under certain conditions. At this stage, if the total production continues to grow in the form of diminishing returns, higher yields can be obtained. The third stage is that the total output reaches the maximum point, that is, the marginal yield turns negative. At this stage, increasing the input of resources causes a decrease in total production, which is called an absolutely unreasonable stage of production.

表163 苜蓿边际分析成果表

处理号	序号	耗水量 (m^3/亩)	总产量 (kg/亩)	平均产量 A	水量增量 ΔX(m^3/亩)	产量增量 ΔY(kg/亩)	边际产量 H	生产弹性 E_p
1	4	556.10	869.41	1.56				
2	3	566.70	897.91	1.58	103.30	28.5	0.27	0.17
3	2	578.70	883.38	1.53	12.00	−14.5	−1.21	−0.79
4	1	589.23	881.67	1.50	10.53	−1.7	0.16	0.11

Table 163 Results Table for Marginal Analysis of Alfalfa

Treatment No.	S/N	Water Consumption (m³/mu)	Total Yield (kg/mu)	Average Product A	Increase in Water Requirement ΔX (m³/mu)	Increase in Yield ΔY (kg/mu)	Marginal Yield H	Production Flexibility E_p
1	4	556.10	869.41	1.56				
2	3	566.70	897.91	1.58	103.30	28.5	0.27	0.17
3	2	578.70	883.38	1.53	12.00	−14.5	−1.21	−0.79
4	1	589.23	881.67	1.50	10.53	−1.7	0.16	0.11

六、价值生产函数及最大纯收益
3.3.6 Value Production Function and Maximum Net Income

由表164、图41中可看出，边际产值曲线与边际成本曲线交于一点，该点表示边际产值等于边际成本，此点即为最大纯收益点，它对应的产值点是137.4元/亩，相应的耗水量为560.0 m³/亩。

From Table 164 and Figure 41, it can be seen that, marginal production value curve and marginal cost curve intersect at one point which means the marginal production value is equal to marginal cost. This point is the maximum net income point, its corresponding production value point is 137.4 yuan/mu, and the corresponding water consumption is 560.0 m³/mu.

表164 苜蓿边际分析统计表
($P_X = 0.04$　　$P_Y = 0.23$)

处理号	序号	耗水量 (m³/亩)	产量 Y (kg/亩)	平均产值 $A_p = A \cdot P_Y$ (元/m³)	边际产值 $M_p = M \cdot P_Y$ (元/m³)	耗水成本 $T_c = P_X \cdot X$ (元/亩)	边际成本 $M_c = P_X$ (元/亩)	资源投入 K (元/亩)	产值 $T_p = Y \cdot P_Y$ (元/亩)	纯收益 $Z = T_p - T_c - K$ (元/亩)
1	4	556.1	869.41	0.360		22.24	0.04	46.0	199.96	131.72
2	3	566.7	897.91	0.363	0.062	22.67	0.04	46.0	206.52	137.85
3	2	578.7	883.38	0.352	−0.278	23.15	0.04	46.0	203.18	134.03
4	1	589.2	881.67	0.345	0.037	23.57	0.04	46.0	202.78	133.21

Table 164　　　　　　　Statistics for Marginal Analysis of Alfalfa

($P_X = 0.04$　　$P_Y = 0.23$)

Treatment No.	S/N	Water Consumption (m^3/mu)	Yield Y (kg/mu)	Average Production Value $A_p = A \cdot P_Y$ (yuan/m^3)	Marginal Production Value $M_p = M \cdot P_Y$ (yuan/m^3)	Water Consumption Cost $T_c = P_X \cdot X$ (yuan/mu)	Marginal Cost $M_c = P_X$ (yuan/mu)	Resources Input K (yuan/mu)	Production Value $T_p = Y \cdot P_Y$ (yuan/mu)	Net Income $Z = T_p - T_c - K$ (yuan/mu)
1	4	556.1	869.41	0.360		22.24	0.04	46.0	199.96	131.72
2	3	566.7	897.91	0.363	0.062	22.67	0.04	46.0	206.52	137.85
3	2	578.7	883.38	0.352	-0.278	23.15	0.04	46.0	203.18	134.03
4	1	589.2	881.67	0.345	0.037	23.57	0.04	46.0	202.78	133.21

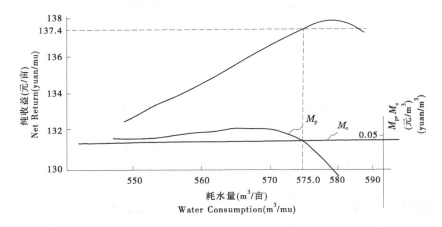

图 41　苜蓿生产函数与纯收益关系图

Figure 41　Relation of Production Function and Net Income of Alfalfa

七、苜蓿产量方差分析
3.3.7　Variance Analysis of Alfalfa Yield

自由度分解：总自由度：$f = nk - 1 = 3 \times 4 - 1 = 11$

Division of degree of freedom：Total degree of freedom：$f = nk - 1 = 3 \times 4 - 1 = 11$

区间自由度：$f = n - 1 = 2$

Degree of freedom in zones：$f = n - 1 = 2$

处理间自由度：$f = k - 1 = 3$

Degree of freedom within treatments: $f = k - 1 = 3$

误差自由度: $f = 11 - 2 - 3 = 6$
Error's degree of freedom: $f = 11 - 2 - 3 = 6$

平方和分解: $C = T_t^2 / nk = 9\ 358\ 228.37$
Division of sum of squares: $C = T_t^2 / nk = 9\ 358\ 228.37$

总平方和: $S_{总} = \sum_1^n X^2 - C = 9\ 380\ 331.44 - C = 22\ 103.07$

Total sum of squares: $S_{total} = \sum_1^n X^2 - C = 9\ 380\ 331.44 - C = 22\ 103.07$

$$S_{区} = \sum T_r^2 / K - C = 9\ 720.86$$
$$S_{zones} = \sum T_r^2 / K - C = 9\ 720.86$$

$$S_{处} = \sum T_t^2 / n - C = 1\ 226.62$$
$$S_{treatments} = \sum T_t^2 / n - C = 1\ 226.62$$

$$S_{误} = S_{总} - S_{区} - S_{处} = 11\ 155.59$$
$$S_{error} = S_{total} - S_{zones} - S_{treatments} = 11\ 155.59$$

将上述数据列入方差分析表(见表165)进行 F 检验,见表166。

Put the data above into the following variance analysis table (see Table 165) and check the value of F (see Table 166).

表165　　　　　　　　　苜蓿方差分析表　　　　　　　　　单位:kg/亩

处理	区组			T_t	X_t
	I	II	III		
1	836.13	887.21	884.89	2 608.23	869.41
2	890.35	920.13	883.25	2 693.73	897.91
3	782.91	897.26	969.97	2 650.14	883.38
4	865.29	878.47	901.25	2 645.01	881.67
T_r	3 374.68	3 583.07	3 639.36	10 597.11	
X_r	843.67	895.77	909.84		883.09

第三章 作物灌溉制度研究
3 Study on Irrigation System of Crops

Table 165　　　　　　　　　　Variance Analysis of Alfalfa　　　　　　　　　　Unit: kg/mu

Treatment	Zone			T_1	X_1
	I	II	III		
1	836.13	887.21	884.89	2 608.23	869.41
2	890.35	920.13	883.25	2 693.73	897.91
3	782.91	897.26	969.97	2 650.14	883.38
4	865.29	878.47	901.25	2 645.01	881.67
T_r	3 374.68	3 583.07	3 639.36	10 597.11	
X_r	843.67	895.77	909.84		883.09

表166　　　　　　　　　方差分析表成果表

方差来源	平方和 SS	自由度 f	均方 Ms	F	$F_{0.05}$	$F_{0.01}$
区组间	9 720.86	2	4 860.43	2.61	5.14	
处理间	1 226.62	3	408.87	0.22	4.76	
误差	1 155.59	6	1 859.27			
总和	22 103.07	11				

Table 166　　　　　　　　Results for Variance Analysis

Variance Sources	Sum of Squares SS	Degree of Freedom f	Mean Squares Ms	F	$F_{0.05}$	$F_{0.01}$
Between Zones	9 720.86	2	4 860.43	2.61	5.14	
Within Treatments	1 226.62	3	408.87	0.22	4.76	
Error	1 155.59	6	1 859.27			
Total	22 103.07	11				

从表166中可以看出,无论是区组间还是处理间的差异都不显著。分析其中的原因,从耗水量统计表中也可发现,4个处理的耗水量相差不大,也就是说这4个处理的差异不明显。

From Table 166, it can be seen that the differences between zones or between periods are not significant. From the statistics table of water consumption, it can be seen that the difference in water consumption of the 4 treatments is not significant, that is, the difference between the 4 treatments is not obvious.

从生产函数曲线和生产收益图中可看出,苜蓿耗水量为578.8 m³/亩,对应的产量最大为 889.9 kg/亩。这一结论与处理 2 相近,即控制水平为:幼苗期 45%,蔓枝伸长期 70%,开花成熟期 65%。而苜蓿的经济灌溉定额为 575.0 m³/亩,其对应的最大纯收益为 137.4 元/亩,最优产量为 885.1 kg/亩。

From the production function curve and production income diagram, it can be seen that the water consumption of alfalfa is 578.8 m³/mu, and the maximum corresponding output is 889.9 kg/mu. This conclusion is similar to treatment 2, that is, the control level is: 45% in seedling period, 70% in branch extension period, and 65% in flowering and maturing period. The economic irrigation quota of alfalfa is 575.0 m³/mu, the corresponding maximum net income is 137.4 yuan/mu, and the optimal yield is 885.1 kg/mu.

八、苜蓿优化灌溉制度
3.3.8 Optimized Irrigation Scheduling of Alfalfa

采用动态规划的方法,按时间顺序、对苜蓿整个生育期划分为 3 个生育阶段,每个阶段根据土壤的含水量和降水量做出决策,即灌水时间、灌水次数和灌水定额,各阶段所组成的最优决策就是苜蓿的最优灌溉制度。

With the dynamic planning method, the growth and development period of alfalfa can be divided into 3 stages according to the chronological order. At each stage, according to the water content of soil and precipitation, the irrigation time, irrigation frequency and irrigation water quota can be determined. The optimal decision of each stage is the optimal irrigation scheduling.

(一)变量设置
3.3.8.1 Variable Setting

阶段变量 i:根据作物生育阶段划分为 i 个阶段;

Stage variable i: the growth and development stage of crops can be divided into i stages;

状态变量:为各生育阶段土壤含水量 Q_i;

State variable: Q_i is the soil water content at every growth and development stage;

决策变量:为各生育阶段灌水量 X_i。

Decision variable: X_i is the irrigation amount at every growth and development period.

(二)约束条件
3.3.8.2 Constraint Condition

作物耗水量约束:

Constraints on crops water consumption:

$$W_{mini} \leq W_i \leq W_{maxi}$$

式中　W_{mini}——各生育阶段适宜下限耗水量；

　　　W_{maxi}——各生育阶段供水充足条件下最大作物耗水量。

Where　W_{mini} = the proper minimum water consumption in every growth and development period

　　　W_{maxi} = the maximum water consumption in every growth and development period when water is sufficient.

土壤含水量约束：

Constraints on soil water content:

$$\theta_{min} \leq \theta_i \leq \theta_{max}$$

式中　θ_{min}——土壤适宜含水率下限；

　　　θ_{max}——田间持水率。

Where　θ_{min} = the lower limit of suitable soil water content

　　　θ_{max} = field capacity.

灌溉定额约束：

Constraints on irrigation quota:

$$\sum_{i=1}^{n} X_i = X^* = X*$$

式中　X_i——各生育阶段灌水量；

　　　X^*——生育阶段优化灌溉定额。

Where　X_i = irrigation amount in every period

　　　X^* = optimized irrigation quota in every period.

（三）状态转移方程（用阶段水量平衡方程表示）

3.3.8.3　**State Transition Equation（expressed in the equation of stage water balance）**

$$H_i = M_{i-1} + X_i - W_i$$
$$M_i = 667Hr(\theta_i - \theta_{min})$$

式中　H——土壤计划湿润层深度；

　　　r——计划湿润层内平均土壤容量；

　　　θ_i——土壤含水率状态；

θ_{min}——凋萎含水量。

Where H = the soil wetting layer

r = the average soil capacity in the soil wetting layer

θ_i = the state of soil water content

θ_{min} = the wilting water content.

(四) 目标函数
3.3.8.4 Objective Function

采用 Jehsen:水量与产量模型:
Jehsen model is adopted: model of water amount and yield:

$$Y/Y_{max} = \max \prod_{i=1}^{n} \left(\frac{W_i}{W_{maxi}}\right)^{\lambda_i}$$

式中 Y——作物供水不足时达到的产量;

Y_{max}——作物供水充足时达到的产量;

n——作物生育阶段数;

W_i——作物供水不足时在阶段 i 的实际耗水量;

λ_i——各生育阶段的水分敏感指数。

Where Y = yield when water is insufficient

Y_{max} = yield when water is sufficient

n = the number of crops periods

W_i = the real water consumption in stage i when there is water shortage

λ_i = the water sensitivity index at every growth and development period.

根据苜蓿生产函数得出苜蓿的经济灌溉定额即优化灌溉定额 $X^* = 575.0 \text{ m}^3/$亩,灌水次数为 17 次。

According to the production function of alfalfa, the economic irrigation quota of alfalfa is obtained, that is, the optimal irrigation quota is $X^* = 575.0 \text{ m}^3/\text{mu}$, and the irrigation frequency is 17 times.

灌溉分配方案为:

The irrigation allocation scheme is stated below:

递推计算过程:采用逆时序递推,即从第三阶段开始依次递推至第一阶段,求出各阶段开始的土壤含水量 θ_{gi}(及相应的土壤供水量 M_i)和可供该阶段及该阶段后各阶段使用的水量 M_i,以及各种组合情况下的最优决策 d_i^*、f_i^* 计算结果见表 167。

Recursive calculation process: reverse chronological recursion, from the third stage to the

3 Study on Irrigation System of Crops

first stage, is adopted to obtain θ_{gi}, the soil water content at every stage (M_i, the corresponding soil water supply amount), M_i, the available water amount during and after that stage, as well as d_i^* and f_i^*, the optimal decision under various combinations. The calculation results are shown in Table 167.

表 167　　　　　　苜蓿第一茬优化灌溉制度递推表

土壤水分状况		第三阶段递推计算 $W_{m3}=86.27\lambda_3=0.3792$ $f^{3*}=d_3^{max}$ $\{f_3\}=d_5^{max}\cdot$ $\{(W_3/86.2)^{0.3792}\}$ 其中：$W_3=\min\{86.27, M_3+37d_3\}$ $0\leqslant d_3 \leqslant I_3/37$				第二阶段递推计算 $W_{m2}=57.59\lambda_2=0.4810$ $f_2^3=d_2^{max}$ $\{f_2\}=d_2^{max}$ $\{W/59.59\}^{0.4810}\cdot f_3^*(M_3,I_3)$ 其中：$W_2=\min\{57.59, M_2+37d_2\}$ $0\leqslant d_4 \leqslant I_4/37$ $M_3=M_2+37d_2-W_2$ $I_3=I_2-37d_2$			第一阶段 $W_{m1}=42.78\lambda_1=0.1519$ $f_1^*=d_1^*$ $\{(\dfrac{W_1}{42.78})^{0.1519\times f_2^*(M_2,I_2)}\}$ 其中：$W_1=\min\{42.78, M_1+37d\}$ $0\leqslant d_1 \leqslant I_1/37$ $M_2=M_1+27d_1-W_1$ $I_2=I_1-37d_1$	
含水量	供水量	d_3^*/f_3^*				d_2^*/f_2^*			$I_1=148$	$I_1=185$
Q3.2	M3.2	$I_3=0$	$I_3=37$	$I_3=74$	$I_3=111$	$I_2=74$	$I_2=111$	$I_2=148$	0/1	0/1
12	44.26	0/0.778	1/0.978	2/1	2/1	2/0.875	2/1	2/1	0/0.982	1/1
11	37.94	0/0.732	1/0.948	2/1	2/1	2/0.839	2/1	2/1	0/0.955	1/1
10	31.62	0/0.683	1/0.917	2/1	2/1	2/0.801	2/0.994	2/1	0/0.938	1/1
9	25.29	0/0.628	1/0.884	2/1	2/1	2/0.759	2/0.966	2/1	0/0.883	1/1
8	18.97	0/0.563	1/0.849	2/1	2/1	2/0.713	2/0.936	2/1	0/0.831	1/1
7	12.65	0/0.483	1/0.811	2/1	2/1	2/0.662	2/0.904	2/1	0/0.749	1/1
6	6.36	0/0.372	1/0.770	2/0.973	3/1	2/0.603	2/0.870	2/1		
5	0									

Table 167　　Recursive Table for Optimized Irrigation Scheduling for the 1st Cropping of Alfalfa

Condition of Soil Water		for the third stage $W_{m3}=86.27\lambda_3=0.3792$ $f^{3*}=d_3^{max}$ $\{f_3\}=d_5^{max}\cdot$ $\{(W_3/86.2)^{0.3792}\}$ Where: $W_3=\min\{86.27, M_3+37d_3\}$ $0\leqslant d_3 \leqslant I_3/37$	for the second stage $W_{m2}=57.59\lambda_2=0.4810$ $f_2^3=d_2^{max}$ $\{f_2\}=d_2^{max}$ $\{W/59.59\}^{0.4810}\cdot$ $f_3^*(M_3,I_3)$ Where: $W_2=\min\{57.59, M_2+37d_2\}$ $0\leqslant d_4 \leqslant I_4/37$ $M_3=M_2+37d_2-W_2$ $I_3=I_2-37d_2$	Stage I $W_{m1}=42.78\lambda_1=0.1519$ $f_1^*=d_1^*$ $\{(\dfrac{W_1}{42.78})^{0.1519\times f_2^*(M_2,I_2)}\}$ Where: $W_1=\min\{42.78, M_1+37d\}$ $0\leqslant d_1 \leqslant I_1/37$ $M_2=M_1+27d_1-W_1$ $I_2=I_1-37d_1$

Continued Table 167

Water Content	Water Supply	d_3^*/f_3^*				d_2^*/f_2^*			$I_1 = 148$	$I_1 = 185$
Q3.2	M3.2	$I_3 = 0$	$I_3 = 37$	$I_3 = 74$	$I_3 = 111$	$I_2 = 74$	$I_2 = 111$	$I_2 = 148$	0/1	0/1
12	44.26	0/0.778	1/0.978	2/1	2/1	2/0.875	2/1	2/1	0/0.982	1/1
11	37.94	0/0.732	1/0.948	2/1	2/1	2/0.839	2/1	2/1	0/0.955	1/1
10	31.62	0/0.683	1/0.917	2/1	2/1	2/0.801	2/0.994	2/1	0/0.938	1/1
9	25.29	0/0.628	1/0.884	2/1	2/1	2/0.759	2/0.966	2/1	0/0.883	1/1
8	18.97	0/0.563	1/0.849	2/1	2/1	2/0.713	2/0.936	2/1	0/0.831	1/1
7	12.65	0/0.483	1/0.811	2/1	2/1	2/0.662	2/0.904	2/1	0/0.749	1/1
6	6.36	0/0.372	1/0.770	2/0.973	3/1	2/0.603	2/0.870	2/1		
5	0									

由表168就可以得出在各种初始条件下的最优灌溉制度。为此,我们将由1阶段进行正向递推可得出灌水决策过程。递推方法如下。

According to Table 168, the optimal irrigation scheduling under various initial conditions can be obtained. Therefore, we will carry out forward recursion to obtain the decision-making process of irrigation by the first stage. The recursive method is as follows.

根据初始土壤含水率在表167中查得对应的 d_1^* 和 f_1^* 及 M_1,由递推方程分别求出第二阶段、第三阶段的 M 和 I 值。再查表167,可得出最优的灌水决策 d_1^*,d_2^*,d_3^*。至此优化灌溉制度就已确定。表168列出初始含水率为9%、7%两种情况。

According to the initial soil water content, the corresponding d_1^*, f_1^* and M_1 can be obtained in Table 167. With the recursive equation, the values of M and I of the second and third stages can be obtained respectively. According to Table 167, the optimal irrigation decisions, d_1^*, d_2^*, and d_3^*, can be obtained, and then the optimized irrigation scheduling is determined. The initial water content is 9% and 7% in Table 168.

表168 不同初始含水率最优决策

初始土壤含水率	阶段	$M_i(m^3/亩)$	$I_i(m^3/亩)$	d^*	f^*	$Q_i(\%)$
9%	1	25.09	185	1	1.00	9.00
	2	19.51	148	2		8.09
	3	35.92	74	2		10.68
7%	1	12.65	185	1	1.00	7.00
	2	6.87	148	2		6.09
	3	23.28	74	2		8.68

Table 168   **Optimal Decision of Different Initial Water Content**

Initial Soil Water Content	Stage	M_i (m³/mu)	I_i (m³/mu)	d^*	f^*	Q_i (%)
9%	1	25.09	185	1	1.00	9.00
	2	19.51	148	2		8.09
	3	35.92	74	2		10.68
7%	1	12.65	185	1	1.00	7.00
	2	6.87	148	2		6.09
	3	23.28	74	2		8.68

当初始含水率为9%时,相应3个阶段在水量上的分配为:37、74、74 m³/亩。

When the initial water content is 9%, the distribution of the water amount in the corresponding 3 stages should be: 37,74 and 74 m³/mu respectively.

$$Y^* = f_1^* \cdot Y_m = 1.00 \times 885.1 = 885.1 \text{ kg/亩}$$
$$Y^* = f_1^* \cdot Y_m = 1.00 \times 885.1 = 885.1 \text{ kg/mu}$$

当初始含水率为7%时,相应3个阶段在水量上的分配为:37、74、74 m³/亩。

When the initial water content is 7%, the distribution of the water amount in corresponding 3 stages should be 37,74,and 74 m³/mu respectively.

$$Y^* = f_1^* \cdot Y_m = 1.00 \times 885.1 = 885.1 \text{ kg/亩}$$
$$Y^* = f_1^* \cdot Y_m = 1.00 \times 885.1 = 885.1 \text{ kg/mu}$$

(五)灌水时间 T 的确定

3.3.8.5 Determination of Irrigation Time T

从水量平衡的原理出发:

According to the water balance principle:

$$ET = P_e + C + \Delta W \text{ 或 } e_t \cdot T = P_e + g \cdot T + \Delta W$$
$$ET = P_e + C + \Delta W \text{ or } e_t \cdot T = P_e + g \cdot T + \Delta W$$

则

$$T = \frac{\Delta W + P_e}{e_t - g}$$

Then
$$T = \frac{\Delta W + P_e}{e_t - g}$$

式中 T——未来灌水的间隔添数；

ΔW——土壤有效潜水量；

P_e——有效降雨量；

e_t——日耗水强度，mm/d，$e_t = Kc \cdot ET$

g——地下水补给强度，mm/d。

Where T = the interval days of future irrigation

ΔW = the amount of the available phreatic water

P_e = effective precipitation

e_t = the diurnal water consumption, mm/d, $e_t = K_c \cdot ET$

g = the groundwater recharge level, mm/d.

本试验区无地下水补给，$g = 0$，计算结果见表169。

There is no groundwater recharge in the plots, so $g = 0$. The calculation results are shown in Table 169.

表169　　　　　　　　　　　　紫花苜蓿灌水时间表

生育阶段		幼苗期	蔓枝伸长期	开花成熟期
9%	ΔW_1(mm)	56.91	51.15	67.53
7%	ΔW_2(mm)	44.26	38.51	54.89
	P_e(mm)	70	3.55	4.57
	e_t(mm/d)	7.17	12.82	12.31
	T_1(d)	8	4	6
	T_2(d)	6	3	5

Table 169　　　　　　　　　　　Irrigation Timetable of Alfalfa

Growth and Development Period		Seedling	Branch Extension	Flowering & Maturing
9%	ΔW_1(mm)	56.91	51.15	67.53
7%	ΔW_2(mm)	44.26	38.51	54.89
	P_e(mm)	70	3.55	4.57
	e_t(mm/d)	7.17	12.82	12.31
	T_1(d)	8	4	6
	T_2(d)	6	3	5

第三章 作物灌溉制度研究
3 Study on Irrigation System of Crops

表中 T_1, T_2 分别代表初始含水率为 9%, 7% 的优化灌溉间隔天数。优化灌溉制度成果见表 169 ~ 表 171。

Table T_1 and T_2 represent the interval days of optimal irrigation, of which the initial water content is 9% and 7%, respectively. The results of optimized irrigation scheduling are shown in Table 169 to 171.

表 170 初始含水率为 9% 和 7% 的优化灌溉制度

初始土壤含水率	阶段	灌水定额(m³/亩)	灌水次数	灌水间隔天数(d)
9%	1		1	8
	2	37	2	4
	3		2	6
7%	1		1	6
	2	37	2	3
	3		2	5

Table 170 Optimized Irrigation Scheduling with Initial Water Content (9% and 7%)

Initial Soil Water content	Stage	Irrigation Water Quota (m³/mu)	Irrigation Frequency	Irrigation Interval (days)
9%	1	37	1	8
	2		2	4
	3		2	6
7%	1	37	1	6
	2		2	3
	3		2	5

表 171 二、三茬优化灌溉制度递推表

土壤水分状况	第三阶段递推计算	第二阶段递推计算	第一阶段递推计算
	$W_{m3} = 56.57$ $\lambda_3 = 0.3792$ $f_3^* = d_5^*$ $\{f_3\} = d_3^*$ $\{(\frac{W_3}{56.54})^{0.3792}\}$ 其中: $W_3 = \min\{56.54, M_3 + 33d_3\}$ $0 \leq d_3 \leq I_5/33$	$W_{m2} = 99.41$ $\lambda_2 = 0.4810$ f_2^* $= d_2^*$ $\{f_3\} = d_2^*$ $\{(\frac{W_2}{99.41})^{0.4810}$, $f_3^*(M_3, I_3)\}$ $0 \leq d_2 \leq I_2/33$ $M_3 = M_4 + 33d_2 - W_4$ $I_3 = I_2 - 33d_2$	$W_{m2} = 40.56$ $\lambda_1 = 0.1519$ $f_1^* = d_1^*$ $\{f_2\} = d_1^*$ $\{(\frac{W_1}{40.56})^{0.1519}$, $f_2^*(M_2, I_2)\}$ 其中: $W_1 = \min\{40.56, M_1 + 33d_1\}$ $0 \leq d_1 \leq I_1/33$ $M_2 = M_1 + 33d_1 - W_1$ $I_2 = I_1 - 33d$

续表 171

含水率 Q3.2	供水量 M3.2	d_3^*/f_3^*			d_2^*/d_2^*			d_1^*/f_1^*		
		$I_3=0$	$I_3=33$	$I_3=66$	$I_3=99$	$I_2=132$	$I_2=165$	$I_1=132$	$I_1=165$	$I_1=188$
12	44.26	0/0.911	1/1	1/1	3/0.892	4/1	4/1	0/0.845	0/1	0/1
11	37.94	0/0.860	1/1	1/1	3/0.856	4/1	4/1	0/0.803	0/1	1/1
10	31.62	0/0.802	1/1	1/1	3/0.788	4/1	4/1	0/0.781	1/1	1/1
9	25.29	0/0.737	1/1	1/1	3/0.733	4/1	4/1	0/0.755	1/0.957	1/1
8	18.97	0/0.661	1/0.969	2/1	3/0.655	4/0.999	4/1	0/0.723	1/0.909	1/1
7	12.65	0/0.566	1/0.922	2/1	3/0.560	4/0.918	4/1	0/0.679	1/0.857	1/1
6	6.36	0/0.437	1/0.872	2/1	3/0.426	4/0.868	4/1	0/0.612	2/0.811	2/1
5	0									

Table 171　Recursive Formulas for Optimized Irrigation Scheduling for the 2nd and 3rd Croppings

	For the Third Stage	For the Second Stage	For the First Stage
Condition of soil water	$W_{m3}=56.57$　$\lambda_3=0.3792$ $f_3^*=d_5^*\{f_3\}=d_3^*$ $\{(\dfrac{W_3}{56.54})^{0.3792}\}$ Where: $W_3=\min\{56.54, M_3+33d_3\}$ $0\leq d_3\leq I_5/33$	$W_{m2}=99.41$　$\lambda_2=0.481$ $0 f_2^*=d_2^*\{f_3\}=d_2^*$ $\{(\dfrac{W_2}{99.41})^{0.4810}$, $f_3^*(M_3,I_3)\}$ $0\leq d_2\leq I_2/33$ $M_3=M_4+33d_2-W_4$ $I_3=I_2-33d_2$	$W_{m2}=40.56$　$\lambda_1=0.1519$ $f_1^*=d_1^*\{f_2\}=d_1^*$ $\{(\dfrac{W_1}{40.56})^{0.1519}f_2^*(M_2,I_2)\}$ Where: $W_1=\min\{40.56, M_1+33d_1\}$ $0\leq d_1\leq I_1/33$ $M_2=M_1+33d_1-W_1$ $I_2=I_1-33d$

Water Content	Water Supply	d_3^*/f_3^*			d_2^*/d_2^*			d_1^*/f_1^*		
Q3.2	M3.2	$I_3=0$	$I_3=33$	$I_3=66$	$I_3=99$	$I_2=132$	$I_2=165$	$I_1=132$	$I_1=165$	$I_1=188$
12	44.26	0/0.911	1/1	1/1	3/0.892	4/1	4/1	0/0.845	0/1	0/1
11	37.94	0/0.860	1/1	1/1	3/0.856	4/1	4/1	0/0.803	0/1	1/1
10	31.62	0/0.802	1/1	1/1	3/0.788	4/1	4/1	0/0.781	1/1	1/1
9	25.29	0/0.737	1/1	1/1	3/0.733	4/1	4/1	0/0.755	1/0.957	1/1
8	18.97	0/0.661	1/0.969	2/1	3/0.655	4/0.999	4/1	0/0.723	1/0.909	1/1
7	12.65	0/0.566	1/0.922	2/1	3/0.560	4/0.918	4/1	0/0.679	1/0.857	1/1
6	6.36	0/0.437	1/0.872	2/1	3/0.426	4/0.868	4/1	0/0.612	2/0.811	2/1
5	0									

3 Study on Irrigation System of Crops

注:二茬和三茬苜蓿的生育期气温因素相近,所以采用平均法求得它们的优化灌溉制度,唯一不同处,在灌水定额中,二茬为33 m³/亩,三茬为32 m³/亩,这样,苜蓿整个生育期的优化灌溉制度划分为两部分,并都推导出,分析计算结果见表172~表174。

Note: the temperature in the growth and development period for the 2nd and 3rd croppings of alfalfa is similar; therefore, the average method is used to obtain the optimized irrigation scheduling. The only difference is that the irrigation water quota of the 2nd cropping is 33 m³/mu, and that of the 3rd cropping is 32 m³/mu. In this way, the optimized irrigation scheduling of the entire growth and development period can be divided into two parts, and its analytical results can be obtained, which can be seen Table 172 to 174.

表172　　　　　　　　　　不同初始含水率最优决策

初始土壤含水率	阶段	M_i	I_i	d^*	f^*	$Q_i(\%)$
9%	1	25.09	188	1	1.00	9.0
	2	17.73	165	3		7.80
	3	17.32	66	2		9.74
7%	1	12.65	188	1	1.00	7.00
	2	5.09	165	3		5.10
	3	4.68	66	2		5.26

Table 172　　　　Optimal Decision for Different Initial Water Content

Initial Soil Water Content	Stage	M_i	I_i	d^*	f^*	$Q_i(\%)$
9%	1	25.09	188	1	1.00	9.0
	2	17.73	165	3		7.80
	3	17.32	66	2		9.74
7%	1	12.65	188	1	1.00	7.00
	2	5.09	165	3		5.10
	3	4.68	66	2		5.26

表173 灌水时间计算表

生育阶段		幼苗期	蔓枝伸长期	开花成熟期
9%	ΔW_1 (mm)	56.91	51.15	7.53
7%	ΔW_2 (mm)	44.26	38.51	54.89
	P_e (mm/d)	0	10.88	8.02
	e_t (mm/d)	127.4	12.74	12.74
	T_1 (d)	5	5	6
	T_2 (天)	4	4	5

Table 173　Calculation Table for Irrigation Time

Growth and Development Period		Seedling	Branch Extension	Flowering & Maturing
9%	ΔW_1 (mm)	56.91	51.15	7.53
7%	ΔW_2 (mm)	44.26	38.51	54.89
	P_e (mm/d)	0	10.88	8.02
	e_t (mm/d)	127.4	12.74	12.74
	T_1 (d)	5	5	6
	T_2 (d)	4	4	5

表174　二茬、三茬苜蓿优化灌溉制度

初始土壤含水率	阶段	灌水定额（m³/亩）	灌水次数	灌水时间(d)
9%	1		1	5
	2	33/32	3	5
	3		2	6
7%	1		1	4
	2	33/32	32	4
	3		2	5

Table 174　Optimized Irrigation Scheduling of Alfalfa's 2nd and 3rd Croppings

Initial Soil Water Content	Stage	Irrigation Water Quota (m³/mu)	Irrigation Frequency	Irrigation Time (d)
9%	1		1	5
	2	33/32	3	5
	3		2	6
7%	1		1	4
	2	33/32	32	4
	3		2	5

九、试验结果讨论
3.3.9 Discussion on Experimental Results

紫花苜蓿优化灌溉定额为 575.0 m³/亩,灌水定额,一茬为 37 m³/亩,灌水次数 5 次,二茬为 33 m³/亩,灌水次数为 6 次,三茬为 32 m³/亩,灌水次数 6 次,最大纯收益为 137.4 元/亩。

The optimized irrigation quota of alfalfa is 575.0 m³/mu. The irrigation water quota of the 1st cropping is 37 m³/mu, and its irrigation frequency is 5. The irrigation water quota of the 2nd cropping is 33 m³/mu, and its irrigation frequency is 6. The irrigation water quota of the 3rd cropping is 32 m³/mu, and its irrigation frequency is 6. The maximum net income is 137.4 yuan/mu.

将苜蓿的整个生育期分为 3 个阶段,建立优化灌溉数学模式,推算出不同土壤初始条件下的优化灌溉制度,本试验最后得出的结果见表 175。

Firstly, the whole growth and development period of alfalfa can be divided into 3 stages. Secondly, the mathematical model of optimized irrigation can be established. Finally, the optimal irrigation scheduling under different initial conditions of soil can be obtained. The final results of this experiment are shown in Table 175.

表 175 紫花苜蓿全生育期优化灌溉制度

生育阶段		幼苗期	蔓枝伸长期	开花成熟期
第一茬	灌水定额(m³/亩)	37	37	37
	灌水次数	1	2	2
	灌水时间(d)	9%/7% 8/6	4/3	6/5
第二、三茬	灌水定额(m³/亩)	33/32	33/32	33/32
	灌水次数	1	3	2
	灌水时间(d)	7%/9% 4/5	4/5	5/6

Table 175 Optimized Irrigation Scheduling of Alfalfa in the whole Growth and Development Period

Growth and Development Period		Seedling	Branch Extension	Flowering & Maturing
1st Cropping	Irrigation Water Quota (m³/mu)	37	37	37
	Irrigation Frequency	1	2	2
	Irrigation Time (d)	9%/7% 8/6	4/3	6/5

Continued Table 175

Growth and Development Period		Seedling	Branch Extension	Flowering & Maturing
2nd and 3rd Cropping	Irrigation Water Quota (m^3/mu)	33/32	33/32	33/32
	Irrigation Frequency	1	3	2
	Irrigation Time (day)	7%/9% 4/5	4/5	5/6

紫花苜蓿各生育阶段对水分的敏感指数为:幼苗期 0.151 9,蔓枝伸长期 0.481 0,开花成熟期 0.379 2。

The water sensitivity index of alfalfa at every growth stage is: 0.151 9 in seedling period, 0.481 0 in branch extension period, and 0.379 2 in flowering and maturing period.

经边际分析和效益分析,该试验最优处理是 2 号处理,即幼苗期田间持水率控制水平为 45%,蔓枝伸长期为 70%,开花成熟期为 65%。

With the marginal analysis and benefit analysis, No. 2 treatment is the optimal one in this experiment, that is, the field capacity is 45% in seedling period, 70% in branch extension period, and 65% in flowering and maturing period.

附 表
Postform

1   作物生育动态观测表   日期

水平	株数	株高	平均株高(cm)	叶数	平均叶数	备注

Attached Table 1   Dynamic Observation of Crop Growth   Date

Level	Plant Number	Plant Height	Average Plant Height (cm)	Leaf Number	Average Leaf Number	Remarks

观测   校核
Observed by: Verified by:

2   作物生育动态观测表   日期

水平	株数(分蘖数)	平均株高(cm)	叶面积(cm^2)	平均叶面积(cm^2)	备注

Attached Table 2   Dynamic Observation of Crop Growth   Date

Level	Plant Number (tiller number)	Average Plant Height (cm)	Leaf Area (cm^2)	Average Leaf Area (cm^2)	Remarks

观测   校核
Observed by: Verified by:

作物各不同水平的实际浇灌制度统计表

试验水平号	第一次灌水			第二次灌水			第三次灌水			第四次灌水			第 n 次灌水			灌溉定额 (m³/亩)	灌水方法	备注
	生育阶段	日期	灌水量 (m³/亩)	生育阶段	日期	灌水量 (m³/亩)	生育阶段	日期	灌水量 (m³/亩)	生育阶段	日期	灌水量 (m³/亩)	生育阶段	日期	灌水量 (m³/亩)			
3																		

观测：　　　　　　　　　　　　　　　　校核：　　　　　　　　　　　　　　　　日期：

Attached Table 3　Statistics for Actual Irrigation Scheduling of Crops at Different Level

Experiment Level Number	1st Irrigation			2nd Irrigation			3rd Irrigation			4th Irrigation			nth Irrigation			Irrigation Quota (m³/mu)	Irrigation Water Method	Remarks
	Growth and Development Period	Date	Irrigation Amount (m³/mu)	Growth and Development Period	Date	Irrigation Amount (m³/mu)	Growth and Development Period	Date	Irrigation Amount (m³/mu)	Growth and Development Period	Date	Irrigation Amount (m³/mu)	Growth and Development Period	Date	Irrigation Amount (m³/mu)			

Observed by:　　　　　　　　　　　　　　　Verified by:　　　　　　　　　　　　　　　Date

附 表
Postform

4 　　　　　　　　　　作物生育期灌水情况记载表　　　　　　　　　　日期

灌水日期	灌水次数	灌水方法	灌溉面积	水表读数	灌水量	灌水计时		灌水时间	灌水定额	灌水强度
（日/月）	（次）		（亩）	始终	(m³)	始	终	(h)	(m³/亩)	(m³/亩)

观测　　　　　　　　　　　　　　　校核

Attached Table 4　　Record of Crop Irrigation in the Growth and Development Period　　Date

Irrigation Date	Irrigation Frequency	Irrigation Method	Irrigation Area	Meter readings	Irrigation Amount	Irrigation Timing		Irrigation Time	Irrigation Water Quota	Irrigation Intensity
(day/month)	(times)		(mu)	Duration	(m³)	Beginning	Ending	(hour)	(m³/mu)	(m³/mu)

Observed by：　　　　　　　　　Verified by：

5 　　　　　　　　　　作物需水量计算表　　　　　　　　　　日期

试验水平号	发育阶段	起止时间		取土时间		计划层内土壤含水量(%)	计划层内土壤储水量(%)	土壤水分差值(m³/亩)	有效降雨量或灌水量(m³/亩)	阶段需水量(m³/亩)	备注
		月	日	月	日						

观测　　　　　　　　计算　　　　　　　　校核

Attached Table 5　　Calculation of Crop Water Requirement　　Date

Experiment Level No.	Development Stage	Duration		Soil Collection Date		Soil Water Content in Planning Layer (%)	Soil Water Storage Capacity in Planning Layer (%)	Soil Water Difference (m³/mu)	Effective Rainfall or Irrigation Amount (m³/mu)	Staged Water Requirement (m³/mu)	Remarks
		Month	Day	Month	Day						

Observed by：　　　　　　Calculated by：　　　　　　Verified by：

6 　　　　　　　　作物 有/无 底测坑浇水量记录表　　　　　　　日期

测坑编号	浇水时间						水表读数		测浇水量 (m³)	备注
	始			终			始(m³)	终(m³)		
	时	分	秒	时	分	秒				

观测　　　　　　　　　　　　　　校核

Attached Table 6　Record of Watering Quantity for Crops in Test Pit or Bottomless Test Pit　　Date

Test Pit No.	Irrigation Time						Meter Readings		Measured Watering Quantity (m³)	Remarks
	Beginning			Ending			Beginning (m³)	Ending (m³)		
	Hour	Minute	Second	Hour	Minute	Second				

Observed by:　　　　　　　　　　　Verified by:

参考文献
Reference

[1] 王千,曾德超.不同耕作措施对降雨入渗的影响.中国科学院禹城综合试验站年报,北京:气象出版社,1991.
Wang Qian, Zeng Dechao. Effect of Different Soil Surface Treatments on Soil Properties of Rainfall Infiltration; Annual Report of the Yucheng Comprehensive Experimental Station of the Chinese Academy of Sciences[J], Beijing: *China Meteorological Press*, 1991.

[2] 王新元.河北缺水盐渍区冬小麦、夏玉米水分利用率的试验研究.节水农业研究,北京:科学出版社,1992。
Wang Xinyuan et al. Experimental Study on Water Use Efficiency of Winter Wheat and Summer Corn in Water-deficient and Saline-soil Areas of Hebei Province; Water Saving Agriculture Research [J]. Beijing: *Science Press*, 1992.

[3] 吴家燕.冬小麦根系生态生理的初步研究.地理集刊,No.17,81-92.
Wu Jiayan. Preliminary Study on Ecological Physiology of Winter Wheat Roots; Geographical Collection [J], No.17, 81-92.

[4] 许越先.我国节水农业的主要趋势.节水农业研究,北京:科学出版社,1992.
Xu Yuexian. The Development of Water Saving Agriculture in China; Water Saving Agriculture Research [J], Beijing: *Science Press*, 1992.

[5] 由懋正.冬小麦根系发育及吸水图式的研究.作物水分关系研究,北京:中国科学技术出版社,1992.
You Maozheng et al. Study on Root Development and Water Absorption Pattern of Winter Wheat; Studies on the Relationship Between Crop and Water [J], Beijing: *Science and Technology of China Press*, 1992.

[6] 由懋正,王会肖.华北平原冬小麦水分利用率研究.农业用水有效性研究,北京:科学出版社,1992.
You Maozheng, Wang Huixiao. Water Use Efficiency of Winter Wheat in North China Plain. Study on the Effectiveness of Agricultural Water Use [M], Beijing: *Science Press*, 1992.

[7] 袁小良,王会肖.冬小麦产量与耗水量的关系.作物水分关系研究,北京:中国科学技术出版社,1992.
Wang Huixiao, Yuan Xiaoliang et al. Relationship between Winter Wheat Yield and Water Consumption; Studies on the Relationship Between Crop and Water [J], Beijing: *Science and Technology of China Press*, 1992.

[8] 张喜英.冬小麦根系生长规律及土壤环境条件对其影响的厂家.生态农业研究,1994,2(3):62-68.
Zhang Xiying et al. Growth Rule of Root System of Winter Wheat and the Effect of Soil Conditions on Manufacturers. Ecological Agriculture Research [J], 1994, 2(3): 62-68.

[9] 张喜英.夏玉米根系生长发育特点及吸水规律的研究.生态农业实验研究,北京:中国科学技术出版社,1994.
Zhang Xiying et al. Study on Growth and Development Characteristics and Water Absorption of Roots of Summer Corn; Experimental Research on Ecological Agriculture [M], Beijing: *Science and Technology of*

China Press, 1994.

[10] 冯谦诚,王焕榜.土壤水资源评价方法的探索,水文,1990(4):28-32.
 Feng Qiancheng, Wang Huanbang. Study on Evaluation Methods of Soil Water Resources, Hydrology [J], 1990(4): 28-32.

[11] 韩仕峰.黄土高原的土壤水分利用与生态环境的关系,生态学杂志,1993,12(1):25-28.
 Han Shifeng et al. Relationship Between Soil Water Utilization and Ecological Environment in the Loess Plateau; Chinese Journal of Ecology [J], 1993, 12(1): 25-28.

[12] 李开元,李玉山.黄土高原旱地土壤水分的利用和管理.陕西农业科学,1991(5):45-47.
 Li Kaiyuan, Li Yushan. Utilization and management of soil water in dryland of loess plateau, Shaanxi Journal of Agricultural Sciences, 1991(5): 45-47.

[13] 刘昌明,窦清晨.土壤—植物—大气连续体模型中的蒸散发计算,水科学进展,1992,3(4):255-263.
 Dou Qingchen, Liu Changming. Evapotranspiration calculation in soil-plant-atmosphere continuum model; Advances in Water Science, 1992, 3(4): 255-263.

[14] 丘山,程维新.不同植被下的土壤水分动态研究,中国科学院禹城综合试验站年报,北京:气象出版社,1991.
 Cheng Weixin, Qiu Shan. Study on Soil Water Dynamics under Different Vegetations. Annual Report of the Yucheng Comprehensive Experimental Station of the Chinese Academy of Sciences, Beijing: China Meteorological Press, 1991.

[15] 王绍仁.生态农业实验研究,北京:中国科学技术出版社,1994.
 Wang Shaoren. Experimental Research on Ecological Agriculture [M], Beijing: Science and Technology of China Press, 1994.

[16] 王树森.土壤水分模拟与调控,作物与水分关系研究,北京:中国科学技术出版社,1992,70-74 1991.
 Wang Shusen. Soil Water Simulation and Regulation. Studies on the Relationship Between Crop and Water, Beijing: Science and Technology of China Press, 1992, 70-74, 1991.

[17] 张书函.农田潜水蒸发的变化规律及其计算方法研究.西北水资源与水工程,1995(1):9-15.
 Zhang Shuhan et al. Study on the Change Rule and Calculation Method of Phreatic Water Evaporation in Farmland; Journal of Water Resources and Water Engineering [J], 1995(1): 9-15.

[18] 刘昌明.土壤水的资源评价,水量转换实验与计算方法,北京:科学出版社,1988.
 Liu Changming. Resources Evaluation Methods of Soil Water. Transfer Experiment and Calculation Method of Water [M], Beijing: China Meteorological Press, 1988.

[19] 牛文元.自然资源开发原理.开封:河南大学出版社,1989.
 Niu Wenyuan. Principles of Natural Resources Development [M], Kaifeng: Hennan University Press, 1989.

[20] 由懋正.发展节水型农业,提高水资源利用效率,华北地区水资源合理开发利用,北京:水利电力出版社,1990.
 You Maozheng et al. The Development of Water-saving Agriculture and the Improvement of Utilization of Water Resources. Rational Development and Utilization of Water Resources in North China [M], Beijing: Water Resources and Electric Power Press, 1990.

[21] 由懋正,袁小良.土壤水资源及其利用.地理学与国土研究,2(4):31-37(2):76-79.
 You Maozheng, Yuan Xiaoliang. Soil Water Resources and their Utilization; Geography and Geo-

参考文献
Reference

information Science (then Geography and Territorial Research) [J],2 (4): 31-37(2): 76-79.

[22] 张和平,袁小良. 土壤水资源的农业评价,2000 年中国水文展望. 南京:河南大学出版社,1991.
Yuan Xiaoliang, Zhang Heping. Agricultural Evaluation of Soil Water Resources; China's Hydrological Vision in 2000, Nanjing: *Henan University Press*, 1991.

[23] 张利. 作物棵间蒸发及叶面蒸腾量的研究. 生态农业研究,1995,3(1):76-78.
Zhang Li. Study on Crop Evaporation and Leaf Transpiration; Ecological Agriculture Research, 1995, 3 (1): 76-78.

[24] Allen S J. 1990. Measurement and estimation of evaporation from soil under sparse barley crops in northern Syria. Agric. For. Meteorol. ,49:291-309.

[25] Boast C W and Robertson T M. 1982. A micro—Lysimeter method for determining evaporation from bare soil: description and laboratory evaluation. Soil Sci. Soc. Am. J. ,46: 689-696.

[26] Daamen C C. 1993. Evaporation from sandy soils beneath crops in the semiarid zone: a study of the use of micro—lysimeter and numerical simulation. Ph. D Thsis, The University of Reading.

[27] Lascano R J. and van Bavel C H M. 1983. Experimental verification of a model to predict soil moisture and temperature profiles. Soil Sci. Soc. Am. J. ,47:441-448.

[28] Matthias A D. Salehi R and Warrick A W. 1986. Bare soil evaporation near a surface pointsource emitter. Agric. Water Manage. ,11:257-277.

[29] Shawcroft R W and Gardner H R. 1983. Direct evaporation from soil under a row crop canopy. Agric. Meteorol. ,28:229-238.

[30] Walker G K. 1983. Measurement of evaporation from soil beneath crop canopies, Canadian J. of Soil Sci. ,63:137-141.